PLACE/CULTURE/REPRESENTATION

D0209965

Spatial and cultural analysis have recently found much common ground; focusing in particular on the nature of the city. *Place/Culture/Representation* brings together new and established voices involved in the reshaping of cultural geography.

The authors argue that as we write we are not just representing some geographic reality, we are also creating meaning; writing is as much about context as it is about a purported reality. The issue becomes not one of scientific truth as an end but the interpretation of cultural constructions as a means.

Discussing authorial power, discourses of the other, texts and intertextuality, landscape metaphor, the sites of power–knowledge relations, and notions of community and sense of place, the authors explore the ways in which a more fluid and sensitive geographer's art can help us make sense of ourselves and the landscapes and places we inhabit and think about.

James Duncan is Professor of Geography at Syracuse University, USA. **David Ley** is Professor of Geography at the University of British Columbia, Canada.

PLACE/CULTURE/ REPRESENTATION

Edited by
James Duncan and David Ley

London and New York

First published 1993
by Routledge
11 New Fetter Lane, London EC4P 4EE

Simultaneously published in the USA and Canada
by Routledge
29 West 35th Street, New York, NY 10001

Reprinted 1994

© James Duncan and David Ley

Typeset in Scantext September by
Solidus (Bristol) Limited

Printed and bound in Great Britain by
Biddles Ltd, Guildford and King's Lynn

British Library Cataloguing in Publication Data
A catalogue record for this book is available from the British Library

*Library of Congress Cataloging in Publication Data
has been applied for*

ISBN 0-415-09450-X (hbk)   ISBN 0-415-09451-8 (pbk)

CONTENTS

ILLUSTRATIONS

PLATES

FIGURES

CONTRIBUTORS

John Agnew is Professor of Geography at Syracuse University.

Denis Cosgrove is Reader in Geography at Loughborough University.

Stephen Daniels is Lecturer in Geography at the University of Nottingham.

Mona Domosh is Assistant Professor of Geography at Florida Atlantic University.

James Duncan is Professor of Geography at Syracuse University.

Derek Gregory is Professor of Geography at the University of British Columbia.

Deryck Holdsworth is Associate Professor of Geography at Pennsylvania State University.

Paul Kariya is Director of Economic Programmes in the Department of Indian Affairs and Northern Development, Ottawa.

Audrey Kobayashi is Associate Professor of Geography at McGill University.

David Ley is Professor of Geography at the University of British Columbia.

Caroline Mills is Lecturer in Geography at Cheltenham and Gloucester College of Higher Education.

Don Mitchell is Assistant Professor of Geography at the University of Colorado.

Jonathan Smith is Assistant Professor of Geography at Texas A&M University.

Brian Stock is Professor of History at the University of Toronto.

Stacy Warren is Assistant Professor of Geography at Eastern Washington University, Cheney.

Topographical Survey

1

INTRODUCTION
Representing the place of culture

James Duncan and David Ley

TOPOGRAPHICAL SURVEY

The drawing by Joanne Sharp entitled *Topographical Survey*, which appears on the facing page, parodies and calls into question a traditional form of geographical representation, the map. Sharp has created a space between two discourses – that of the cartographic surveyor on the one hand, and that of the Foucauldian critique of surveillance on the other. Not only does the drawing expose the concealed values behind the map but so too does its title, for if one performs a close reading of the dictionary definitions of 'topography' and 'survey' one can see that running alongside a language of 'objective' science is another language of power.

> *Topography*: The science or practice of describing a particular place, city, town, manor, parish, or tract of land.
>
> (*The Oxford Universal Dictionary*)

Topography claims to be a 'science', that is, a discourse of technical, objective, rational, Enlightenment knowledge. Such knowledge is often claimed to be universal in its scope and free of cultural or political interests. But topography (like cartography)[1] also is a 'practice', knowledge put to use, knowledge in the service of power that is deeply intertwined in the cultural, social and political webs of a society. Such knowledge is intended to describe the way in which a social formation is made visible on the face of the earth. It is a practice which describes boundaries, including property relations, and thereby objectifies them, rationalizes them and makes them seem like objects of nature through the legitimizing tropes of the discourse of science. Topography is a system of knowledge which can encompass the world; its purview (as the artist shows) takes in the rural and the urban, the secular and the religious. Its theoretical reach even extends beyond the edge of settlement to empty tracts of land. Topography is also therefore a science of domination – confirming boundaries, securing norms and treating questionable social conventions as unquestioned social facts.

Survey: The act of looking at something as a whole, or from a commanding position.

(The Oxford Universal Dictionary)

Here we have the power of observation – the very process that produces the science of topography. It claims to be a totalizing gaze, rational and universal, which sees the whole and orders it. In practice it is usually a white, male, elite, Eurocentric observer who orders the world that he looks upon, one whose observations and classifications provide the rules of representation, of inclusion and exclusion, of precedent and antecedent, of inferior and superior.

Topographical Survey speaks not simply about mapping but more broadly about the nature of representation in cultural geography. The picture and its title also raise the question of mimesis, of an accurate copy. By adopting the discourse of science, thus claiming to grasp the whole, to assume a commanding position, the surveyor 'captures' the world in miniature. The world is mimetically reduced to the text. *Topographical Survey* attacks the surveyor's imperial gaze by returning it; locating this surveyor within the landscape fractures the illusion of absence and reveals, and thus problematizes, the claims to power through mimesis. One might also read in this picture a new cultural geography no longer surveyed exclusively from the authoritative perspective of the Berkeley School and its inspirational leader, Carl Sauer. One of the major tasks of this volume is to begin to map out a series of alternative ways of seeing and interpreting within the field of cultural geography.

ON REPRESENTATION

The task of scholars is to represent the world to others in speech and print. Yet until recently representation has not been widely problematized either within geography or within Anglo-American social science more broadly. Perhaps because it is so central to our whole enterprise, the question of how we should represent the world has usually been taken for granted. This fundamental level of agreement concerns the issue of mimesis, the belief that we should strive to produce as accurate a reflection of the world as possible. Within this broad area of consensus, debate has swirled around the issue of how best to achieve an accurate copy of reality in our writings.

Within twentieth-century Anglo-American human geography we can discern four major modes of representation, two of which operate within the framework of mimetic representation and two of which pose varying degrees of challenge to mimesis. The first and dominant mode in traditional cultural geography, and indeed most of human geography until the 1950s, was 'descriptive fieldwork' based upon observation. The assumption underlying this position is that trained observation transcribed into clear prose and unencumbered by abstract theorizing produces an accurate understanding of the world.

The second mode is a form of mimesis loosely based upon positivist science.

Although popular within geography more generally since the 1950s it has had almost no impact upon the practice of cultural geography because in its drive to produce abstract, reductionist descriptions of the world there is little or no room for the differences between places produced by cultural variation. Although these two modes of representation are thought to be opposed because the former values the concrete and the particular and the latter the abstract and the general, the representational claims of these two types of practice are basically similar in that they both accept the goal of achieving mimesis. Their rhetorical claims differ primarily in that the former draws upon nineteenth-century models of science and is more concerned with description and classification, while the latter draws upon early twentieth-century models and is more concerned with the construction of spatial theory.

The third type of practice is a postmodernism which represents a radical attack upon the mimetic theory of representation and the search for truth.[2] In this sense it is anti-foundational in that it explicitly rejects the totalizing ambitions of modern social science. A postmodern orientation distrusts and interrogates all meta-narratives including those of the researcher. Such an epistemology, if taken seriously, is inescapably and radically relativist. Although there has been great interest within geography in postmodernism of late, few geographers, as we shall see, have been willing to embrace its epistemology to this extent.

The fourth type of practice is interpretative and its basis is hermeneutics. Unlike the first two positions, it acknowledges the role of the interpreter and therefore rules out mimesis in the strict sense of the term. Rather than setting up a model of a universal, value-neutral researcher whose task is to proceed in such a manner that s/he is converted into a cipher, this approach recognizes that interpretation is a dialogue between one's data – other places and other people – and the researcher who is embedded within a particular intellectual and institutional context. It is precisely the interpersonal – and intercultural – nature of the hermeneutic method which poses a challenge to mimesis, since a 'perfect copy' of the world clearly is not possible if the interpreter is present in that textual copy. A number of theoretical positions within geography can be accommodated by adopting a broad definition of hermeneutics. For example, certain types of humanistic geography (Buttimer, 1974; Ley and Samuels 1978), including Marxian humanism (Cosgrove 1983), as well as a new cultural geography influenced by the work of Clifford Geertz and post-structural anthropologists (Cosgrove and Daniels 1988; Duncan 1990; Duncan and Duncan 1988, forthcoming; Gregory 1989; Ley 1987, 1989), all adopt hermeneutic modes of representation.

THE CRITIQUE OF MIMETIC THEORIES OF REPRESENTATION

Of late there have been calls by Rabinow (1986) and others for the abandonment of theories of representation that are mimetic and posit universal validity. The 'crisis of representation' in ethnography, as Marcus and Fischer (1986) have termed it, is part of a broader attack within a number of fields upon mimesis and the 'natural attitude' which underlies it. This 'natural attitude' stems from the philosophers of the Enlightenment, for whom language and imagery appeared to be perfect, transparent media through which reality could be represented to understanding. However, in modern literary and art criticism they are thought of as 'a prison house' (Jameson 1972) which locks us into particular modes of understanding and separates us from the world. The opacity of these media, it is argued, stems from the fact that there is no neutral, univocal, 'visible world' out there to match our vision against. Put slightly differently, 'there is no vision without purpose ... the innocent eye is blind' for the 'world is already clothed in our systems of representation' (Mitchell 1986: 38).[3] As such our representations of the world cannot be other than 'partial truths' (Clifford 1986a).

Bryson (1983: 13–15) defines realism as 'the coincidence between a representation and that which a society assumes as its reality'. However, for a society to maintain the illusion that its representations are natural representations it must conceal their historic specificity. Drawing upon Bourdieu's (1977) notion of habitus, a coherent set of values and orientations, Bryson argues that 'culture produces itself around a "habitus" which, though discontinuous with the natural world, merges into it as an order whose join with Nature is nowhere visible'. Here Bryson uses the term 'nature' to refer to external reality. While we may not see the difference between the world and our own representations of it, he is saying, others, whose cultural site or point of view differs from ours, may see our discrepancies (though perhaps not their own) much more clearly. This 'invisible join' between reality and cultural representations of it, which is so essential to the 'natural attitude' of objectivism and cultural reproduction, might not be acknowledged, he argues, were it not for modes of representation such as painting which, at different historical periods under the influence of varying conventions of genre, and under different social formations, have produced dissimilar essential copies of the world. Any appeal to the natural attitude in painting, then, is a mystification.

In literature the natural attitude takes the form of literary realism and is equally suspect for it conceals the social construction of language. As Eagleton points out, 'it helps to confirm the prejudice that there is a form of "ordinary language" which is somehow natural. This natural language gives us reality "as it is"' (1983: 135). Realism is ideological according to Eagleton, for it passes off as natural that which is in fact cultural. It is such a transformation of cultural productions that are historically specific that Barthes (1986: 129, 143) terms myth. Myths, he says, naturalize; they turn history into nature. The myth of realism creates 'a world wide open and wallowing in the evident, it establishes a blissful clarity: things appear to mean something by themselves'. This mythology of mimesis, Jay (1986: 182) argues, must be overcome by the realization that 'what is "seen" is not a given, objective reality ... but an epistemological field constructed as much linguistically as visually'. Perhaps Foucault (1970: 251) best

4

captures the enormity of the task facing those who wish to traverse the treacherous terrain of representation when he writes that 'the visible order, with its permanent grid of distinctions, is now only a superficial glitter above an abyss'.

White (1978: 130) points out that historians are also guilty of thinking that language can serve as a perfectly transparent medium of representation if only they can find the right words to allow the meaning of an event to '*display itself* to consciousness'. LaCapra (1983: 74–5) claims that historians who hold this view combine the paradigm of late nineteenth-century science with that of mid-nineteenth-century literature. Both share the pre-critical conception of facts as the ultimate 'givens' of an account.

Within geography the humanistic critique (Ley and Samuels 1978) was one of the first sustained attacks upon the positivist variant of mimesis discussed earlier, arguing for the centrality of values and valuing in the construction of knowledge. It was only later with the rise of interest in cultural theory that older, descriptive perspectives on mimesis (such as narrative) were subjected to scrutiny. A critical view of the mimetic claims of description builds a more serious divide between a traditional cultural geography committed to theory-free empiricism and a recent cultural geography committed to theory-laden inter-pretation. The critique of mimesis has recently been extended into cartography by Harley (1988, 1992) using the methods of deconstruction and by Pickles (1992) using the insights of hermeneutics. Both authors demonstrate how, in spite of rhetorical claims to objectivity, maps are texts within discourses of power.

A tenacious set of cultural conventions constantly subverts the achievement of mimesis. A powerful criticism of Western representation has been provided by feminists who have argued that Enlightenment epistemologies are profoundly gendered by patriarchal presuppositions (Harding 1986). The quest for abstraction, for a detached objectivity which disavows any distraction by values in its practice, is said to be characteristic of a Western male. In such a conceptual space there seems to be no room for knowledge as encounter, as 'one of the regions of my care' (Buttimer 1974). Marxism, particularly its abstract, economic variants, is derived from a similar tradition of Western rationality and has also collided with the criticism of a gender blindness which in essentializing the female form has not escaped a pervasively patriarchal view of the world (Deutsche 1991; Massey 1991). In mainstream social science, gendered presuppositions, for example of home and 'work', not only colour representations of men and women but also reflect an unequal distribution of power which constrains what female identities may yet become. The division between the private realm of the home and the public realm of civil society has been a gendered dualism, admitting men to the world of politics and economics while limiting women to the domestic round. Although a cultural convention, the private–public dualism naturalized separation and exclusion and has been a persistent target of feminist resistance (Pateman 1983; Rose 1990).

The feminist critique has been joined by an equally vigorous interrogation of

European representations of former colonial territories. The typification of the other as an abstraction, a collective social fact, is a denial of the complexity of other cultures. Elite, including intellectual, representations share these distortions and have been described as allegories, that is, stories we choose to tell ourselves about ourselves (Clifford 1986b). Moreover, intrinsic to these stories is a Eurocentric view of superiority and inferiority. European representations of the other have been far from benign, for as the interrogation of the Western construction of 'the Orient' (or indeed 'darkest Africa') has indicated, while mimetically false such constructs as Orientalism are significant buttresses of colonial and post-colonial power (Said 1979). That power is also solidified, literally and politically, in architecture and urban planning (King 1976). While there are powerful insights in such a deconstruction, one must be cautious to avoid a tendency towards reification where dominant representations are posited as unduly totalizing. Even such Western abstractions as 'the Orient' were contested among Europeans, and showed variation through time. Nor is it at all clear that non-Western societies do not share in the construction of equivalent cultural fictions. In places, Said's polemic against the myth of the Orient is propelled by a representation of Europe that is almost as uncompromising. The fundamental task, as he reflected elsewhere (Said 1989), is to take the other seriously. While our own cultural categories are tenacious, we can be aware of both their partiality and also their effects (Kahn 1989).

These observations are no less pertinent when directed to the cultural pluralism within nations in many parts of the world. Whether in Canada, or Yugoslavia, or Nigeria, or Malaysia, cultural pluralism is invariably accompanied by a mythology of the other (Smith and Thrift 1990). The boundaries of race provide particularly conducive conditions for the construction of distorted cultural representations. In the United States, for example, a pernicious white mythology about black culture was historically consolidated by the media, legitimated by the social sciences and institutionalized by the state (Ley 1974; cf. Jackson 1987; Smith 1988). Its consequences, grounded in the cultural fiction of 'the black ghetto', included the creation of unequal life-chances for households confined within the stigmatized collectivity of the inner city. The manipulation of space was a standard strategy confirming the yet more explicit apartheid confronting Chinese immigrants in nineteenth-century North American cities (Anderson 1988), or racialized communities in South Africa after the 1950 Group Areas Act. In each instance, space not only contained the other, but was also a label of status or stigma, so that Western (1978: 297) noted that in Cape Town even within the 'Coloured' population the varying destinations of displaced and resettled households became cultural signifiers: 'Stigma or respectability is now conferred by address.'

Gender, imperial power, nation or race provide only some of the lineaments around which discourses of exclusion arise. To them could be added representations shaped by the categories of class, religion, political ideology, age and even physical or mental disability, and a literature exploding in quantity is disclosing their arbitrary constitution and often prejudicial effects. The social construction of knowledge is pervasive; values and valuing are integral to knowing, making any claim to objectivity untenable. How does the scholar engage such a contingent reality? One conclusion is to accept a radical relativism where knowing is culturally and socially contained. More hopeful is the view of representations as partial truths, the outcome of a relation between an empirical world and a historical subject. This was the view of Karl Mannheim, one of the early contributors to the sociology of knowledge, who contrasted an epistemology of radical relativism with what he called relationalism, a hermeneutic endeavour where interpretation problematizes the relations between the empirical world and historically situated subjects (Mannheim 1952).

POSTMODERNISM

Let us now briefly examine the challenge posed to traditional forms of representation by a radical postmodernism. While postmodernists share the hermeneuticians' critique of mimetic representation, they also undertake to decentre the privileged sites from which representations emanate, notably Western, male intellectuals. This decentring takes a number of forms. One strategy attempts to undermine the tropes of representation known as the Enlightenment project, or modernism. This revolt against objectivism, rationality and the meta-narratives which have guided enquiry has been one of the central themes in postmodern representation. Some ethnographers have argued that cultures are composed of fragmented and contested codes of meaning and that ethnographers should acknowledge and participate in this by experimenting with writing. Such experimentation would further challenge the Enlightenment project by adopting a fragmentary writing style that is purposefully ambiguous, incomplete and open-ended. One of the major problems with this view of representation, as Mascia-Lees *et al.* (1989) point out, is that

> once one articulates an epistemology of free play in which there is no inevitable relationship between signifier and signified, how is it possible to write an ethnography that has descriptive force? Once one has no meta-narratives into which the experience of difference can be translated, how is it possible to write any ethnography?

There is a further inconsistency in this writing strategy. In the history of artistic representations, use of the fragment, decontextualized and re-presented as collage, was a standard method of the modern movement in the early twentieth century (Kern 1983). Ironically, in upholding the fragment, the escape from modernism turns back into modernism.

In human geography support for anti-foundationalism and the epistemology of the radical postmodern project is to be found in several of the contributions in the volume edited by Barnes and Duncan (1992)[4] Experiments in writing within geography are most closely associated with Olsson (1980, 1992) and Pred

(1990), both of whom have been closely influenced by post-structuralist authors. Olsson's bold textual experiments, including poetry, are intended to shake loose creative and plural readings of a text, to challenge the tyranny of the word, through ambiguity and language games (Olsson 1978). They do not (indeed, cannot), however, remove the control of the author, whose presence is ever more evident in strongly stylized forms of writing.

A second form of decentring is a shift away from exclusive representational control by the Western male academic. It is argued that the introduction of a polyphony of voices, with the redistribution of representational authority into multiple sites (women in academia, non-academics within our own culture, people from other cultures), not only undermines political authority, and the domination of a single author, but also further undermines the myth of mimesis. Rabinow (1986: 239–41), drawing upon Foucault, argues that academics must begin to think in terms of multiple systems of representation that are produced within discourses tied to social practices. Poster (1982: 142) claims that it is an Enlightenment myth that reason is continuous with reality and that intellectuals are the privileged caretakers of this process. While important in their intent, such dialogical and polyphonic ethnographies have not removed the authority of the author, who has, after all, defined the project in the first place. It appears that a truly radical relativism cannot be secured in practice.

HERMENEUTICS

Hermeneutics offers a less radical break with the rational tradition of the Enlightenment, and has such it offers an important avenue of escape from the bad faith of an objectivism which renders research ironic. The irony lies in the fact that objectivism knows itself to be based on a fiction: that the researcher is an objective recorder of that which s/he sees and hears. The hermeneutician, on the other hand, explicitly recognizes and theorises the site of his or her representation. What is seen as avoidable bias by the positivist is acknowledged by the hermeneutician as an inescapable part in the formation of knowledge. Rather than attempting to banish the historically situated observer, hermeneutics acknowledges the collision between the data and the interpreter (Gadamer 1986: 273). One consequence of this is the recognition that one cannot completely escape from ethnocentrism, for by definition all representations are inextricably intertwined with the theory-laden categories of the research. In the slogan of the phenomenologists, every object of knowledge is an object for a subject. But while hermeneutics breaks with objectivism in some important respects, it shares with it the same relationship of power over objects of study, for the site from which hermeneutic representation usually emanates is an intellectual elite in the West. Meanings are recuperated from 'at home' and 'abroad' and placed within the interpretative frame of the Western intellectual.

In practice the attempts of postmodern ethnographies to undermine this power have not been (we would argue cannot be) entirely successful (Khan

1989). They have argued that by allowing informants their own voice in scholarly monographs, as opposed to having the ethnographer speak for them as has traditionally been the case, a dialogue is established within the text and representational control is shifted away from the academic. While there is some truth to this, the loss of representational control by the academic is more illusory than real for the following reasons. First, this whole project is theorized within a discourse of postmodernism (a hyper-intellectual western academic project). Second, even in cases where informants are co-authors with academics, organizational control over representation remains in the hands of the academic. There are in turn several reasons for this, including the most material: most academic books in the West are produced for other academics to read, and their publication is predicated upon their acceptability to this speech community. We would argue, therefore, that hermeneutics provides a more satisfactory form of representation than that offered by postmodern ethnography because it allows dialogue between the researcher and his or her subject and yet does not misrepresent the power relations that are structured into the Western academy.

Within hermeneutics one may distinguish three major components of a representation (Iser 1989). The first is the text which the academic produces (whether it be a journal article, a book or simply an idea which has yet to be committed to print). The second is the *extra-textual field of reference*, the 'data' used in the production of the text. The third is the *inter-textual field of reference*, elements culled from other texts (both theoretical and empirical) which are used in the production of the text. What this model of academic work suggests is not a mirroring of the extra-textual within the text, but rather *re-presentation*, the production of something which did not exist before outside the text. This process of academic production is essentially disruptive of the extra-textual world. The text disrupts the extra-textual field of reference, by highlighting some elements within the field and deleting others. Elements are reshuffled within the text, thereby splitting up the fields of reference through an act of selection. The basis for this selection and reshuffling lies largely within the inter-textual field of reference. Both the inter-textual and extra-textual fields serve as contexts for each other within the text. Each plays off the other and helps define the possibilities of interpretation. As such, the world within the text is a partial truth, a transformation of the extra-textual world, rather than something wholly different from it.

But this view considers only the relationship between the producer of the text and the extra-textual and inter-textual fields of reference; one must also theorize the relationship of the reader to the text.[5] This relationship is also complex, for if one rules out the possibility of mimesis in the production of a text, then surely there is no reason to apply it to the reading of that text. The same processes are at work in the consumption of texts as in their production. The reader understands a text by situating it within two interpenetrating fields of reference – the extra-textual, the reader's experiences in the world, and the inter-textual, the context of other texts. The world in the text is continually compared to the worlds outside the text in order to see what the former reveals about the latter and as a 'test' of the plausibility of the former. Some readers will object to a text because they do not share its extra-textual field of reference (they think that unsuitable evidence has been chosen to illustrate a particular argument). Others may object because of the author's inter-textual field (they do not agree with the author's theoretical position, or with what is considered to be an important research question). In either case the reader (by reordering the relationship between the text, the extra-textual and the inter-textual) will produce a different interpretation of the text than that which the author intends, thereby extending the hermeneutic cycle. As long as there are readers for a text its reproduction will continue. In this sense, representation is not only a collective but also an iterative process.

RE-PRESENTING CULTURAL GEOGRAPHY

We have raised in some detail the question of different modes of representation and have provided alternatives to mimetic modes and in particular to realism, an atheoretical type of representation which has dominated traditional cultural geography in North America. We have also argued that certain postmodern types of representation currently considered in American ethnography are also problematic, although for very different reasons than mimetic modes. The position which we as editors and most of the contributors to this volume have adopted is closer to that which we have identified as hermeneutics.

But what if we subject this volume to the classificatory hermeneutic framework that we outlined in the preceding section? What can we identify as its extratextual and inter-textual frames? Before attempting to address these issues we must briefly answer a question which no doubt has occurred to some readers: 'Does this book represent the scope of North American cultural geography?' It does not, and the rather complex reason lies in our earlier discussion of the nature of representation. We wish to make no claims to offering a mimetic representation of cultural geography as it has been practised. Rather we see our representation as a creative process, the production of something new – a *re-presentation* of cultural geography by means of a strategic shift in both the extra-textual and the inter-textual field of our text. Certainly a part of what makes this volume different from traditional cultural geography is its extra-textual field; unaccustomed problem areas such as the modern city and the state. To many cultural geographers, however, its inter-textual field will mark even greater difference, for the authors show an interest in broad questions of methodology and theory drawn from other social sciences and from the humanities as well. Together these extra- and inter-textual selections begin to decentre an existing mode of seeing, by fragmenting it. Such fragmentation is a creative endeavour, making room for new theoretical perspectives.

In what ways, then, do the essays in this book constitute a re-presentation of cultural geography? For sure, the view of knowledge as socially constituted is far from a new insight in the social sciences. In the twentieth century, social phenomenologists like Alfred Shutz, neo-Kantians and some materialist scholars have argued the point eloquently. Significant contributions to the sociology of knowledge, including those of Mannheim (1952), Berger and Luckmann (1967) and Raymond Williams (1973, 1977), have sought to integrate some of these disparate strands. Within geography, the centrality of intersubjective values and valuing in the production of knowledge was a message disseminated in early humanistic writing (e.g. Buttimer 1974; Lowenthal 1961).

But despite the intervention of influential authors like Brookfield (1969) and Mikesell (1978) who urged the mediation of environmental perception in the study of cultural geography, much work continued on in the tradition of material culture studies, observing, describing, classifying and mapping such cultural artefacts as houses, barns or fences in order to identify cultural regions or paths of cultural transmission; a remarkable compilation of such work is to be found in the atlas of American and Canadian cultural forms (Rooney *et al.*, 1982). A principal criticism of this work is its preoccupation with geographical description, patterns on the map, which provide an incomplete intellectual project. First, it is a perspective addressed primarily to a stable, rural and often pre-industrial world located ever further in the past. Second, its cartographic patterns are not problematized in terms of social, economic, political or even cultural contexts. The description is thin rather than thick, and treats the landscape in a purely phenomenal manner. In contrast current work applies such interpretative metaphors as spectacle or text to landscapes implicating a web of social processes and intersubjective meanings. For example, Kay Anderson's (1988) interpretation of the meanings of North American Chinatowns has treated the landscape as a text expressing European representations of 'the Orient', a text which imposed limits on the status of what its residents were, and of what they might become. Such an interpretation of landscape incorporates interlocking cultural processes (Eurocentric values), political processes (their institutionalization by the state) and economic processes (competing positions in the labour market). In this interpretation, landscape is anything but a neutral element. Chinatown, as a district of minimum entitlement, both projected and reinforced a status of existing and ongoing marginality for Chinese immigrants.

Attention to theory which problematizes description, and to interpretation which queries pattern, is then an important element of what some have called a new cultural geography (Cosgrove and Jackson 1987; Gregory and Ley 1988). Not less important is the theorization of culture itself. The Berkeley School treated culture as a totality which imprinted its messages mechanically upon the residents of a culture area (Duncan 1980). Such a view is flawed in several ways. Contemporary work, particular in popular culture, sees society as constituted by a plurality of cultures, some dominant, some marginal (see Chapter 10 by Stacy Warren). A dominant or hegemonic culture is rarely passively internalized; commonly it is negotiated, resisted or selectively appropriated by people in everyday life. So too, cultural representations (like landscapes) invoke both ideology and power, a power which is often instutionalized by dominant groups in legal discourse (see Chapters 11 and 12 by Paul Kariya and Audrey Kobayashi respectively). In advanced societies law has become a major site of contestation between interest groups who belong to textual communities with different normative values.

It follows that culture is not a residual element in social life, a secondary realm of variation left over after more powerful explanations have run their course. Cultural explanations have frequently had to defend themselves from such derivative status before an all-pervading economism. But the counter-position of culturalism is equally mistaken. The issue more properly is the nature of the relations between the economic and the cultural rather than a preordained pecking order which sets up superior and inferior realms (McRobbie 1991). This mediated position is amply illustrated by studies in the culture of consumption, arguably the dominant public culture of Western societies today. For a commodity is the site where production and consumption relations meet. On the one hand, a commodity is the outcome of a set of economic relations between people, and between people and nature. But no less it is a site of meanings, of values and valuing, a magical realm where materiality is infused with symbolic meaning. However, the blurring of boundaries does not stop there. The examples of Japan and Hong Kong as centres of production underscore how economic relations are culturally (and politically) mediated. So too any interpretation of an art object passes quickly beyond the aesthetic realm to an international art market which defines the trajectories through which art objects pass. But, then, at a further remove, art collection and the associated act of economic valuing are themselves products of a particular, not universal, Western cultural tradition (Clifford 1988c).

or "reality"

A realization of the constant imbrication of, among others, cultural, economic and political processes is necessary for any interpretation of landscape. Moreover, landscape and place have assumed a remarkably central position in current interdisciplinary interpretations of our times. For a number of authors, postmodern landscapes have been an entry point to a discussion of 'the postmodern condition' – though, like the posing of a single architecture, this is a dangerously monolithic representation. Similarly, Jackson (1989) has noted the geographical metaphors which pervade cultural studies in Britain. The same point could be made in North America, where the 'cognitive mapping' of Frederic Jameson, the 'chronotope' of Mikhail Bakhtin and the 'travelling theory' of Edward Said sustain the view of the literary critic Caren Kaplan (1990: 25) that 'there are many terms circulating today that signify an attention to geography in cultural studies'. It is the same in the new ethnography, where anthropologists speak of 'local knowledge' (Geertz 1983), of a 'poetics of displacement' (Clifford 1988d), of the need for a new geographic strategy of multi-local ethnographies (Marcus 1989). In part, of course, the sheer visible presence of landscape provides a convenient grounding and point of departure for discussion of less fixed and visible cultural domains. But there is a good deal more to it than that. An awareness of the profound Eurocentricity of cultural representations has led to an urgent reappraisal of the specificity of the other, and an acknowledgement that geography is centrally implicated in the constitution of difference. This has been an insistent refrain of critical, post-colonial studies. Edward Said (1990), for example, argues for the fundamental importance of colonial Algeria in the work of Albert Camus. Said challenges the readings of Camus which extract only a discourse on the universal (dis-placed) human condition, where Algeria is a passive backdrop for a placeless saga. Instead, Said claims that the place and the time – the chronotope – have a far more active presence in the constitution of the novel: 'Camus' narratives lay absolutely severe and ontologically prior claims to Algeria's geography' (Said 1990: 82). And they are colonial claims, his fiction 'an element in the methodically constructed French political geography of Algeria that took many generations to complete' (ibid.: 88).

Written texts contain an often hidden geography and an examination of that geography may clarify the unreflected ideologies of the text. In this manner geography is not only duplicitous in its confirmation of existing ideologies (Daniels 1989; Duncan and Duncan 1988). From another perspective it may also undermine dominant cultural representations of the other, by revealing that they are less regularities of nature than conventions of a situated – geographic – imagination.

RE-PRESENTING CULTURAL GEOGRAPHY: AN INTRODUCTION TO THE ESSAYS

In Part I the specific challenges of representation in cultural geography are addressed in the context of a broader discussion in the humanities and social sciences. Denis Cosgrove and Mona Domosh (Chapter 2) examine the post-modern challenge to the authority of authorship. Authorship passes far beyond reason to the embrace of unrevealed political and moral options. They note the gender bias in the Western tradition of science, and review attempts to demystify that bias and its power relations. They conclude with the relativist position that our representations are 'tales of how we have understood the world' which are not to be judged by a theory of correspondence. In 'Sites of representation' (Chapter 3) James Duncan focuses his argument more concretely around the making of cultural geographies. He notes two tropes or rhetorical devices used to represent the place of the other. First, through an empiricist theory of observation, the authority claimed as Geertz (1988) has put it by 'being there', culturally contingent methods have been employed to claim objective representation, methods which include observation itself, the collection and taxonomy of cultural forms, photography and also mapping (cf. Orlove 1991). In a second trope, the other is incorporated into the history of the self, but referred to an earlier, and more primitive, period, thus reinforcing the superior site of the modern viewer. These tropes are illustrated at work in nineteenth-century European representations of Africa.

Discarding the metaphors of machine and system, recent work in cultural studies and cultural geography has made much use of such metaphors as spectacle, theatre and text to represent landscape. In Chapter 4, Stephen Daniels and Denis Cosgrove interrogate these metaphors. Despite their present currency, all three have a long history in the Western intellectual tradition. Often their coexistence has been uneasy and conflictual, as the visual images of spectacle and theatre are brought against the linguistic predilection for the word, the text. As these tensions in sixteenth-century Venice and in Regency England are explored, it is clear not only that they are shaped by specific societal conditions, but also that they may often be played out across the discursive terrain of landscape. Jonathan Smith returns to the metaphor of landscape as text in Chapter 5. Like Daniels and Cosgrove he finds landscape anything but innocent. Landscape is of the *longue durée*, seemingly fixed and natural when experience is not, expressing actors long past and intents no longer relevant. Yet landscape may also have the capacity to reproduce and confirm social relations, such as the status gradients of unequal societies. Again, we can conceive of landscape as irony, where an incongruous emplacement destabilizes meaning and provokes greater reflexivity challenging the taken-for-granted world. In these different moments, the semantic pluralism of landscape as text adequately repays the struggle of a careful reading.

The four essays in Part II all address specific residential landscapes, a major preoccupation of cultural geography. But all approach the house with a critical view of the descriptive, morphological and folk orientation of earlier work. In his essay 'Revaluing the house' (Chapter 6) Deryck Holdsworth observes how research in the 1970s and 1980s failed to advance Fred Kniffen's seminal, if limited, project of the mid-1960s. As a result 'house-type geography' is essentially a descriptive and taxonomic undertaking which raises more questions than it answers. The theoretical developments of human geography more generally have failed to impact this subfield. More promising directions come from material cultural studies and social history which contextualize built form in a range of social and economic conditions, including class relations and the status of dominant and minority cultures in a modernizing world. These contexts are explored in Don Mitchell's case study of workers' housing in Johnstown, Pennsylvania, and Pittsburg, California (Chapter 7). Both towns were dominated by a large steel corporation which had established a social contract with its work-force. In return for greater control over the workers' interests and loyalties the company offered a range of public services including housing. Its paternalism was thus double-edged, including both moral responsibility and a search for domin-ation. The hegemony of the corporation was less than complete, observes Mitchell, as both parties displayed degrees of dependence upon the other. Increasingly company responsibilities devolved to the welfare state, including the wartime construction of employee housing. This housing sustained lines of occupational and racial segregation in the mill. The landscape of company housing thus represents particular social and economic conditions, a social contract negotiated by management and by labour, contested and always subject to change.

In the post-war period the moral responsibility of the state to provide housing expanded considerably. In most Western societies, the public housing programmes of the 1950s and 1960s drew their inspiration from the cultural categories of the modern movement, producing standardized, high-density units. David Ley notes in Chapter 8 that as the limitations of modernism became apparent by 1970, the state looked for new innovative forms and practices of housing delivery. The housing co-operative was one such innovation, the product of a distinctive cultural politics which promised a more powerful voice to residents in the design of housing and which promoted more personalized, humane and communal living arrangements. In Vancouver, postmodern design solutions provided the built forms expressive of this liberal ideology. A second face of postmodernism is interpreted by Caroline Mills in her closely argued paper on the fictional geographies of gentrification (Chapter 9). The makers of these geographies are the planners, architects and developers who shape the site; the sales people and advertisers who code these built forms into consumption codes; and the purchasers who reinterpret these codes against the repertoire of needs and desires which defines their self-identity. Among this textual community a dominant symbol is that of urbanity. The gentrified inner city offers a landscape for self-conscious 'city people' who define themselves in antipathy to the perceived banalities of the suburbs and find an urbane landscape which reinforces that identity. In all there is a complex interchange between rational production and cultural symbolism, between place and identity. Mills's ethnography shows that culture is the medium which relates landscape texture to social text.

In Part III the argument turns more squarely to the institutional context of cultural production. In her study of landscapes of leisure Stacy Warren (Chapter 10) notes how research in cultural studies has always divided around the issue of the source of cultural production, the extent to which it is derived from above (the culture industry) or below (popular culture). In an important sense this is a false dichotomy since both positions have their merit. Most cultural products are undoubtedly produced as commodities, but this overlooks the key point that popular culture 'is primarily about the circulation of meanings'. How are these meanings produced and how are they received? Following Benjamin and Bloch, Warren argues that even fantasy has its possibilities for transcendence; like the fairy story it can problematize the real by reconstituting it in whimsical, even Utopian, forms.

The next two essays examine the construction of, and conflict around, Eurocentric representations of the other in Canada, representations which coded into law have become part of a national institutional culture. Paul Kariya (Chapter 11) shows how, empowered by the Indian Act, the Department of Indian and Northern Affairs appears to be a Foucauldian total institution in its capacity for control over native peoples. But it is rife with contradiction: intended to serve, it is criticized for policing; mandated to improve native conditions, it is challenged with worsening them. Kariya traces these contradictions in part to opposed Eurocentric representations of the native, needing protection but also inevitably fated for assimilation. Two further points emerge from the discussion: first, that the exercise of the hegemonic powers of the Indian Act is continually contested by an informed and astute native body, and, second, that in the impasse over implementation, the bureaucrats are as much captives of the contradictions as are the natives. The power of resistance from below is well illustrated in the conflict over Canada's critical Meech Lake Accord, where ratification in the Manitoba legislature in 1990 was blocked by the action of a single native member. Manitoba was one of two provinces refusing ratification (Newfoundland was the other), precipitating the constitutional crisis which threatens the future of the Canadian confederation in the early 1990s. Audrey Kobayashi (Chapter 12) interprets Canadian multiculturalism as a progressive rupture with the founding model of two dominant charter groups, British and French. Outside Quebec this model produced the cultural hegemony of Anglo-conformity which translated into selective immigration statutes and other entitlement privileges. Anglo-conformity has been contested by a growing cultural pluralism that was initially demographic, and then became institutionalized into a multiculturalism directorate whose actions have moved from symbolic celebrations of diversity to structural demands for human rights. In Quebec these tendencies run against desires to preserve the founding model of privileged francophone status, presented at Meech Lake as recognition of Quebec as a distinct society.

The conjunction between culture and power in Western Societies in legal texts and institutional contexts finds its counterpart in a preindustrial and non-European setting examined by James Duncan in Chapter 13. As institutional cultures not only reflect existing conditions but perpetuate them, so Duncan argues that the built environment of the Kandyan Kingdom in nineteenth-century Sri Lanka was not merely reflective of society but actively constituted it. For landscape was a text connected inter-textually to other cultural elements sustaining the status and power of the king. The vast rebuilding programme of the king of Kandy was densely suffused with iconographic reference to authoritative religious texts. The king, as landscape author, thus claimed their authority, attempting to secure his contested power by building a miniature of the cosmic order with himself at its centre. However, the king's construction project was turned against him by conflicting readings of opposing groups, themselves textualized in poems and allegorical performances. Differences in textual interpretations were indicative of a *realpolitik* of struggles for power; an errant author had no legitimate claim to ritualized authority.

A reassessment of representation has occurred across a broad front of disciplines in the arts, humanities and social sciences. In Part IV some of the cross-disciplinary currents are elaborated in the intellectual positioning of cultural geography today. John Agnew (Chapter 14) examines the spatial presuppositions which underlie the widespread conceptual use of the nation state and core–periphery models in the broad fields of political sociology and international relations. Invariably they display an impoverished view of both place and culture. In contrast recent work in cultural geography and elsewhere sees culture as actively mediating between structural conditions and local circumstances to create distinctive meanings of place at different spatial scales. By way of illustration, Agnew reviews the changing electoral geography of Italy as a product of shifting and locally mediated political cultures.

In the multidisciplinary debate over modernity, late nineteenth-century Paris has often been regarded as exemplary, a seedbed of the criss-crossing practices of modernity in art, culture, politics and economics. But, observes Derek Gregory (Chapter 15), interpretations of Paris are representations not only from other times but, more fundamentally, from other cultures and other politics. David Harvey's reconstruction of Second Empire Paris is guided by a Marxist meta-narrative which isolates the transformative role of finance capital, but as a totalization fails to represent adequately the cultural life of the city. Gregory contrasts Harvey's account with that of Walter Benjamin, who, as he interrogates the nexus of cultural meanings around the Parisian arcades, presents a view of modernity which, unlike Harvey's, is tragic not progressive. His view of the commodity, the artefact, leads to a discourse not of Enlightenment rationality but of a mythic dream world requiring a distinctive textual strategy in representation. Like Benjamin, Allan Pred's account of modernizing Stockholm between 1880 and 1900 is concerned with the practices and meanings of everyday life and similarly employs an innovative textual strategy. Pred recovers local glossaries, key words of popular geographies, which represent the practices and meanings of popular life-worlds. His excavation of popular languages is a sophisticated rendering of social theory which draws upon Foucault's work on discourse as well as Bakhtin's concept of heteroglossia.

In the final essay, (Chapter 16) Brian Stock positions a discussion engaging text, reading, community and sense of place within a literature which includes the contributions of anthropology, history and literary studies. He returns to issues raised earlier on the relations between visual and textual strategies, of seeing and reading. Landscape is amenable to both strategies; it is visual and unproblematic, but it is also readable as a text. That reading is culturally informed, and may indeed be shaped by explicit written conventions (compare the stylized reading and sometimes bizzare representations of the 'wild terrain' of mountains by the nineteenth-century Romantics). Thus metaphors of reading may be extended to visual experiences, including landscapes. As Duncan's essay on Kandy illustrated, even pre-literate socities offer informed landscape readings empowered by the memory of oral tradition. Indeed places may provide cues to memory and signification. To this extent we may even say that landscapes consolidate shared meanings; they act as community builders.

NOTES

1   See Harley (1988, 1992).
2   Postmodernism is one of the most elastic of terms. We use it here to refer specifically

and narrowly to a radical attempt to decentre the authority of the writer, and often to implement experimental styles of writing. In ethnography, see Tyler (1986) and the discussion of dialogical and polyphonic modes of authoring in Clifford (1988a). In geography the key figure is Gunnar Olsson (1980, 1992).

3 One of the most interesting ethnographic examples of this is to be found in the work of Lee (1950) among the Trobriand Islanders. She shows that because the Trobriand Islanders do not codify space and time lineally as we do, their 'common-sense', 'natural' definitions of objects and events do not correspond to our own 'common-sense' ones.

4 For a detailed review and critique of postmodernism in geography see N. Duncan (forthcoming).

5 It is important to note that while we have drawn upon Iser (1989) for our discussion of the producer of the text and his or her various fields of reference, Iser is best known for his theorization of the relationship between the reader and the text.

ACKNOWLEDGEMENTS

We are grateful to Joanne Sharp for preparing the drawing *Topographical Survey* specifically for this volume.

REFERENCES

Anderson, K. (1988) 'Cultural hegemony and the race-definition process in Chinatown, Vancouver: 1880–1980', *Environment and Planning D: Society and Space* 6: 127–49.

Barnes, T. and Duncan, J. (eds) (1992) *Writing Worlds: Discourse, Text and Metaphor in the Representation of Landscape*, London: Routledge.

Barthes, R. (1986) 'Myth today', in *Mythologies*, trans. A. Lavers, New York: Hill & Wang, 109–59.

Berger, P. and Luckmann, T. (1967) *The Social Construction of Reality*, Garden City, NY: Doubleday.

Bourdieu, P. (1977) *Outline of a Theory of Practice*, trans. R. Nice, Cambridge: Cambridge University Press.

Brookfield, H. (1969) 'On the environment as perceived', *Progress in Geography* 1: 51–80.

Bryson, N. (1983) *Vision and Painting: The Logic of the Gaze*, New Haven: Yale University Press.

Buttimer, A. (1974) *Values in Geography*, Resource Paper No. 24, Washington, DC: Association of American Geographers.

Clifford, J. (1986a) 'Introduction: partial truths', in J. Clifford and G. Marcus (eds), *Writing Culture: The Poetics and Politics of Ethnography*, Berkeley: University of California Press, 1–26.

——— (1986b) 'On ethnographic allegory', in J. Clifford and G. Marcus (eds), *Writing Culture: The Poetics and Politics of Ethnography*, Berkeley: University of California Press, 98–121.

——— (1988a) 'On ethnographic authority', in J. Clifford *The Predicament of Culture*, Cambridge, Mass.: Harvard University Press, 21–54.

——— (1988b) 'Histories of the tribal and the modern', in J. Clifford *The Predicament of Culture*, Cambridge, Mass.: Harvard University Press, 189–214.

——— (1988c) 'On collecting art and culture', in J. Clifford *The Predicament of Culture*, Cambridge, Mass.: Harvard University Press, 215–51.

——— (1988d) 'A poetics of displacement', in J. Clifford *The Predicament of Culture*, Cambridge, Mass.: Harvard University Press, 152–63.

Clifford, J. and Marcus, G. (eds) (1986) *Writing Culture: The Poetics and Politics of Ethnography*, Berkeley: University of California Press.

Cosgrove, D. (1983) 'Towards a radical cultural geography: problems of theory', *Antipode* 15: 1–11.

Cosgrove, D. and Daniels, S. (eds) (1988) *The Iconography of Landscape: Essays on the Symbolic Representation, Design and Use of Past Environments*, Cambridge: Cambridge University Press.

Cosgrove, D. and Jackson, P. (1987) 'New directions in cultural geography', *Area* 19: 95–101.

Daniels, S. (1989) 'Marxism, culture and the duplicity of landscape', in R. Peet and N. Thrift (eds) *New Models in Geography* Vol. 2, London: Unwin Hyman, 196–220.

Deutsche, R. (1991) 'Boys' town', *Environment and Planning D: Society and Space* 9: 5–30.

Duncan, J. (1980) 'The superorganic in American cultural geography', *Annals of the Association of American Geographers* 70: 181–90.

—— (1990) *The City as Text: The Politics of Landscape Interpretation in the Kandyan Kingdom*, Cambridge: Cambridge University Press.

Duncan, J. and Duncan, N. (1988) '(Re)reading the landscape', *Environment and Planning D: Society and Space* 6: 117–26.

—— and —— (forthcoming) *The Text in the World: Towards a Poststructural Theory of Landscape Interpretation*, Baltimore: Johns Hopkins University Press.

Duncan, N. (forthcoming) 'Postmodernism: a critique', in C. Earle and M. Kenzer (eds) *Conceptual Thinking in Geography*, New York: Rowman & Littlefield.

Eagleton, T. (1983) *Literary Theory: An Introduction*, Minneapolis: University of Minnesota Press.

Foucault, M. (1970) *The Order of Things*, New York: Random House.

Gadamer, H.G. (1986) *Truth and Method*, New York: Crossroads.

Geertz, C. (1983) *Local Knowledge: Further Essays on Intepretive Anthropology*, New York: Basic Books.

—— (1988) *Works and Lives*, Stanford: Stanford University Press.

Gregory, D. (1989) 'Areal differentiation and postmodern human geography', in D. Gregory and R. Walford (eds) *New Horizons in Human Geography*, London: Macmillan, 67–96.

Gregory, D. and Ley, D. (eds) (1988) 'Culture's geographies', *Environment and Planning D: Society and Space* 6(2): 115–227.

Harding, S. (1986) *The Science Question in Feminism*, Ithaca: Cornell University Press.

Harley, B. (1988) 'Maps, knowledge and power', in D. Cosgrove and S. Daniels (eds), *The Iconography of Landscape*, Cambridge: Cambridge University Press, 277–312.

—— (1992) 'Deconstructing the map', in T. Barnes and J. Duncan (eds) *Writing Worlds: Discourse, Text and Metaphor in the Representation of Landscape*, London: Routledge, 229–45.

Iser, W. (1989) *Prospecting: From Reader Response to Literary Anthropology*, Baltimore: Johns Hopkins University Press.

Jackson, P. (ed.) (1987) *Race and Racism*, London: Allen & Unwin.

—— (1989) *Maps of Meaning: An Introduction to Cultural Geography*, London: Unwin Hyman.

Jameson, F. (1972) *The Prison-House of Language: A Critical Account of Structuralism and Russian Formalism*, Princeton: Princeton University Press.

Jay, M. (1986) 'In the empire of the gaze: Foucault and the denigration of vision in twentieth century French thought', in D.C. Hoy (ed.) *Foucault: A Critical Reader*, New York: Blackwell, 175–204.

Kahn, J. (1989) 'Culture: demise or resurrection?', *Critique of Anthropology* 9(2): 5–25.

Kaplan, C. (1990) 'Reconfigurations of geography and historical narrative: a review essay', *Public Culture* 3: 25–32.

Kern, C. (1983) *The Culture of Time and Space 1880–1918*, Cambridge, Mass.: Harvard University Press.

King, A. (1976) *Colonial Urban Development*, London: Routledge & Kegan Paul.

LaCapra, D. (1983) 'Rethinking intellectual history and reading texts', in D. LaCapra *Rethinking Intellectual History: Texts, Contexts and Language*, Ithaca: Cornell University Press, 23–71.

Lee, D. (1950) 'Lineal and nonlineal codifications of reality', *Psychosomatic Medicine* 12: 89–97.

Ley, D. (1974) *The Black Inner City as Frontier Outpost: Images and Behavior of a Philadelphian Neighborhood*, Monograph Series No. 7, Washington, DC: Association of American Geographers.

—— (1987) 'Styles of the times: liberal and neoconservative landscapes in inner Vancouver, 1968–1986', *Journal of Historical Geography* 13: 40–56.

—— (1989) 'Modernism, postmodernism and the struggle for place', in J. Agnew and J. Duncan (eds) *The Power of Place*, London: Unwin Hyman, 44–65.

Ley, D. and Samuels, M. (eds) (1978) *Humanistic Geography*, Chicago: Maaroufa.

Lowenthal, D. (1961) 'Geography, experience and imagination: toward a geographical epistemology', *Annals of the Association of American Geographers* 51: 241–60.

McRobbie, A. (1991) 'New times in cultural studies', *New Formations* 13: 1–17.

Mannheim, K. (1952) *Essays on the Sociology of Knowledge*, London: Oxford University Press.

Marcus, G. (1989) 'Imagining the whole: ethnography's contemporary efforts to situate itself', *Critique of Anthropology* 9(3): 7–30.

Marcus, G. and Fischer, M.M.J. (1986) *Anthropology as Cultural Critique: An Experimental Moment in the Human Sciences*, Chicago: University of Chicago Press.

Mascia-Lees, F., Sharpe, P. and Cohen, C. (1989) 'The postmodernist turn in anthropology: cautions from a feminist perspective', *Signs: Journal of Women in Culture and Society* 11: 7–33.

Massey, D. (1991) 'Flexible sexism', *Environment and Planning D: Society and Space* 9: 31–57.

Mikesell, M. (1978) 'Tradition and innovation in cultural geography', *Annals of the Association of American Geographers* 68: 1–16.

Mitchell, W.J.T. (1986) *Iconology: Image, Text, Ideology*, Chicago: University of Chicago Press.

Olsson, G. (1978) 'Of ambiguity or far cries from a memorializing mamafesta', in D. Ley and M. Samuels (eds) *Humanistic Geography*, Chicago: Maaroufa, 109–20.

—— (1980) *Birds in Egg: Eggs in Bird*, London: Pion.

—— (1992) 'Lines of power', in T. Barnes and J. Duncan (eds), *Writing Worlds: Discourse, Text and Metaphor in the Representation of Landscape*, London: Routledge, 86–96.

Orlove, B. (1991) 'Mapping reeds and reading maps: the politics of representation in Lake Titicaca', *American Ethnologist* 18: 3–38.

Pateman, C. (1983) 'Feminist critiques of the public/private dichotomy', in S. Benn and G. Gauss (eds) *Public and Private in Social Life*, Beckenham: Croom Helm, 281–303.

Pickles, J. (1992) 'Texts, hermeneutics and propaganda maps', in T. Barnes and J. Duncan (eds) *Writing Worlds: Discourse, Text and Metaphor in the Representation of Landscape*, London: Routledge, 191–228.

Poster, M. (1982) 'The future according to Foucault: the *Archaeology of Knowledge* and intellectual history', in D. LaCapra and S. Kaplan (eds) *Modern European Intellectual History: Reappraisals*, Ithaca: Cornell University Press, 137–52.

Pred, A. (1990) *Lost Words and Lost Worlds: Modernity and the Language of Everyday Life in Late Nineteenth-Century Stockholm*, Cambridge: Cambridge University Press.

Rabinow, P. (1986) 'Representations are social facts: modernity and postmodernity in anthropology', in J. Clifford and G. Marcus (eds) *Writing Culture: The Poetics and Politics of Ethnography*, Berkeley: University of California Press, 234–61.

Rooney, J.F., Zelinksy, W. and Louder, D.R. (eds) (1982) *This Remarkable Continent: An Atlas of United States and Canadian Society and Cultures*, College Station: Texas A & M University.

Rose, G. (1990) 'The struggle for political democracy: emancipation, gender, and geography', *Environment and Planning D: Society and Space* 8: 395–408.

Said, E. (1979) *Orientalism*, New York: Vintage.

—— (1989) 'Representing the colonized: anthropology's interlocutors', *Critical Inquiry* 15: 205–25.

—— (1990) 'Narrative, geography and interpretation', *New Left Review* 180: 81–97.

Smith, S. (1988) 'Political interpretations of "racial segregation" in Britain', *Environment and Planning D: Society and Space* 6: 423–44.

Smith, S. and Thrift, N. (eds) (1990) 'Oppressions and entitlements', *Environment and Planning D: Society and Space* 8: 375–458.

Tyler, S. (1986) 'Post-modern ethnography: from document of the occult to occult document', in J. Clifford and G. Marcus (eds) *Writing Culture: The Poetics and Politics of Ethnography*, Berkeley: University of California Press, 122–40.

Western, J. (1978) 'Knowing one's place: the "Coloured people" and the Group Areas Act in Cape Town', in D. Ley and M. Samuels (eds) *Humanistic Geography*, Chicago: Maaroufa, 297–318.

White, H. (1978) *Tropics of Discourse: Essays in Cultural Criticism*, Baltimore: Johns Hopkins University Press.

Williams, R. (1973) *The Country and the City*, New York: Oxford University Press.

—— (1977) *Marxism and Literature*, New York: Oxford University Press.

Part I

ON REPRESENTATION IN CULTURAL GEOGRAPHY

2

AUTHOR AND AUTHORITY
Writing the new cultural geography
Denis Cosgrove and Mona Domosh

It is simply the wrong approach to package the historical mind into the methods of science. To do so is to squeeze into a fixed shape what is not an explicit methodology.

> (Cole Harris, 'The historical mind and the practice of geography', 1978: 131–2)

True, a keen sense of problem ... is a hard won skill (or art). But it is an absolutely necessary one, because without the ability to see how a usually small, manageable research question emerges from ... a set of theoretical concerns, there is no progress. Certainly there is no theory building, no accumulation of a coherent body of knowledge.

> (Susan Hanson, 'Soaring', 1988: 5)

We will make a revolution ... against history. ... History is the intoxicant, the creation and possession of the Devil ... the greatest of lies – progress, science, rights – against which the Imam has set his face. History is the deviation from the Path, knowledge is an illusion, because the sum of knowledge was complete on the day Allah finished his revelation to Mahound.

> (Salman Rushdie, *The Satanic Verses*, 1988: 210)

The first two of these statements concern 'packaging': the ways in which the understanding we develop of the world in the practice of our geography is articulated, both to ourselves and to others. Both therefore raise questions about representation and thus about communication. For most geographers today the issue of communication is one of writing, since writing has become the single most important form of communicating geographical understanding. This of course was not always so. When geography was dominated by the search for regional identity and chorology the map was often regarded as the central embodiment of its findings. Questions of representation and communication are dependent upon prior questions of ontology (what constitutes reality), epistemology (how we come to know that reality) and science (the formal construction of such knowledge). Implicit in the two statements quoted are significantly different responses to these questions and thus to the issues of representation and writing in geography. Our intention here is not to adjudicate between the responses, but rather to unpack the assumptions upon which they rest in the light of what today passes for a 'new' cultural geography.

If we take Hanson's statement to be indicative of mainstream views of the practice of geography, and she writes as the retiring editor of the most widely circulated geographical journal in the world, then it would appear that Harris's observation of fifteen years ago has been widely ignored. By the 'historical mind' Harris was referring to a habit of thinking, a way of creating a rational account of the world through contextual and synthetic explanations. The historical mind relies upon a degree of openness to evidence and interpretation, and is always subject to re-evaluation in the light of new evidence or changes in perspective. Above all it eschews strict methodology, regarding the production of knowledge as a dialogue between an active subject in the form of the individuality of the scholar and an external reality, but a reality not necessarily possessed of a prior order or pattern. While Harris did not directly confront the issue of how the scholar represents the knowledge from this dialogue, his celebration of Fernand Braudel's historical narrative of the sixteenth-century Mediterranean world indicates his enthusiasm for carefully crafted, imaginative writing, akin to convincing story-telling. Harris's position lies firmly in the mainstream of Western rationalism as it has been articulated in the humanities since the Renaissance.

The set of assumptions implicated in Hanson's statement by contrast reveals a perspective on the nature of geographical enquiry which could indeed be said to 'squeeze' it into a fixed shape, a shape whose outline is drawn by a specific model of 'science'. Ontologically the world is assumed not only to exist independently of its observer and to have empirical warranty in the same forms as Harris would recognize, but also to be possessed of an inherent order or logic. The scholar, or 'scientist' in this formulation, applies a set of tools: theoretical laws, carefully formulated hypotheses, methodological devices, to analyse this reality in ways which reveal its inner order and logic. While considerable emphasis is placed upon the skill of the individual scientist in choosing and applying the appropriate tools, the scholar as active participant in the knowing process is here removed from its central position awarded by Harris. Imagination and creativity are accorded a less important place than definition of the research question and skill in manipulating the methodological tools of formal science and in positioning the analysis within a broader frame of scientific advance. The contribution of the scientist is located within a progressive construction which leads us from ignorance to enlightenment. A clear distinction is established between the production and communication of knowledge, a distinction articulated by the phrase 'writing up' research so familiar to us from graduate thesis production. Having established scientific truth in methodologically acceptable ways, the results are to be communicated to a broader world in the most precise and objective language, preferably that of mathematics or statistics. Under this prescription the form of scholarly writing itself becomes squeezed: literature review, hypothesis, data collection and analysis, empirical results and theoretical implications determine a pattern of representation that reflects and reproduces 'scientific' ontology and epistemology.

For all their differences, Harris and Hanson share a belief in the primacy of intellectual discourse, in reason, in the cumulative progression of knowledge, in the sceptical openness of science and criticism and in a linear history. The third quotation with which we open serves to remind us that these values are not shared by all cultures or by all people. The present fate of its author and his work bears strongly on our theme. Rushdie's is a typically modern work in its doubt and its resistance to the concept of revealed truth, and also postmodern in its treatment of fundamental human ideas in a deconstructionist way. Multiple narratives, shifts in temporal chronology, a resistance to any form of revealed truth or fixed moral order and above all its amused subversion of any idea of authority in words all testify to an attitude to the world which separates it from the positions of both Hanson and Harris. At the same time the reaction to *The Satanic Verses* within parts of Islam springs from a cultural attitude to words and meaning equally removed from any which Harris or Hanson would recognize. Words and text do not, in this Islamic prescription, represent and define meaning, they signify because they are. Who speaks the word speaks Allah. Both Harris and Hanson lie somewhere between this position and that of radical cultural relativism, a complete rejection of authority, which appears to be the direction which some postmodern thinking is taking. In so far as the cultural turn within geography is part of the postmodern sensibility in refusing to privilege one cultural discourse over another, to see each as an equally meaningful representation, it finds itself equally removed from Harris's and Hanson's positions. Issues of authority in making and representing meaning in space and human landscapes are inevitably of central concern to the new cultural geography.

The issues of knowledge and authority raised by these contrasting views go beyond words and meaning alone. Therefore our initial concerns focus on the construction of scientific knowledge and the claim that such knowledge is cumulative and progressive. Deconstructing that claim leads us directly to questions of authorship and authority in representing scientific knowledge, specifically in writing. While we do not question the possibility of constructing and communicating knowledge, we do resist the naïvety of representation as reflection of a separate reality. This forces us to recast the crisis of representation raised within contemporary cultural geography as a crisis of authority and therefore of power. Here, we suggest that the writing of our geographies is a process of creating and inscribing meanings about our places and spaces, and we will explore the implications of such a process within the context of power relations, supporting our argument with constructs developed among feminist thinkers.

PROGRESSIVE AND CUMULATIVE KNOWLEDGE

It has become almost a truism that the scientific revolution of the seventeenth century and its philosophical underpinnings in Cartesian thinking signify the origin of our belief in a cumulative advance of scientific knowledge. In fact the history of Western thought is much more complex and nuanced than such simple interpretations of an epistemological break between a pre-modern and a modern mode of conceptualizing subject–object relations would imply. Many feminist writers would argue that the very process of 'conceptualization' and the privileged place it affords to rational cognition are an indication of an androcentric world-view. We shall deal with this in due course, but for now let us acknowledge that as well as to Cartesianism we are heirs to the beliefs of the eighteenth-century Enlightenment, the self-confidence of Victorian positivism and the early twentieth-century symbiosis of science and technology, as well as, of course, to Romanticism and transcendentalism which over the past century have challenged orthodox constructions of knowledge. Recent writers have packaged the positivist aspects of this diverse inheritance as 'modernism' and sought to identify a critical break into postmodernity during the years between 1960 and 1975. The reasons for this break have been widely discussed and need not detain us. But it has ignited a bonfire of the certainties, yielding a crisis in the production and representation of scientific knowledge which reflects a deeper postmodern crisis of ontology and epistemology.

The deconstruction of a received history of scientific thought is to be found within geography, for example in the re-evaluation of magic and myth in the formulation of past geographical understanding, promoted within the new cultural geography. The interpretation of nature as vital and universally organic, an interpretation widely respected within 'science' until well into the seventeenth century, and much later within 'lay' discourse and among a minority strand of thinkers, stands in opposition to progressivism. Its characteristic models are organic, its logic one of correspondence, sympathy and analogy, and its language one of pictorial symbols: signs, words and numbers. Its revival in popular and even 'scientific' literature in recent years is one dimension of changing cultural assumptions associated with postmodernism. The scientific way of knowing is no longer regarded as a privileged discourse linking us to truth but rather one discourse among many, which constructs both the object of its enquiry and the modes of studying and representing that object.

Postmodern deconstruction of the idea of a progressive historiography of scientific knowledge produces a relativism which rejects all forms of totalizing discourse and denies any possibility of constructing a meta-language for intellectual communication. It stands in radical opposition not only to the assumptions that Hanson's argument rests upon, but also to the pre-modern mentality just outlined. Yet it has similarities to the latter in so far as words are taken to construct rather than reflect the world. But here signs are self-referential. If the world has no inherent logic or order then no criteria of truth exist. We are obliged to accept the magical/mystical interpretation of the world as equally valid or invalid as the scientific, for each is a construction of meaning. Since the creation of meaning is a distinctively human activity, the turn to culture in the construction of geographical knowledge becomes understandable, for culture signifies the characteristically human capacity to shape and share meaning. In geography we trace out the production and communication of cultural meanings in spatial organization, conduct and the landscape. But cultural studies of landscape are no longer regarded as part of a 'coherent body of knowledge', slowly assembling, growing and developing like an architectural structure. Rather they seem disassociated fragments, shards of reflecting glass which at once illuminate, reflect and distort – in sum, re-present – the world of individual and intersubjective experience. It may be that those shards imply a potential single sheet of transparent glass through which a single reality might be visible, but the task of reassembling them is fruitless and we are obliged to deceive as we perceive. Surrealist geography may assume the same intellectual validity as a study which obeys all the rules of traditional scientific method and presentation. We seem obliged to abandon all rules constraining and delimiting the field of scholarly geographical endeavour. Little wonder that editors like Hanson are presented with an acute dilemma. Even Harris would be less than content with the more radical consequences of such an atheoretical stance. Even less wonder that those like the Imam, who believe that the power of the word derives from its role as revelation of a transcendental truth, wish to destroy those who also seek knowledge and progress as illusions but who seek to wrest the power of the word for themselves or, worse, simply play with the word, disdainful of both progress and authority.

THE IDEOLOGICAL DIMENSION OF SCIENCE

This reconstruction of the historiography of modern Western science and the idea of progress have emphasized that the modernist tenets of value neutrality, the uniformity of nature and the experimental method are historical creations of specific time periods, cultures and social formations. We would not go so far as to say that their construction is arbitrary, but rather that discourses of meaning are implicated in struggles for power and dominance between humans. Theories of hegemony suggest that this occurs through a process of naturalizing specific discourses, suppressing others and thus legitimizing uneven distributions of power. The judgements that we make of this process are not based on any concept of empirical truth but rather lie in a realm of moral discourse, a central element in the struggle for meaning. Victory in that struggle reconstitutes the world, our knowledge of it and thus the modes of its representation. Neo-Marxists have narrated this struggle within Western social formations in terms of competing class groups. Cultural geographers have indicated the role that cultural interpretations of space and the concept of landscape have played in this history. Feminism is now forcing awareness of the place that gender plays within the culture and ideology of science. Feminist interpretations add yet greater depth of deconstruction. The objective mode of enquiry may itself be read as a male-gendered activity:

> Science affirms the unique contributions to culture to be made by trans-historical egos that reflect a reality only of abstract entities; by the administrative mode of interacting with nature and other enquirers; by impersonal and universal forms of communication; and by an ethic of elaborating rules for absolute adjudications of competing rights between socially autonomous – that is, value-free – pieces of evidence. These are exactly the social characteristics necessary to become gendered as a man in our society.

> (Harding 1986: 238)

The complete endeavour of science, then, may be read as ideological and male gendered, from the subject-matter that we choose to study, to the structures of enquiry modes, the discourse of verification and the claims of authority in scientific representation. Once again the issue of writing science, or in our case geography, is thrown into crisis, for the very structures of representation are implicated in moral and political discourse.

THE CRISIS OF REPRESENTATION AND THE COMMUNICATION OF MEANING

The new cultural geography is part of the response to the crisis of geographical representation. It has promoted a significant change in the metaphors character-istically employed in geographical explanation. Two features of this shift in metaphor may be noted. The first is the recognition that many traditional geographical and spatial metaphors are heavily gendered. Nature has been seen as characteristically female; housing changes in cities are characterized by 'invasion' and 'succession'; capital 'penetrates' peripheral areas. Indeed the very language of conceptualization in 'hard science' is male gendered. The French feminist Luce Irigaray provocatively points out that truth has been conceptu-alized as analogous to the male organ; it must be singular and unified if it is to stand up to scientific enquiry (Franklin 1985). Female sexuality suggests a different morphology and thus may generate a different mode of representation, one which is potential, ambiguous, enclosing and containing, resolving contra-dictions in the embrace of opposition. It is not without significance that much of the interpretative and humanist work in geography is regarded as 'soft' and that this term is often compared in a pejorative way to 'hard' (read male) physical or quantitative geography.

A second feature of the change introduced by cultural geographers is that metaphors of system and machine have given way to others drawn, significantly, from the realms of culture and the arts: text, theatre, map and painting. Traditional metaphors were taken from the sciences and technology and thus lent themselves to the language of the laboratory report and quantification. The new metaphors, by contrast, are drawn from areas of human activity that are very consciously representational. Thus, in shifting the metaphor from landscape as system to landscape as theatre for example, we become conscious precisely of the metaphor as metaphor. The metaphorical quality of traditional concepts was not acknowledged, rather they were called models and their value was assessed in terms of their approximation to 'truth' or 'reality'. Their efficacy was thus determined by empirical observations which could themselves be expressed in the same neutral language as the metaphor. With the shift to more cultural metaphors we have been forced to abandon that innocence of representation, for we know that our metaphors are themselves drawn from the arena of human meaning creation. We are only too aware that all knowledge is culturally constructed and may as readily be deconstructed.

These two aspects of metaphorical change may not be unrelated, for metaphors do more than serve as heuristic devices that disappear as soon as the new theory they were used to elucidate becomes accepted: they are instrumental to knowledge creation and in fact may become the theory or idea they are intended to explain. Nature itself becomes a system rather than simply being represented as such. And metaphors are not randomly chosen. They reflect the struggle for dominance via social and cultural norms; they actively shape a world-view.

> Men are seen to be more like wolves after the wolf metaphor is used, and wolves seem to be more human. Nature becomes more like a machine in the mechanical philosophy, and actual, concrete machines are seen as if stripped down to their essential qualities of mass in motion.
>
> (Hesse 1966, quoted in Harding 1986: 235)

In becoming conscious of the escape of signs from their referents in this way we are forced to abandon any search for a meta-metaphor.

This lack of a single, unifying metaphor has major consequences for producing and communicating geographical understanding. Mechanistic models, for example those of spatial theory, drew upon widely shared languages of geometry and mathematics and yielded broad agreement of method and reportage. Our present cultural metaphors do not yield such agreement. Not only are there a large number of cultural metaphors but within the arts from which they are drawn there is little left of convention in form or technique. Take the example of landscape as theatre. There are obvious advantages in thinking of the world as a stage on which various players perform roles which are scripted but whose interpretation is the responsibility of the individual actor and a great deal of ad-libbing takes place. The analysis of real places in terms of the spatial structures and interaction of stage set, scenery, choreography, script, actors, performances and audience offers fertile ways of recasting the vexed traditional questions of agency and structure, environment and human conduct. But the metaphor relies upon a very conventional reading of theatre as something which is highly scripted and takes place before an auditorium, contained within the space of representation behind the proscenium arch. As contemporary theatre deconstructs the difference between itself and life so the metaphor becomes simultaneously richer and of less value in representing landscape. Above all its use places strain upon traditional forms of writing and representing our geographical knowledge.

The metaphor of text produces different problems and possibilities. Many of these have been discussed by the Duncans, who point to the consequences of post-structuralist theory for the realist treatment of landscape texts (Duncan and Duncan 1988). According to this perspective the text (landscape) is 'no more than a chain of organized black marks on a page until it is actively produced by a reader' (Eagleton 1983 in Duncan and Duncan 1988: 120). The Duncans regard landscapes as readings of texts on the part of dominant individuals or groups who inscribe those readings into their transformations of the natural world and then naturalize such reading-writings through ideological hegemony. The Duncans, however, do not problematize their own rereading and rewriting of the text in the creation of geography, believing that 'the ideological aspects of landscapes as texts can be unmasked', the bastion against relativism lying in 'the realist recognition that there is an empirical reality to which explanations are accountable' (Duncan and Duncan 1988: 125). But if we accept that the formal qualities of language make their own statement, the problem is rather deeper than such protest allows. The shape and appearance of text relay their own meanings as is made clear in Quoniam's essay on Arizona, one of the most adventurous works in representational terms to emerge in recent cultural geography (Quoniam 1988).

REPRESENTING ARIZONA

Quoniam's approach is to present his notes from a trip through Arizona as reproduced figures placed like conventional illustrations within a written essay. The notes are in fact illegible, except for the occasional stimulus word like 'Coronado', so that the text for the most part is meaningless as language. Its meaning derives from its formal qualities as blocks of inscribed signs which we recognize as written notes and thus assume to have recorded and produced meaning. They therefore store geographical knowledge even if that knowledge is unavailable to us. Their form at times takes on a further signification in, for example, altering alignment on the page so that the text blocks resemble the conventional mode of representing geological strata – a powerful image for the Arizona landscape and one actively suggested by the appearance elsewhere of a geological section of a volcano. The structure of writing thus becomes the structure of language, its meanings layered like those of earth itself.

These lines of manuscript text are also wedded to line drawings, maps and colour sketches in collages which have qualities of formal composition (see Plates 2.1 and 2.2). The work at once suggests the possibility of geographical knowledge and the impossibility of communicating personal geographical

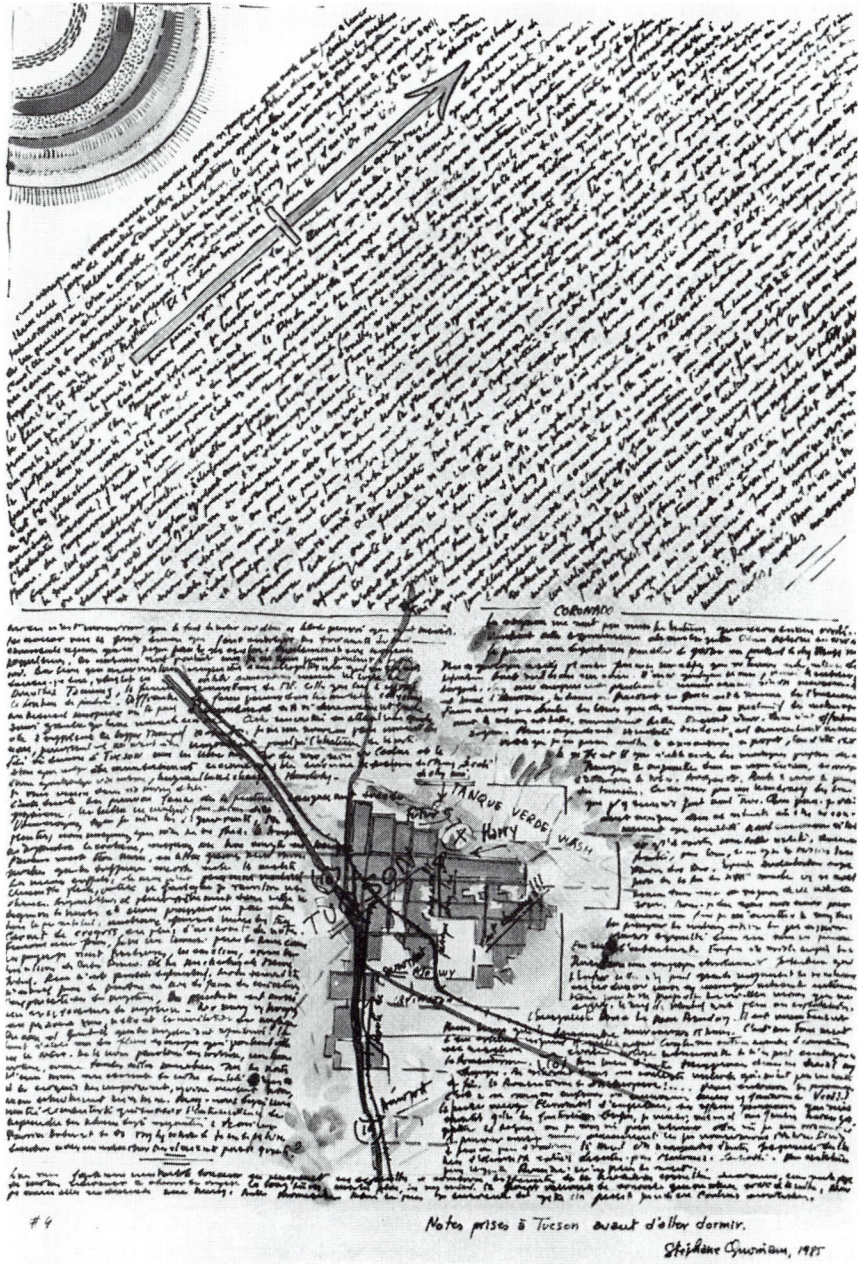

Plate 2.1 Notes prises à Tuscon avant d'aller dormir

Source: Quoniam 1988: 5. Reprinted with permission from Pion Limited.

Plate 2.2 Soirée près du Coronado

Source: Quoniam 1988: 11. Reprinted with permission from Pion Limited.

understanding, indicated by the illegibility of the notes. It also points to the redundancy of traditional textual methods, since much of the printed commentary that accompanies the illustrations is irrelevant.

Quoniam's images directly challenge our notions of language and communication by totally divorcing words from their 'referents'. The words speak for themselves, just as the landscape of the south-west is often said to speak for itself. Quoniam recognizes that the landscape of Arizona can be decoded, read like a text, but that 'it is also the place par excellence where the stones themselves have something to tell' (1988: 4). Similarly his own text can be decoded in the notebook, but in their reproduction those words 'have something to tell' (ibid.: 5). By subverting the roles of picture and text, illustration and word, word and world, Quoniam recasts the patterns of authority in geographical representation. In his art, the text becomes picture, and the picture becomes text: 'painting becomes the text of my ideas and the text becomes the descriptive picture of the landscape'. By revealing the interchangeability of word and picture, he forces us to recognize words as simply another surface. But that surface, like a painting or an illustrated geological section, is not superficial; it contains and creates meaning. Like images, words remake the world. Language is opaque, refracting and perhaps distorting the world it describes, and yet powerful in that its distortions act to create meaning.

The extent to which the meaning of landscape can be recast in words and images is revealed in another work on the cultural geography of the arid south-west. The collection of essays titled *The Desert Is No Lady* examines landscape meaning through various media of art: novels, painting, basketry, pottery, sculpture; all of them produced by women from the three ethnic groups who have humanized the desert (Monk and Norwood 1987). Consistently the essays subvert the dominant male meanings that have been inscribed into this landscape. For example, they present alternatives to the reading of the desert as the stage set for the epic battles of cowboys and villains in the drama of American expansion across the continent. The essays consistently feminize the landscape, reading the naked geology and deep canyons as metaphors of the female body, seeking out and representing characteristics of nurturing domesticity and abandoned sexuality:

The desert is no lady.
She screams at the spring sky
dances with her skirt high,
kicks sand, flings tumbleweeds
digs her nails into all flesh.
Her unveiled lust fascinates the sun.

(Pat Mora, in Monk and Norwood 1987: v)

These recent geographical essays on the south-west remind us that when we write our geographies we are not just representing some reality, we are creating meaning. The problem of representation is only a 'crisis' if we somehow think that we are conveying some independent truth about the world, that we are relaying an authentic representation. Yet we have learned from the rewritings of the history of science that our methods of analysis are morally and socially implicated, and from feminist analysis that the structures of our language are ideologically informed and that the attempt to understand some objective truth is itself part of an androcentric world-view. One of the elements in the deeply constructed and contested battles to create a world that naturalizes male dominance is the cognitive separation of the subjective knower from the objective reality to be known. If we accept that the construction and representation of knowledge are ideologically informed, we must recognize that the problem of representation is in fact a problem of what and who constructs meaning. Debates over interpretations are not about which is the most 'truthful' or 'authentic', but instead are part of a 'social and political struggle for the production of meaning' (Somekawa and Smith 1988: 155). The geographies that we create in our writing enter the realm of moral and political discourse. What we should be asking is not what is the most 'authentic' way to communicate truth, but instead what is the best way to represent and communicate specific and conscious meaning. The crisis becomes one of authority.

How then can we demystify the claims of the objectivity and authority entailed by the act of geographical writing? The types of writing advocated by Harris and Hanson do in fact serve to naturalize and therefore mystify the role of the geographic author in the former and geographic text in the latter. We somehow must let our readers know that what we are creating are themselves cultural, gendered and political products, that our writing is as much about ourselves and our conditions as it is about some purported geographic reality, and that our methodologies and techniques are not ways of establishing ground truth but rather are conventions devised to make our meanings intelligible. We are obliged to share authority with both subject and reader, but equally cannot evade the authority of authorship. Not surprisingly, it is this type of writing, this way of knowing and communicating, that has been identified as 'women's way of knowing', a way of knowing that is contextual and connected, that does not submerge the subjective into the objective, and that questions the notion of single authority and one truth (Belenky *et al.* 1986).

The forms that such writing takes remain an issue. When we problematize the issue of writing, we run the risk of making our writing unintelligible to the people we want to read it. The work of Gunnar Olsson, like that of the French deconstructionists and feminists, testifies to the problem. Is it possible to communicate in text while avoiding the trap of naturalizing and mystifying our role as 'scholars'? Even historical narrative, as Felix Driver has pointed out, has 'been robbed of its old innocence' (Driver 1988: 501). What we seek is a way of writing that is explicit about its political and moral nature, demystifies claims to scientific truth, shares authority with its audience and yet remains understandable.

RATIONAL DISCOURSE AND THE POWER OF LANGUAGE

> It is within moral and political discourses that we should expect to find paradigms of rational discourse, not in scientific discourses claiming to have disavowed morals and politics.
>
> (Harding 1986: 251)

> Where language and naming are power, silence is oppression, is violence.
> (Adrienne Rich 1971; quoted in Belenky *et al.* 1986: 23)

What we have been suggesting in this essay is not necessarily that we need to discover alternative forms of writing, or representing our geographies. A map, a picture, a word, a text – all of them are forms of creation, all are in large measure personal expressions. We write to create sense from the world, but in so doing we authorize that sense at the moment of conveying its importance to our reader. In this way, our writing is like that of any other author, literary or academic, except that we emphasize place and space in our stories. This is not to say that our writing becomes irrational. By acknowledging that our story is related to and constitutive of our social experience, by admitting that our story is part of moral and political discourse, we become more fully rational. We make sense out of the world in the only 'rational' way: from our own experiences. Such an admission makes evident the power of our words, opens the creativity of meaning and honestly faces the assumption of authority. As Adrienne Rich realizes, not to participate in this discourse is to decline power, to court oppression. And given the social construction of the language of science and the academy, the voiceless are the oppressed, the powerless. Thus the constrictive language implied by Hanson's statement is not only irrational in its failure to recognize its internal connections of social and cognitive structure, but potentially oppressive in its silencing of alternative voices.

The problems of writing in a post-Kuhnian world are not that it leaves us without any 'objective' reality from which to base our work, but that it forces us to explicitly recognize our personal and cultural agendas, and the power that words give to those agendas. We must recognize our commitments as authors, and treat them seriously. The postmodern world leaves us nothing behind which we can hide. When we write our geographies, we are creating artefacts that impose meaning on the world. The moral claims implicit in our descriptions and explanations of landscapes and places are what have determined their choice as subject-matter, controlled the mode of study, produced the story we tell and structured the mode of its telling. Our stories add to a growing list of other stories, not listed in a logic of linearity to fit into a coherent body of knowledge, but as a series of cultural constructions, each representing a particular view of the world, to be consulted together to help us make sense of ourselves and our relation to the landscapes and places we inhabit and think about. These stories are to be read not as approximations to a reality, but as tales of how we have understood the world; to be judged not according to a theory of correspondence, but in terms of their internal consistency and their value as moral and political discourse.

REFERENCES

Belenky, M.F., Clinchy, B.M., Goldberger, N.R. and Tarrle, J.M. (1986) *Women's Way of Knowing: The Development of Self, Voice and Mind*, New York: Basic Books.

Driver, F. (1988) 'Moral geographies: social science and the urban environment in mid-nineteenth century England', *Transactions of the Institute of British Geographers* 13: 275–87.

Duncan, J. and Duncan, N. (1988) '(Re)reading the landscape', *Environment and Planning D: Society and Space* 6: 117–26.

Eagleton, T. (1983) *Literary Theory: An Introduction*, Minneapolis: University of Minnesota Press.

Franklin, S. (1985) *Luce Irigaray and the Feminist Critique of Language*. Women's Studies Occasional Papers No. 6, Canterbury: University of Kent.

Hanson, S. (1988) 'Soaring', *The Professional Geographer* 40: 4–7.

Harding, S. (1986) *The Science Question in Feminism*, Ithaca: Cornell University Press.

Harris, R.C. (1978) 'The historical mind and the practice of geography', in D. Ley and M. Samuels (eds) *Humanistic Geography: Prospects and Problems*, Chicago: Maaroufa, 123–37.

Hesse, M. (1966) *Models and Analogies in Science*, Notre Dame: University of Notre Dame Press.

Monk, J. and Norwood, V. (eds) (1987) *The Desert Is No Lady: Southwestern Landscapes in Women's Writing and Art*, New Haven: Yale University Press.

Quoniam, S. (1988) 'A painter, geographer of Arizona', *Environment and Planning D: Society and Space* 6: 3–14.

Rich, A. (1977) 'Conditions for work: the common world of women', in S. Ruddick and P. Daniels (eds) *Working It Out*, New York: Pantheon, xiv–xxiv.

Rushdie, S. (1988) *The Satanic Verses*, New York: Viking.

Somekawa, E. and Smith, E. (1988) 'Theorizing the writing of history or "I can't think why it should be so dull, for a great deal of it must be invention"', *Journal of Social History* 22: 149–62.

3

SITES OF REPRESENTATION
Place, time and the discourse of the Other
James Duncan

INTRODUCTION

The title of this essay 'Sites of representation' is intentionally ambiguous. It is intended to suggest to the reader both the site to be represented (a geographical place), and the site (the geographical, cultural, political, theoretical viewpoint) from which that representation emanates. The representation of places and regions, I will argue, necessarily partakes of this dualism. In this I part company with empiricists who deny the theory-ladenness of representations and urge us to simply open our eyes and see an unmediated world that yields its secrets to our gaze.

My object in this essay, then, is to address a number of issues that circulate around the problem of representing places. The first is to situate the representation of places within the more general context of the 'crisis' of representation as posited by literary theorists, art historians and ethnographers. The second is to argue that one way to investigate the duality of the representation of places is by situating the sites within an analysis of discourses about the Other. Such an analysis highlights the duality by showing how difference in the site of the Other is 'recuperated' by appropriating it into a categorical framework that is familiar and useful within the site from which the representation emanates. To claim simply that discourses of the Other 'distort' the nature of other places and peoples by representing them in ways that are alien to the residents of such places, while justified, misses the inescapability of discourses. Any discourse regardless of its claims, cannot create mimesis (reveal the naked truth); rather, through its ideological distortions, it operates in the service of power. By analysing these relations of power, we can more clearly see how interests play a constitutive role in vision and representation.

Although the majority of studies of the Other have been conducted by scholars in comparative literary theory (Said 1979), anthropology (Clifford 1987; McGrane 1989), art history (Hiller 1991) and history (Baudet 1988; Brantlinger 1986), there exists a small but growing literature in geography on the topic. For example, there have been a number of studies of white representations of minority groups in North America such as Ley's (1974) study of representations of the black ghetto, Anderson's (1988) study of Chinatown or Kariya's (this volume) examination of native peoples in Canada. Geographers have also explored such diverse topics as tourist imagery of the Other (Duncan 1978) and American images of the Soviet Union (Dalby 1990).

In this essay I shall focus upon two of the most important sets of rhetorical devices or tropes that discourses of the Other tend to employ in representing places. The first includes the tropes of mimesis which persuasively claim to represent accurately and objectively the nature of a place. I explore the way the tropes of physical presence and expertise are used to achieve this. The second set of tropes is comparative; these tropes attempt to assimilate the site being represented to the site from which the representation emanates by arraying both sites along a temporal continuum. This second set of tropes, I will argue, contains the key to recuperation since it purports to convert geographical difference into temporal difference. These are the most fascinating tropes of all as they reverse the terms of time and space in which we normally locate ourselves in the world. This real world of geographical difference and temporal co-presence is transformed through these tropes into a world in which different places share fundamental characteristics and potentialities but are separated by occupying different temporal locations. A journey in space is a journey in time; 'a foreign country is the past', to reverse the title of a recent book by David Lowenthal (1985).

The second half of this chapter will show how the discourse of the Other emanating from Europe has been used to represent a particular place and time – Africa in the nineteenth century. This example serves to demonstrate both the historical depth and adaptability of the discourse of the Other, how it has served the authorial sites (in Europe), even as the relations between these sites in Europe and Africa have been transformed over the centuries. I explore how this discourse with its attendant tropes of presence and time travel has appropriated Africa for European imperialism.

SITES OF REPRESENTATION

The sites that are of particular concern in this section of the chapter are those from which geographical and anthropological representation has been most prolific and powerful in its geopolitical and global economic consequences: the cores of imperial power first located in the urban centres of Western Europe and, subsequently, of the United States. We will confront this haughty gaze that has surveyed and appropriated the world since before the age of exploration by tracing its genealogy through various forms of practice – art, science, military conquest and social science. We later explore some of the sites upon which this gaze has been fixed.

The type of representation to which I allude and whose genealogy I trace is objectivism and is characterized by what Husserl termed 'the natural attitude'. It is a type of representation whose origins are associated with the changes in the practice of art in Renaissance Italy and in philosophy as it was revolutionized by Descartes. The invention of linear perspective by Brunelleschi and Alberti in the fifteenth century was a key taxonomic moment in the history of representation. The mathematization of space which underpinned this type of representation appeared to promise a systematic means of producing a mimetic reproduction of the material world. So convincing a representation of the world did this artificial perspective produce that its proponents claimed that it was a 'natural' form of representation, a 'transparent window' as Alberti called it. Perspectival painting became one of the first Western projects which, in the name of science, reason and objectivity, denied its artifice and claimed to offer a mimetic reproduction of the world of experience. It became what Jay referred to as 'an allegedly disincarnated, absolute eye'.[1] As we shall see, such claims rapidly spread from the world of art to other knowledge systems within Europe and became a new 'scopic regime' (Jay 1988).

The end of the fifteenth century marked the beginning of the great age of European exploration and imperial expansion. This period from the beginning of the sixteenth through to the end of the eighteenth centuries was a time of decline in the authority of the Aristotelian world system and Thomist thought. An increasing primacy was given to experience, observation and painstaking description; older hermeneutic readings of the world as a divine text declined in the face of this new empiricism (Eliade 1959; Glacken 1967; Jay 1988; Smith 1987). A tradition of objectivism was developed with the use of perspective instruments, camera lucida and other artistic devices for the faithful replication of nature.

The discoveries of the European explorers called into question older classificatory systems. At first explorers believed that new environments could be described by direct analogy with the old. Things that looked alike were described as if they were identical. Pumas *were* lions and jaguars *were* tigers (Pagden 1982: 11). Finally, the strain of these environmental differences became too great on the older taxonomic system as Europeans confronted the fact that they had an inadequate descriptive vocabulary. Faced with this failure of language, natural scientists who accompanied explorers began to rely more heavily on what was seen as a more reliable form of representation – vision. Using both the new techniques of perspectival art and the collection of fauna and flora for museums and curio cabinets, pictures and objects were made co-partners with language in knowledge (see Pagden 1982: 11–12).

Eventually a descriptive language was appended to this visual array, but, it was thought, mimesis was best captured visually, allowing the person in Europe actually to see the object or landscape as the naturalist in the new world saw it. It was believed that the language that came closest to replicating the mimetic qualities of realist art was a dispassionate, objective language of science that could both describe the visual qualities of objects and classify them in relation to other objects. This project brought together the representational achievements of Renaissance art and Enlightenment language. The focus upon vision and the discourse of mimesis marginalized other modes of representation.

It could be argued that the stimulus for these taxonomic reforms was, in part, the acquisitiveness of many Europeans: an excessive desire to possess the objects of the world. The basis for the selection of objects was a taxonomy that classified things into hierarchies and which would serve as the basis of systematic collections. Such an urge to collect not only extended to plants, animals and artefacts, but to human beings as well, who along with other objects of the natural world were brought back to Europe as curiosities. Todorov claims that Columbus captured Indians 'in order to complete a kind of naturalist's collection, in which they took their place alongside plants and animals' (Todorov 1984: 130). But to say that such people and things were curiosities is only to tell half the story, for the new taxonomy also transformed them into witnesses to European military and intellectual control over the world. Cultural description can be understood as part of this process of collecting the world and assimilating it to European taxonomic systems. The eighteenth century witnessed the increasing elaboration of taxonomic systems within which were progressively incorporated new environments and cultures.

During the nineteenth century there developed a particularly malign variant of the taxonomic system: evolutionary or developmental classification. We shall return to a consideration of this politically powerful form of mastery below, with reference both to our general discussion of time and its use in defining difference and to the representation of Africa in the nineteenth century. Suffice it to say here that cultures around the world were classified into a temporal (and moral) hierarchy ranging from primitive to modern. Definitional power, of course, rested with European taxonomists who by the twentieth century had redefined certain cultures as 'dying', their religious objects becoming consequently 'cultural artefacts' to be 'salvaged' by appropriating them for collections in Western ethnological museums, or 'primitive art' to disappear into museums of art or into the hands of wealthy, private collectors (see Clifford 1987, 1988b).

The descriptions of nineteenth-century travellers, colonial administrators and missionaries claimed to be accurate representations of other places because the writers had 'first-hand experience' there; they had seen these places and peoples with their own eyes (see Pratt 1986a). Sketches usually supplemented their narratives to 'prove' the veracity of the accounts. Twentieth-century ethnographers have also believed in the primacy of personal presence, field*work* as it came to be called in order to professionalize it and thereby elevate its products above the representations of amateurs such as colonial administrators and travellers. The professionals' representations were to be even more highly trusted than those of amateurs because the principles of scientific enquiry further validated the claims of personal experience (Pratt 1986a). The professional ethnographers' representations more closely approximated the rational categorization procedures of Enlightenment discourse and therefore, it was argued, by definition were even more likely to be mimetic. But the ethnographer while claiming to practise objective science in which the site of the observer is denied was also unwilling to give up the amateur's claim to personal presence. This tension between rhetorical modes perhaps accounts for the odd mixture in twentieth-century ethnography of the use of the first person in one part of the text (usually the beginning in order to establish personal presence) and the use of impersonal, scientific language elsewhere (see Clifford 1988c: 26, 30; Pratt 1986a).

Photography has also played a large role in twentieth-century ethnological representation. What better way to assert the primacy of the visual, produce a 'true' representation of the place in question and establish presence than through the use of photography? But the mimetic claims of photography can also be called into question. (See the essays in Burgin 1982; and Tagg 1988). A camera is a machine *constructed to produce* an image based upon artificial perspective. Only if one accepts the claims of the naturalness of Renaissance artificial perspective can we accept photography as a mimetic representation of the world (see Mitchell 1986: 37). Such claims can be cast in doubt, for example, by the failure of peoples unfamiliar with photographs to be able to 'read' them (Goodman 1976: 15).

The twentieth-century objectivist ethnography still largely depends upon the rhetorical devices we have just described. It is claimed that these devices (linguistic and photographic) allow peoples being studied 'to speak for themselves' and show themselves as they truly are. This has increasingly been questioned by critics who see such claims in the context of a literary institution servicing larger positivist goals which have fallen into disrepute (Marcus and Cushman 1982: 29–30, 34). It might also be added that the rather easy acknowledgement of the impossibility of perfect replication on a philosophical level may not substantially alter the natural attitude which thoroughly permeates our common-sense approach to the world. The naturalizing tropes of representation that privilege vision, expertise, first-hand experience and descriptive accuracy through careful reporting are so deeply entrenched in our culture that we often fail to question much that is presented as self-evidently true.

TIME AND THE DISCOURSE OF THE OTHER

While in the previous section I discussed in rather abstract terms the sites of representation, I will now demonstrate how these sites of representation have been filled out: socialized, temporalized and, one would have to add, fantasized. For in order to understand how representational sites were articulated, we must understand how they were 'colonized' socially and temporally as sites of desire, power and weakness. The major tool in this 'colonizing' process is the discourse of the Other.

In Europe this discourse can be traced back to classical Greece. Whereas the specific content of this discourse (who was being represented, by whom and what qualities were being imputed) varied over the millennia, the form and general purpose of the discourse remained similar. Difference, rather than being acknowledged as relational, is seen from an unacknowledged site. From the perspective of realism the site from which the representation emanates is occulted, but nevertheless remains as a phantom space, denied but present. It is not seen in its own historical and cultural specificity. Harbsmeier (1985: 273), in his comparative study of xenology, concludes that probably every culture has its Other, its 'own barbarians, heathens, unbelievers, savages, primitives or whatever specific "counter-concepts"'. Such binary oppositions between us and them serve the dual purpose of reinforcing and defining group identity while simultaneously ordering complex difference into a simpler, homogeneous entity which is more easily appropriated. The Other that I wish to examine here is first Europe's and then to a lesser extent America's.

One of the earliest recorded instances of reference to the Other is the Greek term *barbaros*. From the seventh to the sixth centuries BC Greeks used the term to designate foreigners, including the respected Egyptians. But by the fourth century BC Greeks used the term in a more limited way to distinguish themselves from peoples whom they considered to be their inferiors. The Hellenistic Greeks employed the term *barbaros* (babblers, speakers of strange tongues) to convey the idea of subhumanity. As the Greeks linked intelligible speech to Reason, the term designated the non-Greek-speaking Other as irrational and thus less than fully human (Pagden 1982: 15–18). The term 'barbarian' (*barbarus* in Latin) has continued to be used in the discourse of Otherness up to the present, retaining some of the connotations, albeit in modified forms, with which the Hellenistic Greeks endowed it.

However, as Baudet (1988) has demonstrated, the discourse of the Other by Europeans over the centuries is not uniformly negative, in its imagery at least, if not in its consequences. Rather it is fraught with ambivalence towards other places and peoples in the world. This is not entirely surprising as the concept of the Other, being relational, is dependent upon two sites. Europeans' ambivalence about themselves, therefore, is necessarily projected into their attitude towards the Other. With the rise of Christianity in Europe this ambivalence took the form of two spatio-temporal representations.

The first was of an image of a pre-cultural, primitive age before the Fall. European culture was seen as a degenerate version of this pure state of grace. As long as Latin Christianity was sealed from the rest of the world, images of paradise or primitive innocence were temporalized by being projected into the past (Baudet 1988: 10–13). However, beginning with the Crusades in the twelfth century and extending through several centuries, Europeans spatialized these representations by projecting them on to the world. Paradise became a geographical site which explorers from Marco Polo to Columbus searched for in vain. From the sixteenth to the eighteenth centuries Europeans projected ideals of a lost, primitive state of happiness upon the worlds that they discovered. Such longing did not, however, stop Europeans from pillaging those other worlds. This can in part be explained by the differences between the representations of European intellectuals and the more pragmatic views of the soldiers and merchants who actually encountered these new worlds.

The destruction was supported intellectually by another European representation of the Other that competed with that of the 'noble savage'. This was the Christian transformation of the Greek notion of barbarian. From the end of the twelfth to the beginning of the sixteenth century the term 'barbarian' referred to people who both were non-Christian and behaved in 'uncivilized ways'. To all intents and purposes this meant non-Europeans. This discourse is exemplified in a statement made by the Bishop of Santa Marta (Colombia) in support of the enslavement of Indians. They were, he said, 'not men with rational souls but wild men of the woods, for which reason they could retain no Christian doctrine, nor virtue nor any kind of learning' (Pagden 1982: 20–4).

During the eighteenth century the discourse of the Other continued to contain contestatory strands. On the one hand there was the non-Christian Other of the Renaissance which became transformed into the unenlightened Other (McGrane 1989: 52). We can see here the beginnings of twentieth-century Euro-American representations of 'Third World' Others as ignorant, traditional or less rational than themselves. Such representations helped provide in the eighteenth and nineteenth centuries, as they have in the twentieth, the ideological superstructure for the spreading European exploitation of the world. On the other hand there continued to be a deeply felt affinity for what was seen by many European intellectuals such as Rousseau as the 'primitive'. Certain influential intellectuals, yearning to free themselves from 'the burden of culture', longed for the 'uncivilized' and judged Europe to be 'spiritually, inferior to the non-European world' (Baudet 1988: 36–8). The Other was thus a therapeutic image for the European (Tyler 1986: 127).

Foreignist sentiment took an aesthetic turn and the houses of the rich were filled with 'cultural artefacts' from around the world (a subject to which we shall return when we discuss the representation of the Other in the twentieth century). As Baudet (1988: 49) points out, however, eighteenth-century representations of the non-European world carelessly jumbled together Turks, Hurons, Hindus and South Sea Islanders. One could argue that, while the motivation for these collections of exotica undoubtedly can be understood in aesthetic terms, it also served to contrast with Europe's idea of itself. The world then was doubly appropriated by Europeans. After first ransacking its material resources, the Europeans' desire to consume still remained unfulfilled, and so they attempted to appropriate the world's myriad identities and fill the ensuing void with the myths of Europe's Other. This type of appropriation included not only the collecting of oriental art objects, but also the capturing of picturesque scenes of a timeless Orient or a mysterious, erotic and brutal Orient by the nineteenth-century school of European 'realist' artists called Orientalists. Said (1979: 40) refers to such Orientalism 'as a mode for defining the presumed cultural inferiority of the Islamic Orient ... part of the vast control mechanism of colonialism, designed to justify and perpetuate European dominance'.

As I have already suggested, time was always implicated in the representation of the Other. In the nineteenth century, however, time assumed heightened importance in this representation. As Johannes Fabian (1983: 15) said, it became the century of '*spatialized* time'. The nineteenth century saw what McGrane (1989: 93–4) has termed a 'double transformation'. Difference was increasingly converted into history and history explained in terms of evolution. The prestige of Darwin's theory of evolution and Spencer's theory of social evolution seemed to lend renewed vigour to the age-old propensity to temporalize the Other. The tremendous interest in origins promulgated by Darwin increased the desire of Europeans to represent other people as shedding light on how Europeans' ancestors had once lived (Ranger 1976: 137). '*Beyond* Europe was henceforth *before* Europe' (McGrane 1989: 94). These taxonomic reforms were instituted within Britain during its age of imperial expansion. Accordingly, the new representational taxonomy had implicit political content and thus political consequences. For example, Europe was seen as the highest stage in evolution, while societies in Asia and Africa were thought to occupy progressively lower stages in the process. Evolutionary theory then provided a racist taxonomy which justified imperialism, for 'backward' places could not be expected to govern themselves. Evolutionary theory was put to creative use in the service of imperialism, allowing Europe at times to claim the 'law' of survival of the fittest, and at other times to proclaim its duty to 'protect the weak' by 'looking after their interests' (Jara and Magana 1982: 116). In the words of Fabian (1983: 144), '*Geopolitics* has its ideological foundations in *chronopolitics*.'

The discipline of ethnography arose during this colonial period and played an important role in providing the taxonomic framework of colonial representation. In the late nineteenth century, the non-European Other was seen either as fit only to be a servant in the employ of his or her colonial masters or as a fallen savage, a mere 'living fossil' exotic and endlessly intriguing as such. Thus another strand of this nineteenth-century discourse of the Other was exoticism (Foster 1982: 21–22). This strand, which continues to be significant up to the present, has, since the middle of the last century, fed into a burgeoning travel industry. Other cultures are often portrayed as occupying remote places that are rare or unique and therefore desirable, places where one can escape the social and psychological pressures of modernity and retreat into a 'simpler', more 'natural' place and time. In other words, the Other in the twentieth century has only partially escaped the taxonomic strait-jacket of nineteenth-century evolutionism. I say only partially because one can still discern romantic Manichaean oppositions within political as well as academic discourse.

Twentieth-century cultural anthropology, which claims to distance itself from the evolutionary ranking of cultures, falls victim, albeit in more subtle ways, to the temporalization of space. For such anthropology places the people it studies in what Clifford has called 'the ethnographic present', which is conceived of as a timeless dimension in which tradition blends past and present together as one (1987: 121–2). Such societies and places are seen as always about to be deluged or corrupted by external forces. As Pratt (1986a: 42, 1986b: 139) points out, such a timeless present dehistoricizes people and makes their behaviour appear to be an instance of pre-given custom. Furthermore, ethnography's silence about the larger agenda of European expansion, its forcibly imposed involvement in these cultures and the role which ethnography plays in this process is not merely a strategic political silence, but also part of the self-delusion of ethnographers that they are studying a 'pure', uncorrupted and uncontaminated traditional society.

Dominguez (1987: 132) claims that the concept of Otherness invites forms of representation that are inherently appropriative. Nowhere is this clearer than in the double appropriation involved in the European removal of 'tribal objects' from various sites around the world. Such objects, as Clifford (1988b: 191–2, 212) points out, have also been redefined from their native meaning as religious objects into 'cultural artefacts' when categorized within the ethnographical mode of representation, or 'primitive art' within the artistic mode. Collections of 'primitive art' containing objects from Africa, Oceania and the Arctic are displayed together, thus occupying a taxonomic grid that may have more to do with the availability of funding for Art exhibits than with any other factor. Within the representational frame of art, the similarity of such displays to the collections of the eighteenth century and the *Wunderkamer* of the nineteenth-century bourgeois is striking. But this is not surprising as such displays serve a similar ideological purpose in both cases: to represent Europe's Other as its past.

REPRESENTING A REGION: AFRICA IN THE NINETEENTH CENTURY

Let us now focus upon a specific region and temporal frame. Africa in the nineteenth century serves our purposes admirably, because it was a site which was dramatically and cruelly appropriated into the European taxonomic system. Here we can witness the impact of changing political and economic conditions on how the European discourse of the Other represented the landscape, the people and the region as a whole.

While in the twelfth century Europeans generally expressed great fear and antipathy towards Muslim North Africa, their attitudes towards sub-Saharan Africa were much less well-formulated and more ambivalent. Very little was known about either Africa or Africans. Africa was seen in the European imagination as a mythical land full of fantastic human and non-human creatures. The attitudes predominant among the more educated classes stemmed from two contradictory representations. The first was a belief, whose roots are to be found in classical Greek geography, that Africa south of the Sahara lies within a Torrid Zone of intense heat and that as a consequence life was impossible there. This was verified, it was thought, by the dark skins of the inhabitants of North Africa, burned by the heat of the sun when they had approached too near the Torrid Zone.

In the second representation of Africa, Ethiopia, which had been known to Christian Europe for centuries, was thought of as a possible location of paradise

(Baudet 1988: 15–19). There was a new-found European interest in geography, stemming in part from the Crusades and from the revival of trade between Europe and Asia, and this was accompanied by changing ideas of paradise which shifted in the European imagination from a location in the distant past to a geographical site in the present. Ethiopia seemed a likely location, not only because it was known to be Christian, but also because it lay adjacent to a great south-eastern river system which in Genesis 2:14 was linked to Eden. Although these Europeans did not think that Ethiopians actually lived in paradise, they thought that to be near it was to be in a state of bliss.

This interest in Ethiopia was not merely religious, but intensely political as well, for the legendary, and to many immortal, Christian king Prester John was thought to be the ruler there, and Europeans were interested in joining forces with him to oppose the demonic 'Saracens' who controlled the Holy Land. The legend of Prester John survived well into the fifteenth century, and this European belief in Ethiopia as paradise had a very positive effect on European attitudes towards Africa.

Once European sailors passed the equator without being incinerated, thereby disproving the theory of the Torrid Zone, Africa was gradually opened up for growing European economic expansion. The notion of a Torrid Zone did not completely disappear, however; it simply took the form of a more refined environmental determinism which was to become one of the ideological props of imperialism.

With the import of black slaves into Portugal in the fifteenth century and their transportation to the Americas by the beginning of the sixteenth century, attitudes towards Africa and Africans began to change and harden (Baudet 1988: 29–30). While Africans generally have been perceived in either a neutral or a fairly benign, and sometimes highly respectful, way before the advent of the European slave trade (Hammond and Jablow 1970; JanMohamed 1986: 80), they soon became portrayed in an extremely negative light (see Brantlinger 1986; Curtin 1964; Jordan 1968). As the slave trade grew, Africa became represented as a place so barbarous and dark that the removal of people from its shores was argued to be to the advantage of those removed (Kiernan 1986: 194). One would be hard pressed to find a clearer example of the manner in which changing economic circumstances influenced the representation of a region and its people.

In fact by the eighteenth century the earlier opposition between the view of Muslims versus African blacks had become reversed. Orientalist literature portrayed Europe as indebted to Islamic civilization, while Africans, because of their association with enslavement for nearly three hundred years and the alleged inhospitality of their continent, began to be seen as the lowest form of human life, slaves by nature, savage beasts of burden, who might or might not be capable even of receiving civilization. Such a portrait lent necessary ideological support to European exploitation and colonizing practices.

During the latter part of the eighteenth and the early nineteenth centuries, pressure grew within England for the abolition of the slave trade. As a result, most of the writing on Africa between 1790 and 1840 portrayed Africans in a much more positive light. The upshot of this 'change of heart' on the part of the English was the abolition of slavery in all British territories in 1833. Not content to renounce it themselves, however, the British vowed to force other nations to abandon the trade as well and used the British navy to enforce their ban. Although the debate was framed by the British in moral terms with many genuine supporters, and some scholarly writing as support, England's economic competitors suspected that the English were simply aiming to deprive them of an advantage which they themselves no longer needed (Kiernan 1986: 203).

The abolitionist literature, as Brantlinger (1986: 189) points out, represented Africans as innocent and noble rather than evil, savages who had been doubly victimized, first by their debilitating physical environment and later by unscrupulous European, American and Arab slavers. Curiously, the abolition of slavery by the British led to the representation of Africa as an even more degraded site. For while British naval power could be used to enforce anti-slavery legislation along the West African coast, Arab slavers continued to ply their trade along the east coast and black Africans themselves who had been formerly employed by Europeans continued to deal in slaves in the interior. Abolition allowed the British increasingly to represent themselves as the saviours of Africa, and portray *Africans* as responsible for the continuation of the slave trade. By the end of the American civil war slavery had been confined to Africa itself and increasingly became represented by the British as further evidence of Africa's degradation and savagery (Brantlinger 1986: 192, 194).

The British saw two possible responses to such a state of affairs. The first was to abandon the continent altogether as 'an area of darkness', and the second was for the British to civilize the Africans. The Burton Expedition to West Africa in 1840 is an example of both of these attitudes at work. The expedition began by attempting to set up 'outposts of light' in the interior; but after many members of the expedition succumbed to disease, the remainder concluded that the continent was unfit for Europeans and should be abandoned. Over the next several decades explorers were divided over whether Africans were capable of progressing and being saved from their ignorance and wickedness. Livingstone, for example, thought that Africans could be enlightened, while Burton thought them beyond redemption. By the mid-century, Victorians could not conceive of Africans (whether redeemable or not) living without European rulers (Brantlinger 1986: 189, 192–3).

But such imperialist dreams could not be brought to fruition as long as Europeans were confined to ports along the coast. This limitation to imperialist expansion was reduced by the mid-nineteenth-century explorers who, aided by the discovery of quinine, 'opened up' the continent. Fabian (1986: 24) sums up the role played by these explorers when he writes:

> Colonial expeditions were not just a form of invasion; nor was their purpose just inspection. They were determined efforts at *in-scription*. By

putting regions on a map and native words on a list, explorers laid the first, and deepest, foundations for colonial power. By giving proof of the 'scientific' nature of their enterprise they exercised power in a pure subtle form – as the power to name, to describe, to classify.

One can find in the writings of these nineteenth-century explorers interesting tropes for codifying difference within the discourse of the Other. Explorer-writers such as Burton, Speke, Livingstone, Barrow and Stanley constructed representations of Africa for Europeans, opening it up conceptually for their readers as they opened it up economically for their governments. Such representations, while built upon the bedrock of older representations, simultaneously constructed an Africa suitable both to the needs of nineteenth-century imperial interests and to a European readership longing for tales of exotic worlds being mastered by heroic European males. Such travel accounts could be divided into two genres, the 'heroic' and the informational or scientific.

The journalist H.M. Stanley's expedition in 1871–2 in search of Livingstone was an example of the former. The expedition was designed not only to find Livingstone, but to sell newspapers. Stanley travelled with a huge retinue which because of its size had to plunder villages in order to feed itself. This 'manufactured excitement' made good newsprint (Kiernan 1986: 226). In such accounts, Africans were represented as bloodthirsty savages; the continent an area of darkness. It could be argued that Africans were placed in the text as a foil to the white man (Plate 3.1).

Paradoxically, accounts simultaneously included both images of the white man surrounded by a sea of black savages and images of explorers 'alone in deepest Africa'. The tropes of solitude used to describe the explorations of Europeans who invariably travelled with and among large numbers of Africans only make sense within a discourse which rhetorically makes Africans absent by representing them as less than human. Such rhetoric of absence could only be believable to a European audience that was deeply racist. The encoded message in such writings was that Africa was a land of disorder and savagery and that the European stood as the sole representative of order, the only rational, civilized presence there.

An example of the informational, scientific account was the writings of the English explorer George Barrow. He produced detailed descriptions of African landscapes which minimized the presence of Africans on the land. Africans, when discussed, were 'relocated' in separate chapters dealing with manners and customs, 'textual homelands', to use Pratt's (1986b: 146) provocative phrase. Such descriptions not only partook of the European aesthetic codes favouring panoramic views, but also portrayed a country that was rich in resources, vast and most importantly *empty* (Pratt 1986b: 143–5): a place, in other words, that was open to European imperialism. Some writers even went so far as to suggest that it was the moral duty of Europeans not to let these resources go to waste (Jara and Magana 1982: 120–1). Such a suggestion, of course, was

Plate 3.1 Stanley cutting his way through the 'dark continent'

Source: Buel 1890

premised on the widely held assumption that Africans were incapable of ruling themselves or profiting from Africa's resources. Occasionally writers drew on the discourse of the pure, noble savage, arguing that Africans, being morally superior to the Europeans, lacked the lust for material goods that infected Europeans. Thus, as MacGrane points out, both the alleged moral superiority of Africans as well as their alleged moral inferiority were employed by Europeans to justify their appropriation of African resources such as gold (McGrane 1989: 26).

Using what JanMohamed (1986: 80) has termed a Manichaean allegory, Europeans transformed racial difference into moral and metaphysical difference. Although the representations changed according to political and economic relations, they were strategies that depended upon what Said (1979) has called 'flexible positional superiority'. JanMohamed has referred to this as a 'representational economy' in which

> the writer's task is to 'administer' the relatively scarce resources of the manichean opposition in order to reproduce the native in a potentially infinite variety of images, the apparent diversity of which is determined by the simple machinery of the manichean allegory.
>
> (JanMohamed 1986: 83)

Africa was depicted as a moral abyss that Europeans could possibly fall into if they succumbed to their own repressed animal instincts. This was frightening, seductive and fraught with sexual connotations. Brantlinger (1986: 215) describes Victorian literature and painting of Africa as depicting landscapes again and again 'with the same tarbrush image of pandemonium'. The darkness of Africa, he argues, was seen to have a dangerously seductive effect on the white race. Missionaries were often afraid of being converted to heathenism before converting others to Christianity. 'Going native' was a common desire among the more adventurous travellers and it was thought that whole societies might succumb to this form of degradation. Years spent on African soil were thought to have led to the Boers' degeneration into savages (Brantlinger 1986: 213). Brantlinger quotes one late nineteenth-century author as remarking on 'the rapidity with which such members of the white race as are not of the best class, can throw over the restraints of civilization and develop into savages of unbridled lust and abominable cruelty' (ibid.: 213).

The discourse of race which we have traced through the period of slavery was given renewed vigour by Darwinism after the middle of the nineteenth century. While Darwinism did not create this discourse, it did give it scientific respectability. Africa now could be scientifically transformed from geographical into temporal space. Africans were placed in the evolutionary past by imperialists such as Rhodes in order to legitimize Europeans' right to possess Africa. Rhodes argued that the Anglo-Saxon race was at the top of the human hierarchy and the Africans were at the bottom. Some thought of Africans as a 'child-race' whose development had been permanently arrested at puberty, while to others they were 'but slightly superior to the lower animals' (Kiernan 1986: 233–4). It was argued by some that such mental underdevelopment was caused by the tropical climate which over the millennia had contributed to the unequal achievement of the races (Jara and Magana 1982: 120). The goals of such taxonomic exercises were both 'scientific' and political, or, as is so often the case, science in the service of politics. Essentially it was argued that as white Europeans represented the climax of racial and social evolution, their relations with Africans should be one of masters to servants for this was the only possible relation between peoples occupying such different mental and moral spaces.

The explorer Speke, an English aristocrat by birth, writing in the late 1860s noted:

> the African must soon either step out of the darkness, or be superseded by a being superior to himself [*sic*]. Could a government be formed for them [*sic*] like ours in India, they would be saved; but without it, I fear there is very little chance.
>
> (Kiernan 1986: 215)

As Kiernan (ibid.: 226) goes on to add, Africa became 'a Dark Continent, but its darkness was one the invaders brought with them, the sombre shadow of the white man'. The taxonomy which posited an inferior race living in darkness appropriated within the representational realm what European capitalists were appropriating within the material realm.

During the nineteenth and early twentieth centuries the fledgling disciplines of ethnography and physical anthropology contributed to the project of imperialist representation of Africa through their participation within the discourse of the Other. Such academic support for imperialist representation was continued, as Ranger (1976: 116–28) points out, well into the twentieth century in Africa. The specifics of the discourse on race and evolution had changed into a more general discourse of control through a social scientific knowledge of both the natives and the geography of the country in which they lived. But such 'knowledge', although portrayed as mimetic by those whose practices it supported, is, as we have argued, a representation of one site emanating from another site and thus far more limited in its perspective than its rhetoric suggests.

CONCLUSION

Studies of the construction of taxonomies and discourses of the Other have an important value in the problematizing and politicizing of the representation process; they point to the fact that this process always involves power relations and is mediated through historically changing institutions, class structures, taken-for-granted historical accounts and scientific assumptions. They help one to recognize the limits of an ethnocentric and usually self-congratulatory vantage-point from which scholarly and journalistic accounts arise. While such a recognition cannot allow one totally to escape ethnocentrism, it aids in the denaturalizing of the researcher's own categories and prepares the way for alternative histories. It is not a question of being culturally sensitive; it is much more than that; it requires the modifying or occasionally the radical undermining of the ground upon which one must necessarily stand. It is by no means necessary, however, to give up all representational responsibility when problematizing one's own authority. Because the process of writing history and representing cultural and historical difference is inevitably a political practice, there are responsibilities involved that should be taken very seriously.

It is important that we examine representations such as those which nineteenth-century Europeans held of Africa, *precisely* because they seem so false, so preposterous, so despicable to us today. We must not forget that such representations in their day were seen, by Europeans at least, as scientifically true, persuasive and moral. What from our late twentieth-century site of representation seems a gross distortion of reality, for them appeared as mimetic representation. Only when we seriously explore those representations which we find self-evidently false can we begin to question the representations which we find self-evidently true. Only then will our own sites of representation become visible to us.

NOTES

1   Jay points out that there are contradictory implications in the position or site of the viewer in Cartesian perspectivalism. The most common assumption is that which I have been describing as the universal or transcendental effaced site. Another is far more relativist and contingent upon the individual viewer with his or her own distinct viewpoint. These different implications are the basis for different scopic regimes. (See Jay 1988).

REFERENCES

Anderson, K. (1988) 'Cultural hegemony and the race-definition process in Chinatown, Vancouver: 1880–1980', *Environment and Planning D: Society and Space* 6: 127–49.

Baudet, H. (1988) *Paradise on Earth: Some Thoughts on European Images of Non-European Man*, Middletown, Ct: Wesleyan University Press.

Brantlinger, P. (1986) 'Victorians and Africans: the genealogy of the myth of the dark continent', in H.L. Gates (ed.) *'Race', Writing and Difference*, Chicago: University of Chicago Press, 185–222.

Buel, J.W. (1890) *Heroes of the Dark Continent*, Philadelphia: Historical Publishing.

Burgin, V. (ed.) (1982) *Thinking Photography*, London: Macmillan.

Clifford, J. (1986) 'Introduction: partial truths', in J. Clifford and G. Marcus (eds) *Writing Culture: The Poetics and Politics of Ethnography*, Berkeley: University of California Press, 1–26.

—— (1987) 'Of other peoples: beyond the salvage paradigm', in H. Foster (ed.) *Discussions in Contemporary Culture*, Dia Art Foundation No. 1, Seattle: Bay Press, 121–30.

—— (1988a) 'On collecting art and culture', in J. Clifford *The Predicament of Culture*, Cambridge: Mass.: Harvard University Press, 215–52.

—— (1988b) 'Histories of the tribal and the modern', in J. Clifford *The Predicament of Culture*, Cambridge, Mass.: Harvard University Press, 189–214.

—— (1988c) 'On ethnographic authority', in J. Clifford *The Predicament of Culture*, Cambridge, Mass.: Harvard University Press, 21–54.

Curtin, P. (1964) *The Image of Africa: British Ideas and Action, 1780–1850*, Madison: University of Wisconsin Press.

Dalby, S. (1990) 'American security discourse: the persistence of geopolitics', *Political Geography Quarterly* 9: 171–88.

Dominguez, V. (1987) 'Of other peoples: beyond the salvage paradigm', in H. Foster (ed.) *Discussions in Contemporary Culture*, Dia Art Foundation No. 1, Seattle: Bay Press, 131–7.

Duncan, J.S. (1978) 'The social construction of unreality: an interactionist approach to the tourist's cognition of environment', in D. Ley and M. Samuels (eds) *Humanistic Geography*, Chicago: Maaroufa, 269–82.

Eliade, M.M. (1959) *Cosmos and History: The Myth of Eternal Return*, New York: Pantheon.

Fabian, J. (1983) *Time and the Other: How Anthropology Makes its Object*, New York: Columbia University Press.

—— (1986) *Language and Colonial Power: The Appropriation of Swahili on the Former Belgian Congo 1880–1938*, Cambridge: Cambridge University Press.

Foster, S. (1982) 'The exotic as a symbolic system', *Dialectical Anthropology* 7: 21–30.

Foucault, M. (1972) *The Archaeology of Knowledge*, London: Tavistock.

—— (1980) *Power/Knowledge: Selected Interviews and Other Writings, 1972–1977*, Brighton: Harvester.

Glacken, C. (1967) *Traces on the Rhodian Shore*, Berkeley: University of California Press.

Goodman, N. (1976) *The Languages of Art*, Indianapolis: Hackett.

Hammond, D. and Jablow, A. (1970) *The Africa that Never Was: Four Centuries of British Writing About Africa*, New York: Library of Social Science.

Harbsmeier, M. (1985) 'On travel accounts and cosmological strategies: some models in comparative xenology', *Ethnos* 50: 273–312.

Hiller, S. (1991) *The Myth of Primitivism: Perspectives on Art*, London: Routledge.

JanMohamed, A.R. (1986) 'The economy of manichean allegory: the function of racial difference in colonialist literature', in H.L. Gates (ed.) *'Race,' Writing, and Difference*, Chicago: University of Chicago Press, 78–106.

Jara, F. and Magana, E. (1982) 'Rules of imperialist method', *Dialectical Anthropology* 7: 115–36.

Jay, M. (1988) 'Scopic regimes of modernity', in H. Foster (ed.) *Vision and Visuality: Discussions in Contemporary Culture*, Dia Art Foundation No. 2, Seattle: Bay Press, 3–28.

Jordan, W. (1968) *White over Black: American Attitudes towards the Negro, 1550–1812*, Chapel Hill: University of North Carolina Press.

Kiernan, V.G. (1986) *The Lords of Human Kind: Black Man, Yellow Man, and White Man in an Age of Empire*, New York: Columbia University Press.

Ley, D. (1974) *The Black Inner City as Frontier Outpost: Images and Behavior of a Philadelphia Neighborhood*, Monograph Series No. 7, Washington DC: Association of American Geographers.

Lowenthal, D. (1985) *The Past is a Foreign Country*, Cambridge: Cambridge University Press.

McGrane, B. (1989) *Beyond Anthropology: Society and the Other*, New York: Columbia University Press.

Marcus, G. and Cushman, D. (1982) 'Ethnographies as texts', *Annual Review of Anthropology* 11: 25–69.

Mitchell, W.J.T. (1986) *Iconology: Image, Text, Ideology*, Chicago: University of Chicago Press.

Pagden, A. (1982) *The Fall of Natural Man: The American Indian and the Origins of Comparative Ethnology*, Cambridge: Cambridge University Press.

Pratt, M.L. (1986a) 'Fieldwork in common places', in J. Clifford and G. Marcus (eds) *Writing Culture: The Poetics and Politics of Ethnography*, Berkeley: University of California Press, 27–50.

—— (1986b) 'Scratches on the face of the country; or, what Mr Barrow saw in the land of the Bushmen', in H.L. Gates (ed.) *'Race,' Writing, and Difference*, Chicago: University of Chicago Press, 138–62.

Ranger, T. (1976) 'From humanism to the science of man: colonialism in Africa and the understanding of alien societies', *Transactions of the Royal Historical Society* (new series 26): 115–41.

Said, E. (1979) *Orientalism*, New York: Vintage.

Shapiro, M. (1988) *The Politics of Representation: Writing Practices in Biography, Photography, and Policy Analysis*, Madison: University of Wisconsin Press.

Smith, J.Z. (1987) *To Take Place: Towards a Theory of Ritual*, Chicago: University of Chicago Press.

Tagg, J. (1988) *The Burden of Representation: Essays on Photographies and Histories*, Amherst: University of Massachusetts Press.

Todorov, T. (1984) *The Conquest of America: The Question of the Other*, trans R. Howard, New York: Harper & Row.

Tyler, S. (1986) 'Post-modern ethnography: from document of the occult to occult document', in J. Clifford and G. Marcus (eds) *Writing Culture: The Poetics and Politics of Ethnography*, Berkeley: University of California Press, 122–40.

4

SPECTACLE AND TEXT
Landscape metaphors in cultural geography
Stephen Daniels and Denis Cosgrove

The present cultural turn in human geography has introduced metaphors and analogies more in keeping with an emphasis on meaning than function, and a consequent abandoning of cybernetic and biological analogies. System and organism give way as metaphors to spectacle, theatre and text. The force of this shift may be connected to changes taking place in the production and reproduction of actual environments. Selling the post-industrial city depends upon 'The projection of a definite image of place ... the organization of spectacle and theatricality [which has] been achieved through an eclectic mixture of styles, historical quotation, ornamentation and diversification of surfaces' (Harvey 1989: 93).

For some years geographers have been using the terms 'theatre' and 'text' in a casual way, referring to spatial conduct as 'role-playing' or likening landscape interpretation to 'reading' a written document. But we are now witnessing a more sustained use of these analogies to formulate a new configuration of geographical enquiry (Bonnett 1989; Cosgrove 1985; Daniels and Cosgrove 1988; Duncan and Duncan 1988; Ley and Olds 1988). This is largely a product of growing familiarity with developments in ethnography, social theory and literary criticism. Here scholars, apparently innocent of geography as a discipline, have begun to use geographical discourse to reconstitute their own interpretative schemes both as a material grounding for meaning and more metaphorically (Foucault 1980: 62–77; Said 1990). Under the influence of Roland Barthes, Jacques Derrida and Michel Foucault it is now customary to conceive of a text (along with a variety of images and artefacts) as a discursive 'terrain' across which 'sites' of power may be 'mapped' (Young 1981).

If the currency of spectacle, theatre and text as analogies in geography owes much to the recent weakening of disciplinary boundaries, that same weakening serves to remind us of the historical recency of the boundaries themselves and of the fact that these analogies have an ancestry in what might broadly be called 'geographical thought'. In sixteenth-century Europe, Renaissance humanism's capacity for analogical reasoning developed close links between all these metaphors, using their semantic complexity to develop holistic understanding through a play of metaphorical meaning (Cosgrove 1990; Vickers 1984).

Spectacle for example could mean simple display, but also something to wonder at, thus touching mystery. It could take on the sense of a mirror through which truth which cannot be stated directly may be seen reflected and perhaps distorted, and it could mean an aid to vision as in the still-uncommon corrective lenses (Boelhower 1989). Spectacle was closely related to theatre, although humanists were beginning to distinguish more clearly than their medieval fore-bears between pageant and scripted drama in the classical sense (McArthur 1984). Theatre itself had the meaning not only of a playhouse, but also a conspectus, a place, region or text in which phenomena are presented together for public understanding. Not surprisingly this sense of theatre was particularly appropriate for unitary forms of scientific knowledge: notably of the greater world of the cosmos or the globe – the name of Shakespeare's own theatre – and the lesser world of the human body. Adam Ortelius's great world atlas of 1570 was titled *Teatrum Urbis Terrarum* and published within a decade or so of Giulio Camillo's famous 'memory theatre' in Venice, constructed as a mnemonic for all human knowledge (Yates 1966). Ortelius's intimate friend, Pieter Bruegel, meanwhile adopted the Ptolemaic perspective to paint landscape as a global stage upon which human life was acted out (Gibson 1989). Local regional descriptions, or chorographies, were conceived in terms of theatrical observation, combining written historical and geographical narrative and graphic illustrations that seem today as much landscape paintings as maps. John Speed's work, *The Theatre of the Empire of Great Britain Presenting an Exact Geography of the Kingdomes of England, Scotland, Ireland and the Iles Adioyning* (1611), is an example. Another favoured title for such work was the *speculum*. Thus at all three levels of the Ptolemaic hierarchy of knowledge – cosmography, geography and chorography – theatre as a glass or mirror to the greater world was a common metaphor for revealing order in the macrocosm. In the case of the microcosm, the human body was examined and displayed as a public spectacle in the anatomy theatre. The sixteenth century saw the construction of the first such structures, in Padua and Bologna. Their rising circles of concentric seating, reminiscent of the structure of the greater world, from which one looked down on the lesser world of the corpse displayed on the dissection table, had by law to be open to the public who came dressed in carnival masque for dissections officially scheduled at carnival season and regarded as one of its spectacles (Ferrari 1987).

These hints alone suggest a sophisticated historical relationship in which geographical knowledge is inserted within a wider sphere of scientific production through overlapping metaphors. The metaphors of spectacle, theatre and text are highly malleable and freighted with ideological implications. Here we shall address one of those implications and the resulting epistemological tension. This is the relationship between linguistic and visual metaphors (word and image), a close but tense relationship in Western thought. Before too readily celebrating a spatial conduct as spectacle or theatre we might consider the anthropological critiques of this analogy as a Western 'visualist' colonization (Clifford and

Marcus 1986: 1–26). Similarly we should ponder such anti-visualist critiques as a dimension of the powerful iconoclastic, anti-theatrical tradition of Western thinking. Yi-Fu Tuan's essay 'Sight and pictures' indicates how European culture simultaneously gives primacy to sight as a pathway to true knowledge while deeply suspecting it (Tuan 1979). Since Plato first condemned the scene painters who deal in phantoms Western thinkers have sought to probe beneath the world's illusory surface to reach a deeper, more authentic understanding expressed in language – in the written word or mathematical symbol. The polemic between Robert Fludd and Johannes Kepler at the inception of the scientific revolution concerned the 'meaning and reality of visual imaginings' (Westmann 1984: 79). John Locke's theory of mind was image based, but he looked to the faculty of 'judgement', whose medium was language, to counter and regulate the 'pleasant pictures of fancy', believing that 'it is probably as impossible to control the "pictures of fancy" as it is to control – who else? – women' (Mitchell 1986: 122). Displaying the same distrust, Karl Marx's metaphor for ideology was the camera obscura (ibid.: 168–72). Recently Harvey has conducted an iconoclastic assault on the culture of postmodernity: 'images dominate narratives, ephemerality and fragmentation take precedence over eternal truths and unified politics' (Harvey 1989: 328). The irreducibly visual aspect of landscape has made it a source of deep distrust (Daniels 1989).

Spectacle and text, image and word have always been dialectically related, not least in theatre itself, and this unity has been the site of an intense struggle for meaning. Here we shall examine this struggle at two historical and geographical junctures in European history: in the urban context of sixteenth-century Venice and the rural world of Regency England, approaching it via representations on both sides of the dialectic, first through painted images of spectacle and then through the written text of a novel. In both cases it is in 'theatre' that word and image come together in highest tension and perhaps find uneasy resolution. While the differences between the representations we examine are vast, we shall argue that in each case *landscape* is the discursive terrain across which the struggle between the different, often hostile, codes of meaning construction has been engaged.

SPECTACLE IN VENICE

Renaissance Venice was the paradigm city of spectacular civic ritual (Muir 1981; Muir and Weissman 1989). The entire Venetian calendar was punctuated by highly choreographed public display which demanded the widest possible participation, incorporating both citizens and visitors through a series of overlapping allegiances: social status, occupation, residence, membership of *scuola* (a lay confraternity). Above all, regular procession and spectacle bound together the entire community, the 'body politic', in rehearsing the political and moral order of the city. They celebrated the Venetian sense of place, of Venice as a world: the unique and divinely ordained 'terra sancta nostra' (Fortini-Brown 1988). The myth of Venice as a single body, a perfectly governed, harmonious polity, an ideal city blessed in its geographical location and sanctified by the actual body of the evangelist Mark was publicly rehearsed in Venice's *campi* (church squares), on its canals and bridges and above all on the great stage setting of the Piazza of San Marco and the Piazzetta, the political centre of the city. The spaces and surrounding buildings were continuously altered and elaborated over Venetian history as the contexts and ideological significance of the Venetian myth demanded altered inscriptions in the townscape (Cosgrove 1982). By the sixteenth century it provided a vast setting in which over a third of the city's entire population could gather and participate in celebrations of the *genius urbis*. In San Marco every genre of public spectacle could and did take place, from ecclesiastical ritual connected with the liturgy of St Mark's Basilica and the political processions of state organized around the doge, to the burlesque of carnival and the antics of popular circus performers.

We can discover much of Venetian ritual from diarists and written archival sources, but the most vivid record exists on the great canvases which decorated the walls of the *alberghi* (boardrooms-cum-shrines) of the *scuola* of Venice. These corporate bodies had a central role to play in the social and ceremonial life of Venice. Although open to the nobility, they were governed by non-nobles and their constitutions and privileges mirrored those of the state, allowing an outlet for the political and status aspirations of those disenfranchised by Venice's impenetrable oligarchy. *Scuole* participation in state ritual emphasized this function and their members would compete for privileged places in the tightly ordered and highly stratified processions where ceremonially robed *scuole* members displayed their precious relics and icons. Competition was also practiced in commissions for decorating the meeting rooms of the *scuola*. The standard form was to commission a cycle of paintings celebrating miraculous events connected with the *scuola*'s patron, or sacred relics. Almost invariably these events were shown to take place either at recognizable locations within the city of Venice or in 'invisible cities' – imagined places constituted by architectural reassembly of Venetian buildings. Thus the most significant events of the *scuola* were integrated with the public meaning of the Republic: painting acted as a spectacular mirror in which the world of Venice was reflected and enhanced. Normally the cycle of paintings corresponded to a written chronicle held, and at times updated, by the confraternity itself: a text which legitimized the *scuole*'s myth but, as we shall see, a text which was itself subordinate to the visual image.

We shall consider only two such paintings, single images taken from different *scuole* cycles and painted some half a century apart. The differences between them reveal something of the ways in which relations between image and text were being reconstituted at this period in Venetian history. The first is Gentile Bellini's *Procession of the Relic of the Cross* in St Mark's Square (Plate 4.1), painted as part of a cycle of nine images for the Scuola Grande di San Giovanni Evangelista in the opening years of the sixteenth century. The second is Jacopo Tintoretto's *Translation of the Body of St Mark* (Plate 4.2), from a cycle produced

Plate 4.1 Gentile Bellini, *Procession of the Relic of the Cross*, *c.* 1505

Source: Accademia dei Belli Arti, Venice. Reprinted courtesy of O. Bohm, Venice.

in the 1550s for the Scuola Grande de San Marco. Both illustrate a miraculous event and locate it recognizably in the heart of Venice: in the Piazza in Bellini's case and in the courtyard of the Doge's Palace in Tintoretto's. Both also display the *body* of Venice, but 'place' the corporeal metaphor differently.

PROCESSION IN THE PIAZZA

Bellini's *Procession* is a perfect example of what Fortini-Brown (1988) has called the 'eyewitness' style of Venetian narrative painting, originally developed in the decorative mosaic cycles of the Basilica, and reaching its peak at the turn of the sixteenth century in the cycles by the Bellini family, Vittore Carpaccio, Giovanni Mansueti and Lazzaro Bastiani. In their works a sacred event recorded by chroniclers is presented in recognizable geographical context and authenticated by the presence of large numbers of ordinary people. In the Bellini *Procession* the specific miraculous event is implied rather than demonstrated and it would pass unnoticed by an uninitiated observer, as it does by most of the hundreds of people shown on the canvas. In the group of men who watch the white-robed brothers of the *scuole* display their most precious relic, a fragment of the true cross, in the doge's procession, one single figure kneels. He is Jacopo de Salis, a merchant whose son is at this moment dying in Brescia from a fractured skull. His father's act of piety will result in the miracle of a complete recovery. Gentile's image is a sophisticated and highly dramatic landscape representation. Clever manipulation of perspective allows him to open to our eyes the full space of the piazza while simultaneously enhancing the elevation of the Basilica of San Marco – the iconographic heart of the Republic. The event occurs during the most significant ritual in the Venetian calendar: the doge's procession on the Feast of St Mark which wound around the piazza *en route* from the palace to mass in the basilica. His composition allows Gentile to give prominence to the commissioning *scuola* while showing them soberly participating in a broader civic ritual, even to the point of displaying the crests of other *scuole* along the fringe of the canopy which protects and enhances their own icon. Thus the corporate body of Venice, in both its political expression of doge, Senate and Great Council, and its more demotic form of the *scuole*, is displayed on the ordered stage of the gridded piazza. In the same space a recognizable cross-section of Venice's cosmopolitan population – *popolari*, Germans, Greeks, Turks, indigents and a dwarf – act to emphasize universal participation in the normality of the event. Flattened like a backdrop, but dominating the scene at the perspective point of the painting, is the basilica, whose portal mosaics (rendered by Gentile with astonishing accuracy) record another miraculous cycle: the ninth-century translation of St Mark's body to Venice and its burial in the basilica: the legitimizing story of the Venetian myth. Although the specific miracle is not seen, Jacopo's act is 'witnessed', both by the entire 'body' of Venice and by us, just as the procession itself is witnessed by, and gives witness to, the place of rest for the evangelist's body. Gentile's emphasis on the basilica as the

Plate 4.2 Tintoretto, *Translation of the Body of St Mark*, 1548

Source: Accademia dei Belli Arti, Venice. Reprinted courtesy of O. Bohm, Venice.

miraculous heart of the holy city and the ritual procession in honour of Venice's patron render the specific miracle of healing more natural – it occurs in a sacred place, while the numbers of witnesses and the arbitrary point that the procession has reached make it singular.

Spectators are a crucial element in these 'eyewitness' paintings. The narrative techniques of the works correspond, as Fortini-Brown has demonstrated, to the chronicle mode of historical writing still favoured in early Renaissance Venice, one in which the authenticity of the text depended upon maximizing both local detail and numbers of testimonials. Indeed pictures themselves, like Gentile's, would be called upon by later chroniclers to authenticate the truth of their written history through verbal description of the event as described in paint. Thus:

> a painting was ... more than an evocation of the past, or a vehicle for inciting religious devotion through the representation of a miraculous event. It was, in fact, a piece of testimony with a status equivalent to a public document or written history: an instrument of proof that such an event had actually happened.
>
> (Fortini-Brown 1988: 79)

The huge cast of individuals in these paintings thus act like the chorus in Greek theatre; their presence mediates between audience and event, establishing the credibility of the latter to the former. In this sense the eyewitness paintings are doubly theatrical, but in the sense of theatre as spectacle, one which privileges image over text.

TRANSLATION OF ST MARK'S BODY

Renaissance humanists distrusted such an appeal to vision alone and indeed the entire mode of chronicle history writing. Lauro Martines (1980) has summarized the critical humanist contributions to historical scholarship as follows: an emphasis on the accuracy of the text through identifying original documents and securing their authenticity through philological techniques; contextual placing of the text to establish a signal 'correct meaning'; challenging historical 'myths' through ascertainable 'facts': names, dates, words and documents (i.e. through text rather than image); establishing a secular history freed from eschatology and appeals to divine intervention; the use of new disciplines like archaeology, numismatics and topography as aids to historical verification. However, in writing their narratives the humanist historians did more than merely record the sources upon which they drew. Their histories were to be *invenzione*, personal syntheses, based upon true sources but composed in such a way as to expose the causal structures behind the events they described, to elicit the hidden order in the body of an apparently haphazard world. The technique for achieving this was rhetoric, the contrived use of elegant language and grammar to gain maximum moral and emotional force in the narration (Kristeller 1990). Humanism thus elevated text over image: *ut pictura poesis*. Leon Battista Alberti, whose *On Painting* (1435–6) became the handbook of the new orthodoxy in the arts, elevated *storia* to the highest point of artistic endeavour. Depicting the 'great deeds of great men' and composing the work according to a clear spatial and narrative hierarchy provided a parallel in painting to the changes in historical narrative, conjoining both while subordinating image to text. Alberti's work indicates another feature of Renaissance humanism: the celebration of the individual, including the stature and dignity of the individual human body. These ideas were current in Florence from the early fifteenth century but only enter the mainstream of Venetian culture at the turn of the sixteenth, exemplified in the writings of Giustiniani, Sabellico and Navagero, and the paintings of Titian, Giorgione and later Tintoretto (Fortini-Brown 1988).

When we turn to Tintoretto's painting we see the full impact of humanism's alternative resolution to the discourse of text and image. It depicts the moment when two Venetian merchants, having removed the body of St Mark from its sarcophagus, carry it through the streets of Alexandria to begin the return voyage to Venice. The presence of the uncorrupted body allows the merchants miraculous passage through the hostile Alexandrian streets, terrifying those who would prevent its removal (Nicol 1988: 24–5). Tintoretto's composition is dominated by the merchants and, above all, the muscular, foreshortened body of the saint. St Mark's body, whose physical presence in Venice legitimated the Venetian myth, is presented to us in anatomical detail, rather as equally dead bodies were publicly displayed in the anatomy theatre. It is placed to the right of the central perspective axis defined by the architectural setting and the gridded markings of the piazza. The axis is deceptive, leading past the centre of the drama to an indistinct white building in the depth of the painting. The drama of the event is increased by a complex set of diagonals: a fallen camel driver attempting to restrain his frightened beast, another man on the ground clutching a violently flapping curtain, and sepulchral figures rushing panic-stricken into a building on the left which is modelled on the Ducal Palace at Venice. The sky is thunderous, the illumination eerie, and ghost-like figures flee from the picture. It is a typical mannerist composition and highly theatrical, not to say melodramatic. Indeed, 'theatre' seems a more appropriate term here than 'spectacle' or 'pageant', which apply to the Bellini. The architectural setting resembles contemporary stage designs for the tragic scene and the sharp perspectives of Palladio's contemporary Olympic Theatre at Vicenza. The central characters tumble precipitously towards us, locked together in tense physical interaction and demanding our emotional involvement, rather than parading in separate, static dignity before us. The event is authenticated rhetorically rather than through factual detail and observation, a textual as much as a visual strategy.

The reasons for this shift in the relative authority of the narrative and pictorial modes are complex. Some result from developments in humanist discourse and changing influences and fashions in painting, notably the influence of exiled Roman painters in mid-sixteenth-century Venice. But they also reflect an altered political and ideological climate. Tintoretto's Venice, the only remaining free republic in an Italy otherwise dominated by Spain and France, severely threatened by Turkish power, its maritime hegemony under strain from the emerging Atlantic economies, and struggling for religious autonomy against the Counter-Reformation papacy, was no longer the secure and serene city state of Gentile's time. Its own internal ideological structures were tightening and the modes of public participation in civic spectacle and theatre were being recast by an increasingly aristocratic patriciate. The Venetian noble Daniele Barbaro's (1569) discussion of *scenografia* applies a patrician humanist concept of rank and hierarchy to theatre. Not only does he accept the Aristotelian hierarchy of tragedy, comedy and satire (with its social, corporeal and gender correlates), but he suggests a threefold ranking of public theatre into *theatre* properly so called, for the performance of scripted plays and musical recital, *amphitheatre* for athletics and feats of physical prowess and *circus* for the performance of gladiators and animals. Each should have its own appropriate building and location in the city, and each clearly appeals to a different social rank. Quite clearly, the scripted performance, based on a text, is here elevated over the purely visual spectacle, and the elevation of tragedy over all other forms tends to tarnish certain aspects of theatre with moral disapproval. The text-based form of theatre is the preserve of those in power, those with authority through *virtu*, founded in reason. It takes place increasingly in the closed playhouse, away from vulgar gaze. More purely visual, spectacular forms of theatre are for the masses – they appeal to sense rather than intellect. By the same token, Tintoretto represents the *body* of Venice no longer as public and corporate, but mystical and individual: the sacred, uncorrupted corpse of St Mark.

THEATRE AND TEXT IN GEORGIAN ENGLAND

The elevation of text over image found in Renaissance humanist writings represents a long conservative tradition of respect for the stable and substantive authority of text and distrust of the mutability, versatility and disguise associated with visual representation. Iconoclasm has been strongly associated with religious and moral zeal for holy writ: in the Islamic prohibition of organic images, the regular destruction of images in the Byzantine Church or the depradations of Calvinism. Such attitudes can be fiercely expressed, tarnishing even theatrical concepts with the negative implications of 'stagey', 'play-acting' or 'making a scene' (Barish 1966). In Georgian England iconoclasm had an intense political charge. In his counter-revolutionary tract, *Reflections on the Revolution in France*, Edmund Burke was intent to spell out the connected etymology and dangers of theatre and theory, of spectating and speculation, which he sums up in his description of a sermon by the radical Richard Price, likening it to revolutionary theatre:

> There must be a great change of scene; there must be a magnificent stage

effect; there must be a grand spectacle to rouse the imagination ... The Preacher found them all in the French Revolution. This inspires a juvenile warmth through his whole frame. His enthusiasm kindles as he advances; and when he arrives at his peroration, it is in full blaze. The viewing from the Pisgah of his pulpit, the free, moral, happy, flourishing, and glorious state of France, as in a bird's eye landscape of a promised land, he breaks out into the following rapture.

(Burke 1968: 156–7)

Whereon Burke quotes a 'mine eyes have seen the salvation' set-piece from Price's speech. Now Burke did not wish to vanquish the viewing of landscape – how could he as a spokesman for a class who, more than many things, valued doing precisely this? – but he did in the *Reflections* imply the reform of landscape taste which many conservative writers, some landscape designers, were earnestly proposing at the time. They proposed subduing the clarity and brilliance of the visual field in a taste for shady views and abandoning panoramas (Burke's 'bird's eye landscape') for short-focus views (Daniels 1988, 1990). It was no less pictorial, or observational, a mode of knowledge; indeed it was in some senses more so. But in condemning surface brilliance and sweeping views it presumed to grasp a more reliable sense of the social and topographical texture of the country-side – a pressing virtue to those who thought this texture was in danger of unravelling. And a more textured countryside was a more textualized one. In the practice of landscape gardening, this reform can be seen in the change from broad, circumspect, designs of Capability Brown, whose main design vehicle was maps, to the more intimate, circumspect, style of Humphry Repton, whose main vehicles were 'Red Books' made up of watercolour sketches qualified by a detailed and often highly moralizing text (Daniels 1982).

MANSFIELD PARK

Jane Austen's novel *Mansfield Park* is representative of this conservative tradition. It was written between 1811 and 1813 when there was acute concern among conservative opinion about pressures on landed culture and its integral role in the culture of the nation. If a revolutionary republic was threatening from across the Channel, at home there were fears about developments the Napoleonic Wars seemed to aggravate, notably financial speculation and plebeian unrest (Daniels 1982). If Jane Austen was anxious about plebeian threats to landed culture they do not impinge upon the discursive world of her novels. Even when action is set beyond the boundaries of polite society, beyond the rural estate and town square (and that is rarely), the novels exclude any colloquial or vernacular language. 'The threats to Jane Austen's society and her language – and they are many and increase throughout her work – are essentially from within' (Tanner 1986: 39).

In *Mansfield Park* the main threat to landed culture is the culture of money. Now in this there is nothing new. The spectre of a moneyed interest had long been an essential galvanizing feature of conservative discourse. Embodied in *parvenus* or located in London it conveniently obscured the powerful financial interests of the gentry themselves (as well as any vulgar pedigree), and the necessity of a moneyed economy to the creation and management of landed estates (Daniels and Seymour 1990). But from the later eighteenth century there develops a chorus of complaint on how the solidity, sobriety and stability of land are being progessively undermined by the culture of money, to produce what Burke called 'a paper circulation and stock jobbing constitution' (Daniels 1988: 68). The radical William Cobbett, while firmly outside Jane Austen's discursive domain, expressed this view in contrasting

> a resident *native* gentry, attached to the soil, known to every farmer and labourer from their childhood ... practising hospitality without ceremony from habit and not on calculation; and a gentry only now-and-then resident at all, having no relish for country delights, foreign in their manners, distant and haughty in their behaviour, looking to the soil only for its rents, viewing it as a mere object of speculation.
>
> (quoted in Tanner 1986: 145)

The poet William Cowper, fondly read by Jane Austen and a discernible influence on *Mansfield Park*, brought out the financial and visual connotations of speculation in putting the whole idea of landscape into question. 'Estates are landscapes', he complained in *The Task*, 'gaz'd upon awhile, then advertised and auctioneered away' (quoted in Daniels 1982: 124).

It is the absence of the owner of *Mansfield Park*, Sir Thomas Bertram, which opens up a space for the malignant glamour of money. Bertram is in many ways a model patriarch, enforcing strict standards at Mansfield, but losses in his sugar plantations in Antigue force him away from Mansfield for a year to supervise this less seemly part of his estate which is pointedly described as a 'business'. The forces of corruption are embodied in two wealthy visitors to Mansfield, Henry Crawford and his sister Maria. Crawford has an estate in Norfolk which he neglects. He and his sister are representatives of London, its moneyed world of mobility and endlessly false appearances. The forces of redemption in the novel are Fanny Price, the plain daughter of Lady Bertram's impecunious sister, and Bertram's second son Edmund, who is destined for the Church. After each warding off the alluring advances of the Crawfords, Fanny and Edmund marry and move into a modest parsonage near Mansfield. Due to be inherited by Bertram's dissolute elder son, the future of the Mansfield estate is uncertain. If the custody of the country (both rural England and the nation at large) cannot be trusted to the aristocracy, it is secure in the care of the minor gentry and clergy.

Conservative opinion in later Georgian England defended property as the condition of social order and stability (and there was a bloody penal code to support them), but no less important is a new defence of good manners and morals among the propertied – in a word, 'propriety'. As Tanner points out, it was precisely in Jane Austen's lifetime that propriety assumed the meaning of conformity with good manners as well as the fact of owning something, and much of her fiction examines the relations between these meanings. Propriety was seen to be no less important than law; indeed Burke argued that 'Manners are of more importance than laws. Upon them, in a great measure, laws depend.' So 'Jane Austen's profound concern with good manners was thus not simply a reflection of cloistered gentility; it was a form of politics' (Tanner 1986: 26). And 'the ultimate propriety, upon which all the other proprieties depended, was true propriety of language' (ibid.: 20). Jane Austen's own language is exemplary in its lack of theatricality, in the sense of its lack of physical action, sudden movement or spectacle. There are few vigorous transitive verbs or striking metaphors, and little direct speech. Her plots have few shocks or surprises. And so when such theatricality does occur, as it does in *Mansfield Park*, it is all the more significant and all the more shocking. If theatre is rare as mode of description in Jane Austen it is arguably her main subject. The prominent social occasions in the novels – balls, dinner parties, excursions and private theatricals in *Mansfield Park* – are those whose drama derives from the fact that certain characters so skilfully deploy and manipulate codes of conduct that it is 'difficult, if not impossible, to distinguish "true" good manner from adroitly simulated ones', to 'tell who is "acting" his acting' (ibid.: 29). And it is the very power of theatricality that demands the control of textuality, a textuality deployed both by the author and by her sympathetic characters.

The sober authority of texts and the propriety of reading and writing are emphasized throughout *Mansfield Park*. Our heroine, Fanny, inhabits the old schoolroom where she had happily 'read and written' until the tutor quit and where she does still in private surrounded by books 'of which she had been a collector, from the first hour of her commanding a shilling'. Here she retreats from 'any thing unpleasant below', notably the rehearsals for the theatricals of which she strongly disapproves and which she holds out from joining in: 'I could not act anything if you were to give me the world.' But her bookish space is not inviolate. The flashy Mary Crawford, in her element in the theatricals, enters to entreat Fanny to help her rehearse, threatening both the sanctity of the space and the bookishness which defines it. 'I have brought my book', she announces, which is the script of the morally dubious play *Lovers' Vows*. And while Mary acts her part, Fanny confesses she 'must *read* the part' she is given, 'for I can *say* very little of it'. The issue of proper delivery of a text recurs in relation to another, more proper play, Shakespeare's *Henry VIII*: Fanny is reading it to the lady of the house, Mrs Bertram, when Henry Crawford enters, seizes the volume and delivers a 'truly dramatic' reading. This is a prelude to a conversation with Edmund about 'the art of reading' in church, in which Crawford likens the preacher to an actor, the pulpit to a stage and the sermon to a script, so subordinating truth, the biblical text, to delivery. (Earlier Crawford had observed that these days 'one scarcely sees a clergyman out of his pulpit', to which Edmund replies that while that may apply to London it did not to 'the nation at large'. He felt that being 'useful in his parish and his neighbourhood' was as important as 'fine preaching' and that the clergy's influence on 'public manners' should not be merely as 'the masters of the ceremonies of life'.) In her notes to Fanny, Mary Crawford's writing is revealed as no less duplicitous than her reading. In contrast, Edmund's writing to her has an integrity that subsists in its signature as well as in what it says. While Crawford had offered her a flashy necklace, 'the scrap of paper on which Edmund had begun writing to her [was] treasured beyond all hopes'. 'To her, the handwriting itself, independent of any thing it may convey, is a blessedness. Never were such characters cut by any other human being, as Edmund's commonest handwriting gave.'

The rehearsals of *Lovers' Vows* are the culmination of the licence Sir Thomas Bertram's absence allowed. The disregard of a stable, ordained order, and the release of a destructive extemporizing energy, is described in terms of changes to the physical fabric of the house to make the theatre in which the play will be acted. The billiard room adjoining Sir Thomas's bedroom is appropriated and comes to be called 'The Theatre'. But in a sense the whole house becomes a theatre. In Fanny's room Mary Crawford gleefully exults, 'we are rehearsing all over the house'. At first the estate carpenter is called in to construct a stage and then he is replaced, at great expense, by a professional scene painter who spoils the floor, ruins all the coachman's sponges and makes 'five of the under-servants idle and dissatisfied'. And the play becomes more of a spectacle, for instead of being guided by Edmund 'as to the privacy of the representation', his dissolute elder brother offers 'an invitation to every family who came his way'. Even Edmund is seduced by the rehearsals. While the Crawfords use their roles to conduct their strategems, he is lost in his: 'between his theatrical and real part [he] was equally unobservant'. Sir Thomas's sudden entry into 'The Theatre', itself a dramatic *coup de théâtre*, smashes this illusory world. What 'struck him especially' was 'the removal of the bookcase from before the billiard room door'. It is not just that texts are displaced but that the bookcase had functioned as a barrier to a domain of games and one which, moreover, was strictly reserved for Bertram and his male friends and relations. Sir Thomas has the stage dismantled, the scene painter dismissed, the bookcase put back into its proper place and all copies of the play burnt. 'Under his government' the house returns to 'its proper state'.

There is as much theatricality outside the house as within. This concerns proposals to alter the landscape of two nearby estates, or, to put it more precisely, to landscape these estates, to refashion their topography according to the latest styles of landscape gardening. The issue, as in the house at Mansfield, is tampering with an established order (Duckworth 1971). Since Burke had used the imagery of excessive estate improvements to illustrate the statecraft of the French Revolution the metaphor had acquired a powerful political charge. Literary complaints about radical or inordinate changes to the fabric of mansions and parks were part of the long tradition criticizing the vanity of money and the deceit of appearance. In *The Task* William Cowper called 'improvement' (landscaping in the style of Capability Brown) 'the idol of the age' (quoted in

Daniels and Seymour 1990: 501). 'Improvement' also comes in for criticism in *Mansfield Park*. But it is not opposed to a static, unchanging status quo which, as in Burke, connotes a dangerous state of atrophy, ripe for radical overhaul. In one episode Edmund opposes 'improvement' with 'progress', by which he means changing an estate in a gradual, piecemeal fashion, attending to the substance of the estate rather than its visible surface, organically repairing the deficiency of the parts with the whole in mind. Humphry Repton is singled out as a figure of the improver, which in hindsight seems a little unfair, now we know Repton in fact shared many of Jane Austen's views of estates as places to be lived in not just looked at (Daniels 1982). But in popular opinion then Repton was seen (as he saw himself) as Brown's successor. And, in any case, Repton is not so much considered for what he in fact thought or did, as a representative of professionalism. Mary Crawford says that if she 'had a place of my own in the country, I should be most thankful to any Mr Repton who would undertake it, and give me as much beauty as he could for my money'. She would not set eyes on it until it was 'complete', whereas Edmund, doing his own gradual refashioning, would attend to the entire, ongoing, unspectacular process. In terms of the private theatricals it is the difference between the work of the estate carpenter and that of the professional scene painter. And yet Jane Austen does not fasten upon something Repton did frequently, making a small estate appear much more grand than in fact it was and so blurring status distinctions for ambition's sake. So Mr Rushworth returns astonished from an estate a seventh the size of his own which has recently been improved by Repton. Compared with this, however, his own, Sotherton Court, an old Elizabethan house surrounded by walled formal gardens, seems horribly confined, 'a prison – quite a dismal old prison'. Repton's landscaping has effected a dramatic change of scene. 'I never saw a place so altered in my life,' says Rushworth, 'I did not know where I was.' Far from unnerving Rushworth, these improvements provoke him to have some for himself, and the company from Mansfield (he is engaged to one of the Bertram daughters) make an excursion to Sotherton to look over the estate and assess its capabilities. Sotherton is heavy with tradition, and Fanny, who is 'delighted to connect any thing with history', is eager to see in it its 'old state'. But she is disappointed, for Sotherton is not just out of date, it has fallen into neglect. She is despondent to hear that the late Mr Rushworth discontinued family prayers in the chapel ('every generation has its improvements', says Miss Crawford, with a smile). And Rushworth has already removed 'two or three fine old trees' from the oak avenue, which provokes Fanny to quote from Cowper 'Ye fallen avenues, once more I mourn your fate unmerited.'

If Sotherton is moribund, Thornton Lacey, the small parsonage which will be Edmund's living, is healthy, its glebe meadows 'finely sprinkled with timber', the church 'within a stone's throw' of the house; which makes Henry Crawford's suggestions on improving it all the more delinquent. He ventures these during a card game, significantly called 'Speculation'. Crawford reckons 'there will be work for five summers at least before the place is live-able' and the work he suggests is designed to cut off the parson from his parish.

> The farm-yard must be cleared away entirely, and planted up to shut out the blacksmith's shop. The house must be turned to front the east instead of the north ... where the view is really very pretty ... The meadows beyond that *will* be the garden, as well as now what it *is* ... [they] must be all laid together of course ... They belong to the living, I suppose. If not you must purchase them. Then the stream – something must be done with the stream; but I could not determine what. I had two or three ideas.

It is a scheme for property without propriety. The indelicacy of Crawford's improvements for the parsonage are such that they are to be 'stamped on it' and 'that house receive such an air as to make its owner be set down as the great land-holder of the parish, by every creature travelling the road'.

Textual knowledge is not categorically opposed to visual knowledge in *Mansfield Park*; indeed it sets out to redeem it. As the novel distinguishes true from false ways of reading, writing and speaking, so it distinguishes true from false ways of seeing and appearing. The falsity of the visual is its surface brilliance, represented by Mary Crawford and all she sees. And if she sees many things – she is restlessly mobile – she sees, in the sense of understanding, nothing. And the brilliance of the Crawfords blinds others, notably Edmund, but not Fanny: 'I was quiet,' she says 'but I was not blind.' While Mary wantonly speculates, Fanny carefully observes. This contrast is nicely brought out on the trip to Sotherton. Mary fights for the highest seat on the carriage (the barouche-box) which will give a fine 'burst of country' (a tellingly abstract and violent phrase). Fanny takes a more lowly position and a more gentle and carefully detailed view 'in observing the appearance of the country, the bearings of the road, the differences of soil, the state of the harvest, the cottages, the cattle, the children'. While Fanny, 'whose rides had never been extensive, was soon beyond her knowledge', she is able here to comprehend the country by deploying a sophisticated pictorial code, one evident in picturesque sketches and paintings but more so in moralizing texts on the countryside which emphasized its utility, or lack of it. Fanny is as short-sighted as she is still. At Thornton Lacey 'her eye was eagerly taking in everything within her reach'. At Mansfield she spends most time in the flower garden 'stooping among the roses' and her walk into 'the hot park' (its heat like that of the play *Lovers' Vows* is a sign of its immodesty) severely fatigues her. In her room Fanny cares for her plants along with her books, and the care of plants was then inseparable from learning from books about them, labelling them, in thought or in fact, with written details of their natural history.

Fanny represents a powerful movement in later Georgian England for the reform of polite society, a reform which became encoded not only in texts but in designs for houses and grounds. If sweeping parks and panoramic ways of seeing became morally suspect, a sign of ambition and too indelicate or overstretched a display of power, gardens and conservatories offered an alternative arena of virtue, one which did not shut out the world at large but reframed it as a set of detailed vignettes (Daniels and Seymour 1990). This might be seen as a response to larger structural forces. For example, Repton's emphasis on gardening rather than landscaping towards the end of his career made, as he acknowledged, a virtue out of the fact that large landowners were unable or unwilling to spend large sums on large-scale improvements, either because of taxes during the Napoleonic Wars or because of the agricultural depression which followed them. But, as he also observed, the largest and most vacuous landscape improvements were being made by those best placed to avoid property taxes, namely financial speculators and industrialists, who were eagerly acquiring estates as badges of status (Daniels 1987). The earnest, domestic code of country life represented in the marriage of Fanny and Edmund seizes the moral high ground (if that is not too panoramic a way of putting it) from both aristocratic grandees and capitalist entrepreneurs, and while it found an emblem in plain old manorial estates, what gave it its vigour was its adoption by the professions, both rural professions (parsons, tenant farmers, millers, landscape gardeners) and those urban professionals who lived in suburbs (Plate 4.3). It was in such homes that the poems of William Cowper were read aloud to the family, and it was in the newspapers to which such homes subscribed that Cowper's poetry was reprinted. As Davidoff and Hall observe, the new domesticity subtly reformulated gender roles. It offered a new model of manliness – bookish, kindly, attentive to women – and, in the care of her home and garden, a new sphere for women, both elevating them and containing them 'like the plant in the pot, limited and domesticated, sexually controlled, not spilling into spheres to which she did not belong, not being overpowered by "weeds of social disorder"' (1987: 191–2). If the dazzling world of London and its landed outliers was difficult to read, the textual, textured world of the suburbs could be comfortably grasped.

WORD AND IMAGE IN GEOGRAPHY

In our contemporary world, where the visual seems not to provide a transparent window to truth but shatters appearance into a set of dazzling surfaces, many reflecting the vanity of the spectator, textuality is upheld as providing an instrument to probe into substance. But, from the other pole of the struggle between word and image, the very power of textuality can be seen as repressive and mystifying, scribbling as it were on the window on the world. And this apprehension lies deep in the drive in Western thought for naturalism. Norman Bryson brings out this apprehension by contrasting a 'textually saturated' medieval window in Canterbury Cathedral showing biblical scenes, whose status is 'that of a relay or a place of transit through which the eye must pass to reach its goal, which is the Word', with Massacio's fifteenth-century fresco of another biblical scene, *The Rendering of the Tribute Money*, which displays 'a marked excess of the image over the text' in 'supplying us with more visual information than we need to grasp its narrative content'. We accept this surplus of visual data because

Plate 4.3 Member of the Pole Family, *Woman Writing at a Desk*, c. 1803–6

Source: Museum and Art Gallery, Bristol

it is carefully controlled, largely through perspective, defused, neutralized to give the impression of the natural, the real, the lifelike. It is the very logic and spaciousness of the image that increase the area of understanding beyond the threshold of the textual function. Where the Text ends, Life – that is imagination – begins (Bryson 1981: 1–14). This naturalistic view is written deep in much geographic thought, whose ancestry is to be found in the Enlightenment.

The most flagrant examples recently occur in some of David Stoddart's essays in *On Geography*, in an idiom which might be described as hairy-chested naturalism. Stoddart seeks to restore the integrity of the visual in geography by emphasizing precisely its innocence, a move which, not incidentally, involves shedding the textual trappings of civilization as the geographer leaves the maligned armchair (locus of the book) and goes into the field, especially in those physically demanding, less sophisticated parts of the world. Stoddart affirms Joseph Conrad's contrast between 'this world of action and discovery, and that of the lecture room and library' (Stoddart 1986: 143). It is Conrad's 'world of action and discovery', we presume, which purifies the writing of the geographer as much as of the novelist. Direct observation makes Stoddart's geography incline to the condition of realism, and rubs the scribbling off the window on the world. In fact, Stoddart takes the camera obscura, specifically that in Patrick Geddes' Outlook Tower in Edinburgh, as a symbol of 'a form of observation and definition of subject matter which has governed field studies' (ibid.: 144). But for Marx, textual author and authority, the camera obscura was a symbol of mystification. We may compare Stoddart's view with that in the paper we referred to at the outset of this essay by Yi-Fu Tuan, which endorses the suspicion of the visual evidence he documents, worrying that it can 'mislead and enthral'. Tuan likens the obsessively visual tradition of geography to cinema in its concern with 'objects and actions'. Words are more accommodating to Tuan's concern with 'the dispositions of the mind' and for this reason he looks to a form of representation which, in contrast to modern cinema, seems to have a low visual register, to be less superficial: pre-Restoration theatre. 'In the cinema we feel that we gain at least some geographical knowledge, whereas in the theatre this claim would be inadmissible' (Tuan 1979: 421).

It should be apparent by now that we do not want to adjudicate between these views, nor do we want to propose, in Mitchell's words, a 'peaceful settlement' between the interests of visual and the verbal representation 'under the terms of some all embracing theory of signs'. Rather, like him, we want to historicize 'a struggle that carries the fundamental contradictions of our culture into the heart of theoretical discourse itself' (Mitchell 1986: 44). Geography is both a key element of our culture and a theoretical discourse. It is hardly surprising that the struggle between the visual and the verbal should find its echoes within our own discipline.

REFERENCES

Alberti, L.B. (1435–6) *On Painting*, trans. J.R. Spencer, 1956, New Haven and London: Yale University Press.

Barbaro, D. (1569) *La practica della perspettiva*, Venice: Camillo & Rutilio Borgominieri.

Barish, J.A. (1966) 'The antitheatrical prejudice', *The Critical Quarterly* 8: 329–48.

Boelhower, W. (1989) 'Inventing America: a model of cartographic semiosis', in G. Zanetto (ed.) *Les langages aes representations geographiques*, Venice: Università degli studi di Venezia, Dipartimento di scienze Economiche, 373–99.

Bonnett, A. (1989) 'Situationism, geography and poststructuralism', *Environment and Planning D: Society and Space* 7: 131–46.

Bryson, N. (1981) *Word and Image: French Painting of the Ancien Régime*, Cambridge: Cambridge University Press.

Burke, E. (1968) *Reflections on the Revolution in France*, Harmondsworth: Penguin.

Clifford, J. and Marcus, G.E. (eds) (1986) *Writing Culture: The Poetics and Politics of Ethnography*, Berkeley: University of California Press.

Cosgrove, D. (1982) 'The myth and the stones of Venice: the historical geography of a symbolic landscape', *Journal of Historical Geography* 8(2): 145–69.

—— (1985) 'Prospect, perspective and the evolution of the landscape idea', *Transactions of the Institute of British Geographers* (new series) 10: 45–62.

—— (1990) 'Environmental thought and action: pre-modern and post-modern', *Transactions of the Institute of British Geographers* (new series) 15: 1–15.

Daniels, S. (1982) 'Humphry Repton and the morality of landscape', in J. Gold and J. Burgess (eds) *Valued Environments*, London: Allen & Unwin, 124–44.

—— (1987) 'Cankerous Blossom: troubles in the later career of Humphry Repton documented in the Repton correspondence in the Huntington Library', *Journal of Garden History* 6: 146–61.

—— (1988) 'The political iconography of woodland in later Georgian England', in D. Cosgrove and S. Daniels (eds) *The Iconography of Landscape: Essays on the Symbolic Representation, Design and Use of Past Environments*, Cambridge: Cambridge University Press, 43–82.

—— (1989) 'Marxism, culture and the duplicity of landscape', in R. Peet and N. Thrift (eds) *New Models in Geography*, vol. 2, London: Unwin Hyman, 196–220.

—— (1990) 'Goodly prospects: English estate portraiture, 1670–1730, in N. Alfrey and S. Daniels (eds) *Mapping the Landscape*, Nottingham: University of Nottingham, 9–12.

Daniels, S. and Cosgrove, D. (1988) 'Iconography and landscape', in D. Cosgrove and S. Daniels (eds) *The Iconography of Landscape: Essays on the Symbolic Representation, Design and Use of Past Environments*, Cambridge: Cambridge University Press, 1–10.

Daniels, S. and Seymour, S. (1990) 'Landscape and the idea of improvement', in R. Dodgshon and R.A. Butlin (eds) *A New Historical Geography of England and Wales*, 2nd edn, London: Academic Press, 487–519.

Davidoff, L. and Hall, C. (1987) *Family Fortunes: Men and Women of the English Middle Class 1780–1850*, London: Hutchinson.

Duckworth, A. (1971) *The Improvement of the Estate: A Study of Jane Austen's Novels*, Baltimore: Johns Hopkins University Press.

Duncan, J. and Duncan, N. (1988) '(Re)reading the landscape', *Environment and Planning D: Society and Space* 6: 117–26.

Ferrari, G. (1987) 'Public anatomy lessons and the carnival: the anatomy theatre of Bologna', *Past and Present* 117: 50–106.

Fortini-Brown, P. (1988) *Venetian Narrative Painting in the Age of Carpaccio*, New Haven and London: Yale University Press.

Foucault, M. (1980) *Power/Knowledge: Selected Interviews and Other Writings, 1972–1977*, Brighton: Harvester.

Gibson, W.S. (1989) *'Mirror of the Earth': The World Landscape in Sixteenth Century Flemish Painting*, Princeton: Princeton University Press.

Harvey, D. (1989) *The Condition of Postmodernity: An Inquiry into the Origins of Cultural Change*, Oxford: Blackwell.

Kristeller, P.O. (1990) 'Rhetoric in medieval and Renaissance culture', in P.O. Kristeller *Renaissance Thought and its Arts*, Princeton: Princeton University Press, 228–46.

Ley, D. and Olds, K. (1988) 'Landscape as spectacle: world's fairs and the culture of heroic consumption', *Environment and Planning D: Society and Space* 6: 191–212.

McArthur, J. (1984) 'The panopticon and the theatre', in Foucault, Tafuri, Utopia: essays in the history and theory of architecture', unpublished master of design studies thesis, Queensland University.

Martines, L. (1980) *Power and Imagination: City States in Renaissance Italy*, New York: Vintage.

Mitchell, W.J.T. (1986) *Iconology: Image, Text, Ideology*, Chicago: University of Chicago Press.

Muir, E. (1981) *Civic Ritual in Renaissance Venice*, Princeton: Princeton University Press.

Muir, E. and Weissman, R.F.E. (1989) 'Social and symbolic places in Renaissance Venice and Florence', in J. Agnew and J.S. Duncan, (eds) *The Power of Place: Bringing Together the Geographical and Sociological Imaginations*, London: Unwin Hyman, 81–104.

Nicol, D.M. (1988) *Byzantium and Venice: A Study in Diplomatic and Cultural Relations*, Cambridge: Cambridge University Press.

Said, E. (1990) 'Narrative, geography and interpretation', *New Left Review* 180: 81–97.

Stoddart, D.R. (1986) *On Geography and its History*, Oxford: Blackwell.

Tanner, T. (1986) *Jane Austen*, Basingstoke: Macmillan.

Tuan, Y.-F. (1979) 'Sight and pictures', *Geographical Review* 69: 413–22.

Vickers, B. (1984) 'Introduction', in B. Vickers (ed.) *Occult and Scientific Mentalities in the Renaissance*, Cambridge: Cambridge University Press, 1–56.

Westmann, R.S. (1984) 'Nature, art and psyche: Jung, Pauli and the Kepler-Fludd polemic', in B. Vickers (ed.) *Occult and Scientific Mentalities in the Renaissance*, Cambridge: Cambridge University Press, 177–230.

Yates, F. (1966) *The Art of Memory*, London: Routledge & Kegan Paul.

Young, R. (ed.) (1981) *Untying the Text: A Post-Structuralist Reader*, London: Routledge & Kegan Paul.

5

THE LIE THAT BLINDS
Destabilizing the text of landscape
Jonathan Smith

Now I appear to myself as one who was under the delusion of being his own while he was the subject of a style.
(Czeslaw Milosz, *Unattainable Earth* 1986: 18)

It is frequently asserted that the distinguishing mark of a geographer's writing is its elaboration of the context of a place. With every increase in the complexity of the description of this context the place undergoes a corresponding simplification. In the end it is supposed to appear almost self-explanatory. It is simply a matter of putting places in place. If this is so, an irony appears, because one very geographical aspect of this context, the landscape, seems bent on contrary projects of displacement and decontextualization, both of its subjects and of its objects, by what I have called a blinding lie.

'Lie' is a strong word, perhaps overly strong as it charges with wilful deception an inanimate thing that at the worst, and then only to the degree necessitated by its nature, misrepresents. Misrepresentation is inherent to landscape, a term used here in the sense of scenery, for reasons I will discuss shortly. What is more important, however, is the complicity of this inherently deceptive text in the communication of social pretensions by privileged persons and groups. Here the word 'lie' may not be intemperate. At the heart of this essay is a discussion of the means by which the text of landscape stabilizes these pretensions through the feat of spatial displacement, and destabilizes these pretensions through irony. In conclusion I will suggest that, in addition to destabilizing pretensions expressed in the landscape, irony destabilizes idealist definitions of the medium in a manner that is altogether salutary for the postmodern conception of landscape.

ABSENT PORTENT

The displacement of the subject occurs because to the eye that has been properly educated a landscape presents itself as a spectacle, a deportment which in turn creates the position of spectator. Whether depicted in paint, or rolled out as a tableau vivant below a scenic overlook, a landscape situates its spectator in an Olympian position, and it rewards its spectator with the pleasures of distance and detachment and the personal inconsequence of all that they survey. Thus, in regarding the landscape as scenery the spectator is transformed into a species of voyeur.

Let me exemplify with a common reverie. I was driving east from the village of Canandaigua towards Geneva, which stands at the foot of Seneca Lake in the state of New York. The road undulated over swells of glacial drift, while the setting sun turned the corn stubble to gold and threw crisp shadows over the fields. Infatuated by the scene, I wished that it could be somehow transfixed. I did not want the sun to drop, or spring to come, or the road to end. I wanted to continue through a silent landscape where everything stayed the same, where there was no substance or consequence, where there was only stylized motion through a stylized scene. Then a truck loaded with crates of cabbage lurched into my path.

Apart from spectacular egoism, this episode illustrates a momentary resemblance of experience to memory. The landscape was, briefly, like a setting for life remembered. Its presence seemed to offer the 'sense of completion, of stability, of permanence' that Lowenthal (1985: 62) identifies as valued attributes of the past. The appearance was deceptive, of course. As Sartre (1964) wrote, one might as well try to catch time by the tail as to try to live a life as a life remembered. But the pleasure of the illusion was not impaired by its brevity or its futility. The pleasures of retrospection are multiple, but among the more prominent of these are the sense of detachment and the luxury of indifference. Reinhold Niebuhr (1951: 19) wrote that, with 'its weaker twin of foresight', it represents a human 'capacity to rise above', even while remaining within, 'the temporal flux'. It is the sense of transcendence provided by the metaphorical elevation of memory that landscape, treated as scenery, unwittingly reproduces and recalls. It is this partial escape from the temporal flux which landscape, treated as scenery, unfailingly represents as an occurrence (Barthes 1979: 17; de Certeau 1984: 92; Harvey 1989: 1).

Considering the reverie I have described, I must admit that it appears to be a rather shameless bit of plagiarism derived from one of those automobile advertisements in which there are no filling stations, pot-holes or traffic lights. But the subtext is more general. There is a scene, for example, in the movie *Betty Blue* (1986) in which a man and a woman drive out on to an open field. The horizon is spacious but, with the exception of a dilapidated cottage, empty; the light is almost preposterously golden on the dry meadow grass; the characters are heroically alone. Some dreamy, pointless things are said, significantly inconsequential things, things that stand outside the accelerating tragedy of the story, things that do not, as everything else so alarmingly does, matter in the least. As a second example, there is a scene near the end of the movie *The Unbearable Lightness of Being* (1988). Thomas and Katrina are returning from an inn, the paradoxes of their weightless existence momentarily resolved. They are driving on a straight dirt road; rain is falling lightly and the windshield wipers are slapping with a homey regularity; to either side are forests of young trees whose leaves are a luminous, tender, succulent green. Again, it is a scene removed from the tragedy that will ensue, that is already in progress. Both scenes are luminous – golden in one case, green in the other – but their real similarity is their resemblance to the placid finality of memory, their quite erroneous suggestion that the characters have escaped the temporal flux.

Both of these scenes, and my own episode, are examples of a sort of nostalgia in the present tense, and each relies heavily on an aestheticized landscape. Each offers a brief reprieve from warranted apprehension, or even terror, of the future. What this suggests is that we believe ourselves to have stepped out of history when we step into an aestheticized landscape, a landscape that seems pregnant with meaning when, and perhaps precisely because, it omits any reference to that which will follow. Its aura of poignant significance derives from the absence of portents of the pain that will come.

INNOCENT INSCRIPTIONS

At the same time that a landscape invites its subject to indulge in the illusion that they have stepped outside of their context of consequential actions, and to assume a position from which the future is obstructed, it also raises a wall that blocks the past. This is not readily apparent since the landscape is, above all else, a legacy of the past. But it is precisely the ability of landscape to outlive that past, its tenacious durability, which causes its objects to pile up in front of history, shielding it from our view and substituting a seemingly greater reality of spotless innocence for its guilty and gritty processes.

To understand the importance of landscape as an inscription of the past we must recognize the critical difference between the evanescent and the enduring, the saying, as Geertz (1983: 19–35) puts it, and the said. As an example of the said, we may consider the landscape in which gestures, which correspond to a sort of saying, are inscribed. Here, as in the transformation from the saying to the said, the speaker often slips away (Stock 1990: 45). The act of will which is concretized, or we might say reified, in a landscape is no longer centred on the person or people who made it. It stays put, while the creative agents move away. Without the aid of deconstructive critics, the author is decentred. At the same time, the historical context decomposes. Over time the enduring element is alienated from both the agency and the scene of its creation, and with this displacement it loses the taint of intention and assumes the purity of nature. One thinks of David Harvey's (1979) discussion of the specious innocence of Sacré Coeur, which was lifted, like the scenes that I have described, from the tragic and terrifying process of history that brought it into being.

In addition to innocence, the memorialized event seems to assume a greater reality than the ephemeral experience. In writing or telling, the event is concretized in the symbols of language, and its meaning is closed, or fixed, and at least partly insulated from the disordering influence of multiple reinterpretations. Following the ontology of Huston Smith (1982), we might say that the recorded or embodied event is more real than the unrecorded event because, transformed into symbols, it is able to exist beyond the time and the place of its original enactment. Just as money allows the dislocation of the points of sale and purchase, symbols allow a dislocation of event and meaning. Like language, landscapes can serve as a device for the chronological extension of ephemeral intentions. Because they are able to endure, which is a function of their tangibility, landscapes are believed to possess a reality surpassing that of the process by which they were created. At the same time, their endurance allows them to be cleansed of the taint of their creators, and to displace themselves from this context into the realm of private memories. And, as a character of Aldous Huxley's (1929: 114–15) observes:

> They are dangerous, these things and places inhabited by memory. It is as though, by a process of metempsychosis, the soul of the dead event goes out and lodges itself in a house, a flower, a landscape, in a group of trees seen from the train against the skyline, an old snap-shot, a broken pen knife, a book, a perfume. In these memory charged places, among these things haunted by the ghosts of dead days, one is tempted to brood too lovingly over the past, to live it again, more elaborately, more consciously, more beautifully and harmoniously, almost as though it were an imagined life in the future. Surrounded by these ghosts one can neglect the present in which one bodily lives.

STYLE

We have seen how the landscape, regarded as a visual text, tends to decon-textualize both its subject and its objects, and become what Roland Barthes would call an 'empty sign' (1979: 3–22). This empty sign is easily appropriated as 'the soul of the dead event goes out and lodges itself' in the object and a landscape becomes the signifier of a reverie. The basic style of these reveries is what Raymond Williams (1973: 18) called the 'enamelled' pastoral. This is one part of a larger cultural style that is founded on a privileging of visual experience and visible material over which 'one is tempted to brood too lovingly' because of the apparent fullness that results from the almost complete evacuation of historical meaning.

The terms of this cultural style of visualization are aesthetic. As in all cases of styling, something ordinary is made extraordinary. Its lure is the promise, as Stuart Ewen (1988: 14) writes, 'that it will lift us out of the dreariness of necessity'. As in the world of commercial style, this aesthetic approach to landscape seems to yield a world without limits, constraints, inhibitions or consequences. This is the illusion that makes the simulacrum more compelling than the reality itself (Eco 1983). In the episodes with which I began the characters are hoisted out of a world of dreary necessity and dropped, albeit temporarily, into an infantile heaven where nothing would seem to impinge on their freedom, where nothing would seem to bound their luxurious desire. They are incorporeal spirits in fields where no one toils, Elysian Fields not Potter's Fields, which are certainly more desirable than the reality.

To illustrate the enhanced desirability of the stylized event, we might turn to three works on the uses of the past. In his magnificent study of why people 'writing out pathology ... use past texts to organize their experience', the historian John Owen King (1983: 7–9) has argued that 'textual expressions ... are in themselves capable of creating a person's character'. As persons 'read their lives against past scripts' they lead their lives 'according to and in fulfillment of scripture'. In so doing they relate their experience, terrifying in itself, to a larger pattern, in this case the Puritan conversion experience as mapped by John Bunyan, which has a predictable and desirable conclusion.

The anthropologist Victor Turner (1982) has found a similar principle in social dramas, eruptions of disorder for which some meaning must be discovered. Enacted by the members of what he calls a 'star group', or those from whom the member's identity is drawn, these dramas begin with a breach in the group, they unfold as a crisis, they climax as an attempted redress, and they conclude as either reintegration or schism. Because these groups constitute what H. Richard Niebuhr (1943) called a value centre, their disruption occasions a cosmic crisis. Like its symbolic situation, the self is thrown into terrifying doubt. This is a liminal stage, a threshold. Crossing it the individual signifies the event as something extraordinary. Containing this event, their own biography becomes extraordinary, aesthetic, desirable.

Hayden White (1980: 11) has argued that historical narrative attempts a similar signification, as it 'strains ... to put an image of continuity, coherency and meaning in place of the fantasies of emptiness, need and frustrated desire that inhabit our nightmares about the destructive powers of time'. This image is 'concluded' when the pattern is revealed, when the moral that is their ultimate referent is identified. This completion, this 'closure' that makes the narrated more full and real than disordered experience, is reached through a simplification of what happened to what what happened 'meant'. The meaning is the symbol in which the event is inscribed, through which the event endures and by which the outcome is made the proper object of collective desire.

Each of these is a means to organize disorganized experience. Each is an example of a decision to regard an event in terms that make its meaning clear and final. In each of these cases trauma is settled by writing the experience out in particular terms. Fear is overcome by styling the event in such a way that it is seen as profoundly significant and desirable. The style of these interpretations is dictated not only by dominant groups and historical precedent, but also by the inherent structures of the written text. This suggests the question of how the visible landscape might structure our regard of elements in that landscape. Particularly, this leads to the question of how it might, when judiciously styled, structure our regard of groups with certain social pretensions, privileged groups with a particular stake in the mode in which they are regarded.

A pretension is a claim to distinction, either from the phenomenal and ordering principles of the natural world, or from other humans (Niebuhr 1932, 1941). It is most convenient when social pretensions are honoured by witnesses who are impressed with their suitability and fitness. Coercive devices can discourage the expression of incredulity, but only at the cost of real credibility. In conventional terminology pretensions that are accepted carry the weight of authority, while pretensions that have overcome the initial setback of disrespect with supplemental threats of violence carry the opprobrium of authoritarianism. I am not concerned with authoritarian pretensions or the modes by which their stability is ensured. It is the power of right rather than the power of might that interests me, mainly because there is something dubious in any claim that appears so securely founded.

The most common pretensions in Western societies are economic. Wealth can, of course, be used to buy and build landscapes; and for secure public regard few investments are better. This is simply because people who live in big houses seem with only rare exception to be the sort of people who deserve to live in big houses. Big houses say so many complimentary things about their owners. F. Scott Fitzgerald (1922: 124) caught the meaning in *The Beautiful and the Damned* when he wrote:

> at the left a great bulk of granite and marble muttered dully a millionaire's chaotic message to whosoever would listen: something about 'I worked and I saved and I was sharper than all Adam and here I sit, by golly, by golly.'

Only somewhat less common are political pretensions. Take, for example, a quotation from John Kennedy (1895: 19), a writer on western New York:

> I have said that the big red barn is an infallible sign of fertility ... but the big white house, with its fresh coat of paint, and its fresh green trimmings, and its well-kept little front of lawn and flowers and shrubbery, and with its large sightly trees casting a grateful shade from their picturesque branches of foliage, is an index of an entirely different matter. It is an index of the population; it is an index of race and civilization.... It is the highest glory of the Genesee Country that you can never get away from the big white house, the house that sends forth judges, generals, governors, senators and presidents.

Kennedy's index consists entirely of landscape elements, and it registers the pretentious claim of the declining rural elite to unusual political skill and virtue (Rossell 1988). The fresh white paint of the house elicits advantageous comparison to sooty urban brick; the tidiness of the lawn suggests the wholesome pastime of gardening; the shade, in which unhurried ruminations on the destiny of the Republic flourish, is a symbol of the cool deliberation prized in statesmen. One has only to appear beneath these trees to take on these unimpeachable qualities.

In both of these cases, a pretension is stylized in a landscape, and extravagant claims are presented as true. The claim to extraordinary civic virtue or extraordinary economic privilege is somewhat preposterous – the necessity of the tie between the signifier and the signified is in a sense a lie – but as readers we are quite frequently blinded by this lie. This favourable appearance, this blinding, is possible in part because the activities of governments and privileged classes are mediated to a degree which is impossible for the ordinary citizen. These groups are in a position to manage impressions, and thus to aestheticize the self or the group – making the self or the group both extraordinary and extraordinarily deserving.

Michel Foucault described this asymmetrical allocation of mediation as panopticism (1979: 195–228). By this he means 'a machine for dissociating the seeing/being seen dyad', a device which employs observation as a discipline. This is the reason that one-way mirrors are the lens of choice for the police. They are allowed information that the unshielded eye is denied, and an asymmetry of power is established. In the landscape the same inequality is enforced by the privacy of privileged persons and the crowding of their subordinates. To use Goffman's ever useful terms, privilege co-opts the right to have all its public existence on a 'front stage', while it is at the very heart of subordination to be caught at all times 'backstage'. Anyone who lives in a shoddy apartment building or tenement knows when their neighbour uses the toilet, showers, copulates, argues, or weeps through the night; by the public aroma of their cooking they know what their neighbour has prepared to eat; by the barely muffled shouts, stumbles and crashes, they know when their neighbour has had too much to drink. This knowledge then adds irony to their neighbour's public display, although this is generally avoided due to a well-founded fear that their neighbour possess similar intelligence.

For the privileged, on the other hand, the estate that surrounds their houses is a mirror, and the unflattering mechanisms that make their life possible are as obscure as the source of their wealth. Thus, the unequal allocation of space which the landscape fixes both expresses and creates a more significant inequality. The privileged are allowed to match their aestheticized ideal, while the unprivileged are condemned to exist for the public in all their grim reality. Privacy is a privilege that supports the pretensions of the privileged; for the unprivileged, the absence of privacy riddles their pretensions with absurdity and irony.

This inequality is given what is perhaps its sharpest focus in the domain of production. The privileged produce their wealth privately, either in a private office or through surrogates in a distant factory. They therefore seem simply to have what most of us must conspicuously get. Producing in private or at a distance gives the consumption that production allows the appearance of an entitlement. In *Tender Is the Night* (1933: 65, 111), Fitzgerald describes Nichole Diver as 'the product of much ingenuity and toil', for whose sake trains ran, factories fumed and men and women laboured. Rosemary, a temporary protégée observes on a shopping trip:

Once again they spent their money in different ways, and again Rosemary admired Nichole's method of spending. Nichole was sure that the money she spent was hers – Rosemary still thought that her money was miraculously lent to her and she must consequently be very careful of it.

The greater confidence with which Nichole spent her money is in part attributable to her more voluminous riches, but Fitzgerald is attempting to say something about the effect of the source as well as the size of her fortune. He is suggesting that Nichole's sense of proprietorship over her money was enhanced by the social and geographical distance that separated her from the place of its production. Meanwhile, Rosemary doubted her desserts. Although she worked for what she had, it seemed miraculous, ironic, almost pretentious that she should have it.

It seems that spatial control and geographical distance may be crucial to the naturalization and the stabilization of pretensions. Claims to extraordinary virtue, an exceptionally coherent personality or extravagant economic rewards all derive some portion of their legitimacy from the geographical cloak that surrounds the claimant. At the same time charges of turpitude, duplicity and rapacity are most easily levelled when immediate visual surveillance destabilizes the pretensions of the accused. Stable pretensions rest on the luxury of denying unacceptable behaviour without fear of exposure; preposterous pretensions collapse under a mounting weight of ironic contradictions. Following the somewhat idiosyncratic definition of Milan Kundera, this luxury of denial implicates these pretensions as kitsch. In *The Unbearable Lightness of Being* he writes: 'kitsch excludes everything from its purview which is essentially unacceptable in human existence' (1984: 248, 256). This is particularly a denial of human excrement, actual and metaphorical. But there is no escape from kitsch, he continues, as 'kitsch is an integral part of the human condition'.

The landscape mediates social communication, and privileged landscapes are designed to hide whatever defiles. Thus they serve as an instrument that maintains what Kundera defines as kitsch. This also satisfies more conventional definitions of the word, since landscapes, particularly the highly stylized landscapes of consumerism, often appear to satisfy desires which they in fact frustrate. This definition is taken from the famous essay on kitsch by Clement Greenberg (1939). Greenberg begins with the argument that abstract, non-representational art is an 'imitation of imitating' (ibid.: 37), which takes nothing but its own form, and formation, for content and therefore has no pretensions. Drawing an analogy with which I cannot agree, he writes that these imitations are 'valid, in the way a landscape – not its picture – is aesthetically valid' (ibid.: 36). Kitsch art on the other hand pretends to mean something, and thereby solves its own problem by furnishing an answer to the simple question that it poses. It does not demand reflection. It 'reflects' for the viewer (ibid.: 44). It interprets itself, and the viewer accepts its interpretation without resistance. Thus kitsch art commits the deception of presenting meaning as something that can be consumed through reception when it is in fact something that can only be created through reflection. While the counterfeit satisfies, it also frustrates, since it hides the fact that society fails to equip most of its members with the faculties to generate meaning in the presence of art.

In Western societies landscapes are also presented as objects for visual consumption, and their meaning – both immediate and as interpreted by guides – is construed as something that is given. Greenberg calls them 'valid'. However, this visual consumption may forestall reflection on the failure of society to furnish its members with the means to consume landscapes in more practical ways, and thus it may frustrate political efforts to enact equitable distribution of its elements.

For individuals who do in fact possess some fragment of a landscape, stylistic emulation fosters a related illusion of participation. In his discussion of the striking emphasis on styling in the consumer products that filled the American landscape after the Second World War, for example, Thomas Hine (1987: 90) elaborates the simple notion that this was merely a means to ensure premature obsolescence and boost consumption. He observes that through styling, which primarily evoked the past or the future, objects that had become ordinary necessities retained the appearance of extravagant luxuries. This appearance of having been included in the world of luxury, which had the odd pretension of exclusivity, was augmented by automobile styles that mimicked the forms of jet fighters, satisfying a corollary desire 'to participate in the technological revolution'. Allen Gowans (1986) makes a similar point in regard to the post-Victorian suburban house. Possession of one of these symbols, he writes, implied possession of the past and participation in an aestheticized America.

IRONY

The reverie that I recounted at the beginning of this essay was terminated by a truck loaded with cabbage. This brought me back to the world of consequential actions. I stepped on the brakes and forgot about landscape. This was an irony which subverted my silly reverie, and exploded my fantastic pretension by hinting at aspects of reality, cabbage farmers and collisions, which it did not include. In the landscape subversion is triggered by the fleck of irony that can make farce of any represented pretension. Irony is a representational discrepancy, a symbol out of place. Irony betrays the existence of representation in the purportedly presentational, and opens social convention to exploration and critique.

It is to avoid irony that the social performer exercises what Erving Goffman (1959: 34–66) called 'synecdochic responsibility' to control expressive 'dissonance'. Irony occurs when there is an incongruity in what Kenneth Burke (1969) called the scene–act–agent ratio: when the scene contains actions that are not assumed implicit to the scene. A pigeon on the head of an august statue; fresh dandelions on a recently trimmed lawn; the infelicitous faeces of fugitive dogs: all are minor ironies. Anything that is swiftly cleaned up or cleared away is an irony that threatens the stability of the scene. Beer bottles on the church steps; 'bums' in the civic centre; an automobile immobilized; a radio blaring in the wilderness. Ironies are also those things it is tactful to ignore. In respectable neighbourhoods the vehicles of extermination companies and collection agencies are all insufferably ironic. When Leo Marx (1964: 25) describes the 'machine in the garden' he is describing an awkward irony. Also ironic are unwelcome reminders of urban alienation that crop up in what Raymond Williams (1973: 298) described as the pastoral fantasy of 'a world in which one is not necessarily a stranger'. As with ironies in social encounters, it is tactful to ignore ironies in the landscape. For example, when Zaring (1977: 410) describes romantic travellers in Wales as 'studiously blind' to the presence of agricultural, social and industrial change, she is describing tact, or the willingness of the audience to 'sustain the quite false impression that they have not absorbed the meaning of what has happened' (Goffman 1959: 232).

Although actions often destabilize the scene of their enactment, it is possible for a scene to repulse the challenge of an act and make the act or the actor ironic. In Sherwood Anderson's *Winesburg, Ohio* (1919), for example, a horrible recognition of the irony of her own self comes to Alice Hindeman after her wild and desperate adventure running naked through the rainy streets of Winesburg. She finds that her impetuous gesture has no place in the scene, and that she has been spurned. Humiliated, she retreats on hands and knees. In this case it is Alice Hindeman who is made ironic; it is her dream of release that is subverted as she comes to realize that 'many people must live and die alone, even in Winesburg'.

When closely observed, every self-image humans have written into the landscape will betray its pretensions with ironic affirmations of an order that is both wider and weirder. This is why it is best to remain sceptical and assume that a gap exists between the thing as it would be taken and the thing as it is, between the synchronic landscape, which is a representation, and the diachronic world that it temporarily represents. It is in this gap that the ironic discrepancies may appear, and through these discrepancies that the 'imperial urban cosmology', of which the built landscape is surely a part, is revealed as incongruous, insufficient, inadequate. The responses to these ironies, the religious historian Jonathan Z. Smith (1978: 289–309) observes, are either a dreadful fright or a joke. Each culture discovers the perversity of its 'structure of limits' and the perversity of the myth that supplanted the creation of these limits for the creation of the world; and each culture develops techniques for escaping these limits, a laugh or a cry as they crumble away. Smith has described this structure of limits as a 'locative tradition'; I have called it a style or a pretension; and it is from this, Smith concludes, that 'a whole language of symbols and social structures will follow'. Concomitantly, as a new society takes place, or existing society takes a new place, it is the symbols of reference that suffer and work change (ibid.: 140–1).

POSTMODERN LANDSCAPE INTERPRETATION AND ITS HAZARDS

In this essay I have suggested three ideas, displacement, pretension and irony, which I find add a useful precision to my understanding of the analogy of landscape as a text. They are useful because they seem to define the places where it is and is not text-like, the places where it can and cannot be made to suffer and work change. This requires further specification, however, since there is a temptation, difficult to decline, to overestimate the destabilizing power of 'rewriting' the world.

Action, Kenneth Burke (1969: 3–20, 38–43) argued, articulates a quality that is ambiguously contained in the scene. Explicating this quality the action becomes a synecdoche of the scene, and the actor participates in the scene. Worship is implicit in the sanctuary; fights are implicit in waterfront bars. These actions realize their scenes. Although scenes call for a particular behaviour, it is quite possible to blunder or consciously rebel, and thus introduce an irony. When we say that such behaviour is uncalled for, it is the scene that we impute as mute. Consistent blundering or rebellion will rewrite the scene and transform the ambiguous quality until the blunder or the rebellion becomes the implicit action and is relieved or stripped of the impact of incongruity. It will also define a new set of ironies.

For example, the action implicit in the rooms of a college dormitory has changed significantly over the last twenty years. With persistent rebellion students rewrote the meaning of that scene. Some universities are attempting to restore the old meaning. Whatever one's opinion on the purpose and perils of dormitory life, this struggle between differently organized power groups over the meaning of an environment is an example close to home of a struggle to signify environments and through this signification to authorize or discredit action.

We are always acting with and against scenes. From the scene our act elicits a reaction, and this reaction defines the initial action, and by extension the actor. The act has an active and a passive phase, it is acted and suffered. Whether it reinforces or discredits the scene it is a definition by the actor which rebounds as a definition of the actor. This 'dialectic of tragedy' is inscribed in the self in the transformation from 'actus to status': from one who does a defining act, to one who has done a defining act; it is inscribed in the scene as the social definition of that scene's quality. In essence this is not unlike the naïve experimentation by which we learn the laws of nature. We try something, see what happens to us and go forth with new understanding. But this is also the way that we explore the rules of our culture. As a subset of this, it is the way that we explore our landscapes. But this notion of exploration requires a further distinction.

Any landscape is, in part, a 'work' consisting in itself as the construction of specific individuals and parties in pursuit of specific technical, political and, sometimes, artistic goals (LaCapra 1982). An individual village, for example, consists of unique spaces organized and divided by real lines manifested in concrete substances like wood, brick and barberry bush. The individual village has a surface, more or less obdurate, and a morphological factuality, more or less secure. At the same time the village is a 'text', and it is as a text that we explore it (Barker and Wright 1954). Unspecified readers, impelled by unspecified intentions, are offered opportunities to act and to have their act defined. It may be a technical goal of any given landscape to control this response, and to limit interpretative play, but few landscapes can scotch every contestatory reading. Thus, for example, a crossroads offers the traveller an opportunity to act. The act it suggests is an act of decision. In fact the crossroads is a metaphor of decision, even if it is only rarely that we treat it as two roads that diverge in a yellow wood.

Viewing landscape as a 'work', we ask what it is and how it got there. The answers are seldom simple. Viewing it as a 'text', we ask how, or by whose authority, is it able to mean what it does? For those parts of the landscape that are examples of what Langer (1951) calls signals and signs the source of authority is seldom recondite, but in the case of what she calls the symbol it is far from clear. For example, a cottonwood tree growing on the prairie signals water; the smell of coffee signals breakfast; the sound of a doorbell signals a visitor. Each of these is limited to their respective natural, cultural and social environments: but the *communiqué* and its message are really a part of the event they announce. The tie that binds the signifier and the signified is stabilized by natural laws. The sign or icon, which operates by some sort of homology, allows greater opportunity for deception or misreading, but the link between signifier and signified is in some sense obvious. The silhouettes of unlucky adversaries with which military aviators adorn their aircraft is iconic. No mystery surrounds its selection. The barber's pole is a chromatic icon recalling blood-letting. Again, no mystery here. The Christian's cross, although it may intrigue with the story of its ascendence over the fish, is ultimately an icon of an instrument of painful execution and through this a synecdoche of suffering. Unlike these examples the symbol is arbitrary and its authority is conventional. Numbers are such a symbol. Letters and phonemes are such symbols. Money is such a symbol. These symbols are stable only so long as a power is in place to penalize misreadings.

Symbolic meanings, then, are held in place by power, and it is only by challenging a definition that we can discover where this power lies. It is only through an interpretative struggle with the text of landscape that we can discover how this power operates. If you want to know what makes a red light mean stop, drive through one. If you are really interested in what makes a brick wall say stop, drive into it. What you will discover is that the red light, unlike the brick wall, is potentially unstable. In other words, the landscape defines our actions in two ways: it tells us what we can do and what we may or ought to do. As another example, consider a graveyard. As stones the markers say do not walk into me, and unless you want to skin your knees you will comply. At the same time, as headstones they say read me and contemplate your own mortality. This is the culturally and historically specific symbolism that is held in place by conventions that are worth exploring and, perhaps, upsetting. This is the part of landscape that is an unstable text, this is the point where the landscape touches the inter-textual.

Near the end of his book *In Defense of Literary Interpretation* (1986), K.M. Newton ventures a parody of Foucault. He describes a beach on which there is a sign bearing the words 'Beware of the sharks'. The vast majority of those who read the sign will take it to indicate that they should not bathe on this particular beach. At the same time a suicidal individual, whose death wish has problematized the assumption of the dominant reading – the desirability of staying alive – could accomplish a deconstructive reading, and take the sign to indicate that this was an ideal place to bathe. As J. Hillis Miller has observed, this reading, like any deconstructive reading, is parasitic on the obvious reading. The sign cannot be an invitation to those who would die unless it is at the same time a warning to those who would live. This leads to the observation that the sign refers to two things, the sharks which no reading can remove, and the desirability of life, which is, to the suicidal individual, highly questionable. It refers to a fact of nature and a fact of culture, and the second part is the text. If a mutilated body washes ashore it is not at all ironic in light of the first reference. Sharks do that sort of thing. But it is highly ironic in light of the second reference, which says that people don't.

The literary critic Harold Bloom (1982) has described reading as an agonism, as a struggle between a reader and a text. This is a struggle of adversaries who strive to usurp each other's meaning, who dodge and squirm to escape each other's dictations, who are simultaneously recalcitrant and imperious, who would define but will not be defined. The progress is undetermined; the settlement is inconclusive. In the end the reader and the text fall back, so to speak, still discrete; fatigued perhaps, but not exhausted or consumed. In metaphor at least, strong reading is faintly erotic. Barthes (1981: 96) has used this same metaphor to describe our relations with and in urban landscapes, 'the place', as he puts it, 'of our meeting with the *other*'. The same construction can be placed on our encounter with *other* meanings in the landscape. The project is alluring: I only hope that it will not be brought to a premature termination by an optimistic plunge into seas that are teeming with real, live and quite implacable sharks.

REFERENCES

Anderson, S. (1919) *Winesburg, Ohio*, New York: B.W. Huebsch.

Barker, R.G. and Wright, H.F. (1954) *Midwest and its Children: The Psychological Ecology of an American Town*, Evanstown, Il.: Row, Peterson.

Barthes, R. (1979) *The Eiffel Tower and Other Mythologies*, trans. R. Howard, New York: Hill & Wang.

—— (1981) 'Semiology and the urban', in M.Gottdiener and A.Ph. Lagopoulos (eds) *The City and the Sign: An Introduction to Urban Semiotics*, New York: Columbia University Press, 87–98.

Bloom, H. (1982) *Agon: Toward a Theory of Revisionism*, New York: Oxford University Press.

Burke, K. (1969) *The Grammar of Motives*, Berkeley: University of California Press.

de Certeau, M. (1984) *The Practice of Everyday Life*, trans. S. Rendall, Berkeley: University of California Press.

Eco, U. (1986) *Travels in Hyperreality*, trans. W. Weaver, New York: Harcourt Brace & Jovanovich.

Ewen, S. (1988) *All Consuming Images: The Politics of Style in Contemporary Culture*, New York: Basic Books.

Fitzgerald, F.S. (1922) *The Beautiful and the Damned*, New York: Scribners.

—— (1933) *Tender Is the Night*, New York: Scribners.

Foucault, M. (1979) *Discipline and Punishment: The Birth of the Prison*, trans. A. Sheridan, New York: Vintage.

Geertz, C. (1983) *Local Knowledge: Further Essays on Interpretive Anthropology*, New York: Basic Books.

Goffman, E. (1959) *The Presentation of Self in Everyday Life*, Garden City, NY: Doubleday Anchor.

Gowans, A. (1986) *The Comfortable House: North American Suburban Architecture: 1890–1930*, Cambridge, Mass.: MIT Press.

Greenberg, C. (1939) 'Avant garde and kitsch', *Partisan Review* 6: 34–49.

Harvey, D. (1979) 'Monument and myth', *Annals of the Association of American Geographers* 69: 362–81.

—— (1989) *The Urban Experience*, Baltimore: Johns Hopkins University Press.

Hine, T. (1987) *Populuxe*, New York: Knopf.

Huxley, A. (1929) *Those Barren Leaves*, London: Chatto & Windus.

Kennedy, J. (1895) *The Genesee Country*, Batavia, NY: Calkins & Lent.

King, J.O. III (1983) *The Iron of Melancholy: Structure of Spiritual Conversion in America from the Puritan Conversion to Victorian Neurosis*, Middletown, Ct: Wesleyan University Press.

Kundera, M. (1984) *The Unbearable Lightness of Being*, New York: Harper & Row.

LaCapra, D. (1982) 'Rethinking intellectual history and reading texts', in D. LaCapra and S.L. Kaplan (eds) *Modern European Intellectual History: Reappraisals and New Perspectives*, Ithaca: Cornell University Press, 47–85.

Langer, S. (1951) *Philosophy in a New Key: A Study in the Symbolism of Reason, Rite and Art*, Cambridge, Mass.: Harvard University Press.

Lowenthal, D. (1985) *The Past is a Foreign Country*, Cambridge: Cambridge University Press.

Marx, L. (1964) *The Machine in the Garden: Technology and the Pastoral Ideal in America*, New York: Oxford University Press.

Milosz, C. (1986) *Unattainable Earth*, trans. C. Milosz and R. Hass, New York: Ecco Press.

Newton, K.M. (1986) *In Defense of Literary Interpretation*, New York: St Martin's Press.

Niebuhr, H.R. (1943) *Radical Monotheism and Western Culture*, New York: Harper & Row.

Niebuhr, R. (1932) *Moral Man and Immoral Society*. New York: Scribners.

—— (1941) *The Nature and Destiny of Man*, Vol. 1, New York: Scribner's.

—— (1951) *Faith and History: A Comparison of Christian and Modern Views of History*, New York: Scribners.

Rossell, D. (1988) 'Tended images: verbal and visual idolatry of rural life in America, 1800–1850', *New York History* 69: 425–40.

Sartre, J.P. (1964) *Nausea*, trans. L. Alexander, New York: New Directions.

Smith, H. (1982) *Beyond the Post-Modern Mind*, New York: Crossroads.

Smith, J.Z. (1978) *Map Is Not Territory*, Leiden: E.J. Brill.

Stock, B. (1990) *Listening for the Text: On the Uses of the Past*, Baltimore: Johns Hopkins University Press.

Turner, V. (1982) *From Ritual to Theater: The Human Seriousness of Play*, New York: Performing Arts Journal Publications.

White, H. (1980) 'The value of narrativity in the representation of reality', in W.J.T. Mitchell (ed.) *On Narrative*, Chicago: University of Chicago Press, 1–24.

Williams, R. (1973) *The Country and the City*, New York: Oxford University Press.

Zaring, J. (1977) 'The romantic face of Wales', *Annals of the Association of American Geographers* 67: 397–418.

Part II

ON REPRESENTING RESIDENTIAL LANDSCAPES

6

REVALUING THE HOUSE

Deryck Holdsworth

For many geographers working in alternative genres of the discipline, the label 'cultural geography' still seems to conjure up studies and descriptions of fence and barn types. More charitable observers have allowed that these artefacts were occasionally woven, along with other elements of material and non-material culture, into eloquent descriptions of places. Typically, material objects have been data points from which something called 'cultural regions' could be discerned. More than barns and fences and gravestones, the key artefacts for delineating such regions have been houses, and in particular folk or vernacular houses. Yet for all the efforts within this tradition, it has appeared to many human geographers to be puzzlingly adrift from the rest of the discipline. The focus on relic features of a largely rural, pre-industrial society has been criticized for demonstrating a nostalgic preference for pre-modern, even 'pioneering' worlds. The tradition has made little attempt to assess industrial or urban societies, a difficult task given that the artefacts that had been privileged were valued because of what they revealed about the individual in an early, implicitly folk society. There was an act-of-faith assumption that a one-to-one correspondence existed between house and individual. Individuals had mental constructs about how dwellings ought to be built and these ideas moved with people as they migrated; thus their new folk houses were 'sure traces' of culture.

This essay first assesses the house-as-key-to-diffusion tradition within American cultural geography, arguing that there has been an undue emphasis on form and type at the expense of other factors that tease out social and economic meanings. It then contrasts the approach of cultural geographers with those of other scholars from related disciplines, who also focus on the house, to suggest the necessity for a more complex political and social analysis. The essay concludes with some suggestions for revaluing the house as an object of analysis for cultural geography for the 1990s.

THE TRADITION OF HOUSE DIFFUSION IN CULTURAL GEOGRAPHY

Although there had been an interest in the house as a geographic form for most of this century, Fred Kniffen's Presidential Address to the Association of

American Geographers in 1965 can be said to offer a watershed (Kniffen 1965). In his essay on folk housing as the key to diffusion, he proposed the concepts of 'initial occupance' and the 'dominance of contemporary fashion', the former to identify distinctive house types brought by European settlers to regions along the Atlantic seaboard and the latter to explain why their westward-moving variants developed a distinctive appearance at different times in different places. It was a suggestive piece, written at a time when geography and geographers were beginning to ride the tide of the quantitative revolution. In this approach more data were required, as precise as possible, and Kniffen was suggesting some potential sources and some fascinating conclusions. In essence he was saying: here are some house types built originally in colonial New England; here is where you find variations in other, more western states at later dates in the eighteenth and nineteenth centuries; here are still other house types that seem to be present in other states; and here are possible lines of connection. Kniffen pointed to county atlases as a data source for old house types in addition to promoting the merits of field survey. To support a map of three dominant diffusion corridors, he presented fifteen photographs and seventeen line drawings of houses and barns as well as five other maps.

Kniffen's paper was a bold advance on his previous, Louisiana-based work and it tried, on a canvas of the eastern United States, to bring the house as an object of research focus into more prominent view within geography. Until then, the tradition of examining housing had been somewhat marginal, at least in American geography. For example, in their review essay on cultural areas and distributions, Wagner and Mikesell include only six sentences and six footnotes on the contribution of house types (Wagner and Mikesell 1962: 58–9) although their widely used reader did reprint Kniffen's study of house types and culture areas in Louisiana (Kniffen 1936).

Kniffen's perspective has had a curious impact. At one level, it was remarkably deep, not the least because of the meticulous work that the folklorist Henry Glassie presented with Kniffen (Kniffen and Glassie 1966) and subsequently by himself as one of the major scholars in material culture studies over the last three decades (see, among many, Glassie 1968, 1974, 1975). Both the detailed structure-by-structure analysis of county-level case studies and the broad generalizations that Glassie offered created a fertile field of enquiry emulated by many in material culture (see, for example, Carter and Herman 1989; Wells 1982, 1986). But within cultural geography, there has been little follow-up. In the series edited by Wagner on Foundations of Cultural Geography, Amos Rapoport's volume on *House Form and Culture* makes no reference to Kniffen's work (Rapoport 1969). By contrast, Walker and Haag (1974) include four essays by geographers on house types in their volume in honour of Kniffen, including one by Estyn Evans, the doyen of Irish folk geography, engaging Kniffen on some of his 1965 arguments. The fascination with log houses (e.g. Jordan 1978, 1985; Jordan and Kaups 1989; Kaups 1981, 1983; Newton and Pulliam-Di Napoli 1977), also lent itself to the diffusion approach.

Within mainstream human geography, Kniffen's ideas largely fell on deaf ears. The rich body of geographical essays that were developing through the translation of Hägerstrand's work on diffusion (Hägerstrand 1967), stimulating considerable research in Anglo-American geography through the 1970s (see Gregory 1985 for an overview), did not consider Kniffen's perspective nor his type of evidence. The types of diffusion studied were those that could be treated in somewhat sharper categories – information and innovation – than those based on vague sets of houses dating over a couple of centuries and spread over side roads in fourteen states. It is interesting to note that for Blaut, writing in the wake of the first tide of innovation diffusion literature in geography, Kniffen's conceptual diffusion theory was regarded as important as Hägerstrand's formal theory. The need for clarity, breadth and depth, central to Kniffen's programme of research, was presented as an important reminder for those in the field of diffusion studies (Blaut 1977).

It is therefore a disappointment that given the possibilities for dialogue and cross-learning in the 1970s, the two traditions did not come together. It was clear that many human geographers were not looking to artefacts for evidence, while traditional cultural geographers did not seem to think of hierarchical diffusion, which might have clarified emerging urban and metropolitan influences. By privileging folk over other agencies, only the arrows of contagious diffusion were considered in this typical framework. As a result, Kniffen's broad conceptualization persists a quarter of a century later as the same broad and impressionistic lines floating on a broad surface.

One geographer who did help to keep the perspective alive was Peirce Lewis (1970, 1975). Even so, the map is still essentially the same, the arrows going in the same directions from the same sources, with Lewis mapping Kniffen's concepts on to his own field observations and adding his own illustrations, but with no detailed 'quantities' of data. For Lewis, 'common' houses had a very elastic definition, embracing massive houses in New Castle, Delaware, and simple one-room cabins in Maysville, Kentucky. There were poor folk and rich folk, some in meaner houses, others in grander houses; there were even rows of company houses in coal country (West Virginia) seen as 'I-houses', thus folk extensions into the upland south; rows of 'pyramidal houses' in coal patches near Carbondale, Illinois, were interpreted as common houses from the south. These common, ordinary houses perhaps were being used to construct 'ordinary landscapes', a theme that Lewis was to develop more rigorously later in the decade. But by that time, the 'Kniffen arrows' are used as proof rather than conjecture (Lewis 1979: 16):

> people in one place see what is happening elsewhere, like it, and imitate it if possible. The timing and location of such innovative changes are governed by various forms of geographic and social diffusion, which are surprisingly predictable, and which tell us a good deal about the way that cultural ideas spread and change.

In a stock-taking of work by cultural geographers on structures (Bastian 1982), the arrow maps appear again, along with a variety of illustrations of generic house types. In his introductory essay, Bastian makes some revealing observations about methodology:

> Field work is the essence of research concerned with traditional rural architecture. Mature scholars have often based their notions about patterns of distribution on impressions derived from extensive travel over the years. Unseasoned observers with less self-confidence have relied more upon systematic sampling techniques. Archival research usually plays a minor role in investigations of rural structures.
>
> (1982: 71)

Thus only decades of travel, and confidence, permit the arrows to be constructed; anyone starting off on their journey has little solid ground, other than conceptual doodles, on which to build. Bastian concludes by speculating on whether different house types were symbols of the social status of initial residents, observing that 'many additional questions can be addressed if scholars are willing to pursue in more than a superficial manner the latent cultural and social messages of North American structures' (ibid.: 73). Whether Bastian saw the scholarship up to that point as superficial, or whether he saw a need to relate form and society more explicitly, is left to the reader.

In the next decade, at least three broad undergraduate texts sanctified the Kniffen map, adding colour and different photographs, but none adding data (Fellman *et al.* 1990: 201–22; Jordan and Rowntree 1982: 231–5; Rubenstein 1989: 208–10). The same disjunctures between 'folk' and 'mass style' are always noted: folk traditions died, and national styles and national taste came in, some time in the second third of the nineteenth century, making it less possible to see the individual and cultural regions. The lines of diffusion end in a desert of national uniformity somewhere near the Mississippi. The image is then fast-forwarded, and today no one lives in houses that can be 'read'.

Has anyone in cultural geography responded to Kniffen's challenge to assemble detailed field evidence? Two books in the 1980s seem to be candidates. Allen Noble (1984) published two volumes on the North America settlement landscape, a twenty-chapter summary of dominant house and barn types. The first volume, devoted to houses, presents the hearth, domain and outliers of distinctive house types. Depending on whether one counts variants and subtypes, there are (a) some 26 or 33 house types from the colonial period, (b) another 13 largely folk or ethnically related house types that fall into the category of modifications or diffusion of the eastern hearth models and then (c) another 12 house styles that were identified from the phases of academic style or fashion that shaped housing in the nineteenth and early twentieth centuries. After at least 50 different house types have been described and illustrated, Noble ends with 23 unanswered questions and 45 research agendas that merely extend the identification of structures and their classification. Examples of the basic research questions still outstanding include: 'What was the process by which shotgun houses migrated across the southeastern United States and the Mississippi Valley?' (p. 166) and 'What is the specific relationship between the Cornbelt cube house and the western bungalow?' (p. 167). Suggestions for research in the future include: 'A study of the camelback house ought to be undertaken, especially a comparative study of the house in, say, Louisville, Kentucky, with that in Louisiana', and 'No scholar has yet produced any study exclusively devoted to the upright and wing house, although this is one of the most important dwellings of the eastern Midwest' (p. 168).

Nowhere in Noble's treatise is there any sense that industrialization, or class, or economy, might be worth including as a perspective on how and why the built environment was produced in specific ways in different times. In this, as in other accounts, houses are remarkably unconnected to the social and economic fabric of a transforming world. The shell of the house, and the plan, seem to substitute for all possible aspects of society. By privileging hearth and domain through the vehicle of the house type alone, space has been reduced to the same 'featureless plain' that the quantitative geographers used for their analyses of spatial pattern and process. If the cultural geography of housing is to be more than a mere typology, ever attentive to smaller and smaller branches of side trees in a diagram of form,[1] then it needs to move on to questions of the political or social meaning of housing. John Jakle, Robert Bastian and Douglas Meyer (1989) attempt to work more directly within the diffusion corridors presented by Kniffen. In a daunting round of field observation, they address the issues of common houses from the Atlantic seaboard to the Mississippi Valley, but make the pragmatic decision at the outset to consider only 20 small towns spread over 19 states. These towns of 2,500–3,000 people were selected because they felt that the historic waves of housing change were absorbed and more fully conserved there, protected from some of the destructive waves of modernization that affect major urban areas. They identified sixty-seven different house types, organized primarily by form:

> cultural geographers have focussed ... on the elements of form that characterize the folk, the vernacular, and the popular in common housing ... [they] also include within their purview the more common dwellings built by tradesmen inspired by the popular media of plan books and catalogies.... When [they] see houses they categorize them first according to form or as structural types.
>
> (Jakle *et al.*, 1989: 3)

Their exhaustive typology allowed them to calibrate variants of and alternatives to Kniffen's folk houses, as well as some of the more 'national' styles of the past century. To assess the validity of the regions associated with housing types, they sampled within eight 'culture regions'. New house types included 'One-Third Double-Pile Cottage', 'L-Shaped House (Yankee House)' and 'Hall and Parlor House (Pre-Classic I-House or Early I-House)'.

By sampling so thinly over a vast area, and covering houses built over a long period of time, it is little wonder that this volume raises more questions than it answers. Although the authors were happy that their 67 dwelling types accounted for 96 per cent of the structures they examined in these 20 small towns, they do not seem to have nailed down the existence of culture regions in today's world, nor the sense that 'folk' in the east gives way to 'mass' or 'popular' housing in the west – there were lots of bungalows in Millersburg, Pennsylvania, for example. They conclude that concern with cultural origins offers only a partial explanation for the similarities, and that

> scholars need to focus more on local builders who, under the influence of national trends, replicated popular dwelling types in town to town.... Thus it was coincidences of economic and population growth that explain many of the similarities observed place to place.
>
> (Jakle *et al.*, 1989: 204)

They also conclude that houses seem to have been built for speculation, and thus their marketability rather than some transcendental cultural meanings of folk identity seems to have been operative. Rather than rely on field evidence alone, they realize that future research might need to look at the role of builders in defining and servicing local housing markets where efforts mirrored private-sector commodification of shelter rather than organic choice. Jakle, Bastian and Meyer's final insights suggest a crisis around the relevance and applicability of a narrow focus on house type and house form as a way of identifying important issues. Further, they seem to indicate some willingness to move beyond the diffusion arrow approach, reintegrating spatio-cultural enquiry with more mainstream human geography and a redefined agenda of issues related to the production of the built environment. Yet their finale stays within the cosy confines of the cultural group: 'what is required is a curiosity about common houses and a striving to understand them in a landscape context' (p. 206).

APPROACHES IN OTHER DISCIPLINES

A critical reading of these texts suggests that the weakness of the house-as-key-to-diffusion approach has been not only the absence of (indeed perhaps the impossibility of) detailed field research, but also the absence of any concerted attempt to locate structures within complex, dynamic economic and social transformations. Studies are necessarily temporally rooted in distinct intellectual niches of the discipline and the broader academy, so that we should not judge a study set in the 1960s through the lenses of the 1990s. But in this case much work in the 1970s and 1980s has perpetuated the ideas of the 1960s without heeding the shifting context of human geography. If studies of the object 'house' are to make an important contribution to questions in cultural geography, rather than be seen as an archaic remnant that is devalued for current enquiry, then those disjunctures have to be addressed. To this end, we should first examine what parallel transformations in the study of the object 'house' have been developing in other fields. Some of the best work on housing and place, incorporating both detailed analysis of social systems and geographical linkages, has been done by scholars in the material culture tradition. Two studies in the 1980s from the eastern seaboard show how difficult it is to have a simple model of a cultural hearth, let alone a simple line of diffusion west over long periods of time. Bernard Herman (1987) has worked on two centuries of architecture and rural life in Delaware. He conveys the sense of housing as undergoing fitful transformation, be it improvement, renewal or rebuilding, with architecture used 'to control meaning in social discourse' (p. 2). There was a dichotomy between durable architecture and impermanent, transient dwelling, the former adopting different forms as some were able to 'purchase social distance in an emerging agrarian class structure' (p. 167). Herman argues that there has been considerable change in one small region, and concludes that changes of form and style were manifestations of rural class articulation: 'what was perfected during the rebuilding period in the middle decades of the 1800s was the architectural ordering of a rural class structure around the concept of an estate' (p. 230). Herman reminds us that time and redevelopment have taken their toll – only half of the brick dwellings surveyed in 1812 remain, and only one-tenth of the wooden structures – so we should be careful of seeing relic structures as somehow typical of the historic reality. Perhaps most distinctive in his meticulous assessment of the social context of housing is the sense of tenancy and labouring:

> the eighteenth century vernacular housing surviving in S. New Castle County is that of a minority of the population who had wealth, land and a preference for tenant or day labour over a chattel workforce. Two thirds of the total taxable population were landless, living in other people's dwellings, and inhabiting houses unworthy of notice.
>
> (p. 113).

What are the implications of such a fraction for a perspective that looks at the westward-moving house as a diffusion corridor from a clear colonial hearth? What mental notion of housing did these two-thirds have?

Equally persuasive on the need to detail change, and to tie form to economy and society, is Thomas Hubka's research on the connected farm buildings of New England that explores why New England houses changed in the way they did (Hubka 1984). The connectedness of house, additions and barn evolved as farmers adjusted to the regional market for their produce and the changed possibilities of craft in an industrializing world; their options were narrowed by 'an increasingly harsh context of historical and economic conditions' (p. 204). Hubka concludes his book with two delightful essays, 'Tobias Walker moves his shed' and 'Why Tobias Walker moved his shed'; over forty pages of text, elevation drawings, plans, axonometric drawings and old photographs chronicle the Walker farm over two centuries. It is a sobering text for any reader who thinks that an analysis of present form alone can unlock the meanings of houses in one place, let alone within or between regions.

Both these works, and others like them (e.g. Steinitz 1989), establish the need to understand the complexities of people and society in an agrarian age. Herman and Hubka remind us that house dwellers in the colonial and post-colonial periods were not all 'folk people' in a 'folk society', and even if some were, the ways in which modernizing processes penetrated their lives need to be measured. The process of industrialization folded different worlds and lifeways together, and in so far as the cottage economy was a critical nexus of proto-industrialization, the ways in which those changes impacted on daily life and built form can be noted. Historian Rudolf Braun (1990: 111–30) offers one such example for Switzerland in the countryside of Zurich. Historian Barrie Trinder (1982: 170–201), for an industrializing Britain, records the range of labourers' cottages, weavers' cottages, squatters' cottages, tenements, barracks, pit-rows and other working-class houses in the early nineteenth century. For the United States, Henry Glassie (1975: 176–93) has argued that the social revolution that accompanied the onset of modernity made people more private, more individual and less community oriented, which had a clear outcome in the form and organization of the house. Most notable in this adjustment was the development of an interior entrance hallway where a stranger could be met, rather than have the outside door open directly on to the intimate family house; the hall and parlour house gave way to a more codified set of interior rooms, matched on the outside by a more consciously designed façade.

In seeking to understand the importance of the house in a modernizing world, it is clear that the house is not just shelter. It is also property. Perhaps the finest exemplification of the house as property is offered by Anthony King's book on the bungalow (King 1984), aptly subtitled 'The production of a global culture'. His empirical work as a historian and a sociologist takes him to India, Britain, North America, Africa and Australia over three centuries, and he defines his project as not a study in diffusion, 'but an exploratory attempt to suggest the emergence of an ever-expanding capitalist world economy in which a particular type of individual and consumer oriented urban development, represented by the bungalow, has developed' (p. 8). For King, culture is not just the anthropological framework developed by Rapoport in his work on housing, but more particularly the culture produced in political economy. Thus the house as the visible articulation of agrarian class consciousness argued by Herman, and the realization of the house as commodity that Jakle *et al.* back into, is seen by King as the most effective way of understanding the meaning of a specific house type. His focus is on

> the social and political symbolism of building and architecture, the influence of economy and culture on dwelling form and of dwelling form on economy and culture, the role of housing, whether shelter or property, in the transformation of economic life and material culture.
>
> (p. 2)

FINDING NEW FOCI

At the risk of over-simplification, studies of vernacular architecture in Europe and of material culture in North America have been at least implicitly interested in social relations because of having to engage both the appearance and the internal plan of the big manor house and the tenants' cottage (Brunskill 1971; Mercer 1975), and even embracing industrial housing (Caffyn 1986; Lowe 1977). By contrast, the American cultural geographers have been so interested in broad cultural regions that they have simplified the house to a mere type to be plotted across space.

Accordingly, it would seem that, for re-presented cultural geographies of the house and housing, rather than dwelling on the pioneering world, one should see a complex mosaic of shelter. It is also unnecessary to persist in the notion of 'folk', which transmutes to 'common' or 'popular' or 'ordinary', since this masks important categories of identity that were present in an industrializing and modernizing world. If class can be a category, as Herman (1987) shows for rural Delaware, and Lemon (1980) argues for colonial America in general, then is it possible to reassess the continuum and disjunctures of houses and household worlds as economies as well as folk streams.

Ethnicity might also be used more centrally than in most American research. Poor Irish immigrants formed 'paddy camps' in the half-acre south of Lowell's textile mills in the 1820s; Mitchell (1988: 26) quotes a contemporary observer who noted that these camps resembled 'an Irish village, with the real Irish cabins and shanties, built of board, sod and mud as can be seen in Ballyshannon'. According to Donna Gabaccia (1984: 44–5, 80–3), Italian immigrants from Sicilian agro-towns brought some sense of the *cantile* to their use of tenements in New York's Lower East Side. In their intended design, tenements were entered through the kitchen with a dining room to one side and a bedroom to the other. As a response to overcrowding that put two families in one tenement unit, the kitchen became the equivalent of the outside courtyard, a middle social space for women as well as a bedroom for extra boarders at night.

However, for this more modern world, it would be just as naïve only to look for culture defined as ethnicity (the romantic echoes of a simpler world), in the increasing uniformity of the industrial, standardized world, as it would be to find culture only in pioneer/pre-industrial/folk houses. For all the Irish acres at early Lowell, the predominant built environment of a more mature textile town involves the development of boarding-house shelter for women and girls (Horwitz 1973). Similarly the 'corporations', or company tenements, provided by the Amoskeag Cotton Company at Manchester, New Hampshire, were at the heart of both family life and industrial life, especially for French Canadians (Haraven 1978). And even though there is currently an important search going on for elements of transferred African folk culture into slave America (Anthony 1976–7; Herman 1984; Upton 1985), the later environments of shelter by and for African-Americans in the urban north are equally significant foci;

Osofsky (1963), for example, has chronicled the role of industrialists and philanthropists in shaping the reform tenements in Harlem, such as Henry Phipp's Model Tenements for Coloured Families, or middlemen such as the Afro-American Realty Company. Similarly, Borchert (1980) explores the alley homes of Washington, DC, as the back-door portals for African-American urbanization.

There is also a fertile debate on the concepts of culture, class and identity that is contributing to a broadening of the notions of culture as used by an earlier generation of cultural geographers. As Peter Jackson (1989) has recently shown, the work of Raymond Williams provides an enormously rich entry into the question of culture in an industrial world. Similarly, Jackson Lears (1985) has offered a persuasive argument, inspired by Gramsci, on the relations between culture and power under capitalism. His essay on the concept of cultural hegemony is useful to American historians trying to grapple with the notion of dominant cultures and the possibilities of autonomy for subordinated groups. In the United States the most powerful examples of cultural hegemony and subordinated groups could not avoid race as a primary category. Ethnicity, along with race, fractures a class perspective (see Palmer 1990: 120–44) but can be an important window for tracing the tensions of class-segmented communities (Zunz 1985: 86–92).

A more explicit concern for class and property encourages a view of housing that spans a much wider range of shelter and encourages an exploration into the associated social relations and structures. For the early twentieth century, Paul Groth (1986) has demonstrated how 'unusual and invisible housing', such as rooming houses and apartment houses, can be characterized and analysed. In so doing, he has grappled with the notion of vernacular building within a clearly commercial property system. For the colonial period, instead of just folk or common, categories and divisions must include occupational and status classes such as yeoman and planters, yeoman and tenants, planters and servants, planters, servants and slaves, or fish merchants and servants. In the post-colonial period, on the trans-Appalachian frontier, there were not just pioneer farm families living in simple single-pen houses, but also men working in logging camps and as seasonal labour while also making a farm. The culture present in logging camps also involves tracing the impact of resource capital and seasonal labour as it worked its way through forests across the eastern continent in the first half of the nineteenth century. Loggers' shelter is not necessarily explained by regional hearth, but rather by a way of organizing gang labour on a temporary, seasonal basis within a world of first mercantile and then industrial and corporate capital skill divisions (Holdsworth 1990; Warf 1988).

A parallel but more stable example of early workers' housing is associated with coal-mining. Historian M.J. Daunton (1980) has argued that the different social meanings of company housing in the Northumberland and Durham and the South Wales coalfields revolved around prior assumptions about tenants' rights in an agricultural world; in the Northumberland and Durham coalfields, agrarian landowners accustomed to providing shelter for their agricultural workers extended tenancy practices and assumptions to their early housing for coal-miners on their estates; coal-miners' housing in the South Wales coalfields, on the other hand, was developed by extra-regional urban merchant capital in Swansea and Cardiff, and none of the tied housing assumptions were present (Jones 1969). With so many distinctive regional articulations of the social meanings of working-class or industrial housing, distinctive housing forms *and* social relations moved with miners and mining companies from parts of Europe to the New World. If this is a cultural diffusion, it is not a question of simply placing arrows on a broad map, but rather of a careful reconstruction of information on capital and labour and specific resource frontiers (e.g. Mulrooney 1989). It is less a question of I-houses, or L-houses, or T-houses, as a form, than the use of letters of the alphabet as blueprint sequences on industrially designed company houses, to then be located on A Street, B Street, etc. in some company patch. Were the model company towns of Hopedale, Massachusetts (Garner 1984), or Vander-grift, Pennsylvania (Mosher 1989), in different culture regions, or a national region of industrial capitalism, as Lears has suggested, that was articulated by certain phases of capital development? Over a decade ago, Harvey (1978) explored the question of struggles between capital and labour around the built environment without needing to delineate house types or distinctive cultural regions. Yet perhaps there are now interesting challenges to link culture, more sharply defined as a set of negotiated relationships between capital and labour, and economy, in the industrial world (Mosher and Holdsworth 1992).

The search for different ways of thinking about housing should creatively intersect with urban and social geography which have been often guilty of placelessness and somehow detached from their human subjects. The struggle to assert some neighbourhood sovereignty (Castells 1983), to negotiate some dignity (Mitchell 1989) or to reshape domestic and household spaces (Hayden 1981; Matrix 1984; Spain 1992; Watson 1988) suggests that 'housing becomes an indicator and a potent symbol of the shifting power relations between classes and within different sectors of capital' (Klausner 1986: 38). If housing has long been a politically charged category in the United Kingdom, then the broad possibility of home ownership in the United States and Canada is equally open to a political interpretation. The achievements of the union movement and the social contract of a Fordist-cum-Keynesian society did indicate very significant gains by North American working people. The 'Cape Cod strawberry boxes' of suburbia, and the bungaloid antecedents (Doucet and Weaver 1985; Holdsworth 1982, 1986), are symbolic of the cultural and political hegemony of capitalism in its industrial phase, just as farm family housing on homestead land in the great plains and prairies was part of the hegemony in a proto-industrial phase (Jackson 1972; Mills and Holdsworth 1976).

Earlier in this essay, the disjuncture between the Kniffen view of diffusion and that developed by Hägerstrand and others was seen as a measure of the degree of separation of questions on the house from those addressed by mainstream geography. The critique of house-type geography argued that it is quite impossible to understand shelter without attention to the social, the cultural, the economic and the political. Ironically both the spatial and the cultural diffusion approaches, at one level, shared a featureless plain on which arrows floated, too divorced from the complex economic and social contours of the society that did or did not absorb phenomena. Just as Derek Gregory's recent critique of the Hägerstrand approach to diffusion argues the need to see agency rather than just structure (Gregory 1985), so too there is a need to go more deeply beyond the shape of houses to examine complexity in society and economy that intersects with the making and using of shelter. The house as key to diffusion seems to have been a vehicle for examining pioneering society. Yet even for colonial and early post-colonial society, research on the house form and related social systems can provide a profitable insight into social and economic processes, contested spaces and political struggles to shape and identify space. By revalorizing the house, in the light of studies that pay attention to the social meanings of property, for producer, owner and consumer, perhaps we can reshape shelter into a useful tool for addressing complex questions on people and their places.

NOTE

1   Frustration at the view of architecture among early practitioners of cultural geography is expressed in Goss's association of the tradition with 'the geography of kitchen utensils' (Goss 1988: 392).

REFERENCES

Anthony, C. (1976–7) 'The big house and the slave quarters, part 1', *Landscape* 20: 8–19; 'part 2', *Landscape* 21: 9–15.

Bastian, R.W. (1982) 'Structures', in J.F. Rooney, W. Zelinsky and D.R. Louder (eds) *This Remarkable Continent: An Atlas of United States and Canadian Society and Cultures*, College Station: Texas A&M Press, 71–95.

Blaut, J. (1977) 'Two views of diffusion', *Annals of the Association of American Geographers* 67: 343–9.

Borchert, J. (1980) *Alley Life in Washington*, Urbana: University of Illinois Press.

Braun, R. (1990) *Industrialisation and Everyday Life*, London: Cambridge University Press.

Brunskill, R.W. (1971) *Illustrated Handbook of Vernacular Architecture*, London: Faber.

Caffyn, L. (1986) *Workers' Housing in West Yorkshire, 1750–1920*, London: HMSO.

Carter, T. and Herman, B.L. (eds) (1989) *Perspectives in Vernacular Architecture, III*, Columbia: University of Missouri Press.

Castells, M. (1983) *The City and the Grassroots: A Cross-Cultural Theory of Urban Social Movements*, Berkeley: University of California Press.

Daunton, M.J. (1980) 'Miners' houses: South Wales and the Great Northern Coalfield, 1880–1914', *International Review of Social History* 25: 143–75.

Doucet, M.J. and Weaver, J.C. (1985) 'Material culture and the North American house: the era of the common man, 1870–1920', *Journal of American History* 72: 560–87.

Fellman, J., Getis, A. and Getis, J. (1990) *Human Geography, Landscapes of Human Activities*, Dubuque, Iowa: Brown.

Gabaccia, D.R. (1984) *From Sicily to Elizabeth Street: Housing and Social Change among Italian Immigrants, 1880–1930*, Albany: SUNY Press.

Garner, J.S. (1984) *The Model Company Town: Urban Design through Private Enterprise in Nineteenth-Century New England*, Amherst: University of Massachusetts Press.

Glassie, H. (1968) *Pattern in the Material Folk Culture of the Eastern United States*, Philadelphia: University of Pennsylvania Press.

—— (1974) 'The variation of concepts within tradition: barn building in Otsego County, New York', in H.J. Walker and W.G. Haag (eds) *Man and Cultural Heritage: Papers in Honor of Fred B. Kniffen*, Geoscience and Man vol. V, Baton Rouge: Louisiana State University Press, 177–235.

—— (1975) *Folk Housing in Virginia: A Structural Analysis of Human Artifacts*, Knoxville: University of Tennessee Press.

Goss, J. (1988) 'The built environment and social theory: towards an architectural geography', *Professional Geographer* 40: 392–403.

Gregory, D. (1985) 'Suspended animation: the stasis of diffusion theory', in D. Gregory and J. Urry (eds) *Social Relations and Spatial Structures*, London: Macmillan, 296–336.

Groth, P. (1986) '"Marketplace" vernacular design: the case of downtown rooming houses', in C. Wells (ed.) *Perspectives in Vernacular Architecture, II*, Columbia: University of Missouri Press, 179–91.

Hägerstrand, T. (1967) *Innovation Diffusion as a Spatial Process*, Chicago: University of Chicago Press.

Haraven, T.K. (1978) *Amoskeag: Life and Work in an American Factory-City*, New York: Pantheon.

Harvey, D. (1978) 'Labor, capital, and class struggle around the built environment in advanced capitalist societies', in K.R. Cox (ed.) *Urbanization and Conflict in Market Societies*, Chicago: Maaroufa, 9–37.

Hasell, M.J. (1990) 'Exploring connections between women's changing roles and house forms', *Environment and Behavior* 22: 3–26.

Hayden, D. (1981) *The Grand Domestic Revolution: A History of Feminist Designs for American Homes, Neighborhoods, and Cities*, Cambridge, Mass.: MIT Press.

Herman, B.L. (1984) 'Slave quarters in Virginia: the persona behind historical artifacts', in D.G. Orr and D.G. Crozier (eds) *The Scope of Historical Anthropology*, Philadelphia: Laboratory of Anthropology, Temple University, 253–83.

—— (1987) *Architecture and Rural Life in Central Delaware, 1700–1900*, Knoxville: University of Tennessee Press.

Holdsworth, D.W. (1982) 'Regional distinctiveness in an industrial age: some California influences on British Columbia housing', *American Review of Canadian Studies* 12: 64–81.

—— (1986) 'Cottages and castles for Vancouver home-seekers', *British Columbia Studies* 69–70: 11–32.

—— (1990) 'The archives and the landscape: texts for the analysis of the built environment', in P. Groth (ed.) *Vision, Culture, and Landscape*, Berkeley: Department of Landscape Architecture, University of California, 187–204.

Horwitz, R.P. (1973) 'Architecture and culture: the meaning of the Lowell Boarding House', *American Quarterly* 25: 64–82.

Hubka, T.C. (1984) *Big House, Little House, Back House, Barn: The Connected Farm Buildings of New England*, Hanover: University Press of New England.

Jackson, J.B. (1972) *American Spaces: The Centennial Years: 1865–1876*, New York: Norton.

Jackson, P. (1989) *Maps of Meaning: An Introduction to Cultural Geography*, London: Unwin Hyman.

Jakle, J.A., Bastian, R.W. and Meyer, D.K. (1989) *Common Houses in America's Small Towns: The Atlantic Seaboard to the Mississippi Valley*, Athens: University of Georgia Press.

Jones, P.H. (1969) 'Colliery settlement in the South Wales Coalfield 1850 to 1926', Occasional Papers in Geography No 14, University of Hull.

Jordan, T.G. (1978) *Texas Log Buildings: A Folk Architecture*, Austin: University of Texas Press.

—— (1985) *American Log Buildings: An Old World Heritage*, Chapel Hill: University of North Carolina Press.

Jordan, T.G. and Kaups, M. (1989) *The American Backwoods Frontier*, Baltimore: Johns Hopkins University Press.

Jordan, T.G. and Rowntree, L. (1982) *The Human Mosaic: A Thematic Introduction to Cultural Geography*, 3rd edn, New York: Harper & Row.

Kaups, M. (1981) 'Log architecture in America: European antecedents in a Finnish context', *Journal of Cultural Geography* 2: 131–53.

—— (1983) 'Finnish log houses in the Upper Middle West: 1890–1920', *Journal of Cultural Geography* 3: 2–26.

King, A.D. (1984) *The Bungalow: The Production of a Global Culture*, London: Routledge & Kegan Paul.

Klausner, D. (1986) 'Beyond separate spheres: linking production with social reproduction and consumption', *Environment and Planning D: Society and Space* 4: 29–40.

Kniffen, F.B. (1936) 'Louisiana house types', *Annals of the Association of American Geographers* 26: 179–93.

—— (1965) 'Folk housing: key to diffusion', *Annals of the Association of American Geographers*, 55: 549–77.

Kniffen, F.B. and Glassie, H. (1966) 'Building in wood in the eastern United States' *Geographical Review* 56: 40–66.

Lears, T.J.J. (1985) 'The concept of cultural hegemony: problems and possibilities', *American Historical Review* 90: 567–93.

Lemon, J.T. (1980) 'Early Americans and their social environment', *Journal of Historical Geography* 6: 115–32.

Lewis, P.F. (1970) 'The geography of old houses', *Earth and Mineral Sciences* 39: 33–7.

—— (1975) 'Common house, cultural spoor', *Landscape* 19: 1–22.

—— (1979) 'Axioms for reading the landscape: some guides to the American scene', in D.W. Meinig (ed.) *The Interpretation of Ordinary Landscapes: Geographical Essays*, New York: Oxford University Press, 11–32

Lowe, J.B. (1977) *Welsh Industrial Workers' Housing, 1775–1875*, Cardiff: National Museum of Wales.

Matrix (1984) *Making Space: Women and the Man-Made Environment*, London: Philo Press.

Mercer, E. (1975) *English Vernacular Houses*, London: HMSO.

Mills, G.E. and Holdsworth, D.W. (1976) 'The BC mills prefabricated system: the emergence of ready-made buildings in western Canada', Occasional Papers in Archaeology and History No. 14, Ottawa: National Historic Sites, 127–69.

Mitchell, B.C. (1988) *The Paddy Camps: The Irish of Lowell, 1821–61*, Urbana: University of Illinois Press.

Mitchell, D.M. (1989) 'A history of homelessness – a geography of control: the production of order and marginality in Johnstown, Pennsylvania', unpublished master's thesis, Pennsylvania State University.

Mosher, A.E. (1989) 'Capital transformation and the restructuring of place: the creation of a model company town', unpublished Ph.D thesis, Pennsylvania State University.

—— and Holdsworth, D.W. (1992) 'The meaning of alley housing in industrial towns',

Journal of Historical Geography, 18: 174–89.

Mulrooney, M.M. (1989) *A Legacy of Coal: The Coal Company Towns of Southwestern Pennsylvania*, Washington, DC: Historic America Building Survey/Historic American Engineering Record, National Parks Service.

Newton, M.B. Jr and Pulliam-Di Napoli, L. (1977) 'Log houses as public occasions: a historical theory', *Annals of the Association of American Geographers* 67: 360–83.

Noble, A.G. (1984) *Wood, Brick, and Stone: The North American Settlement Landscape*, Vol. I: *Houses*; Vol II: *Barns and Farm Structures*. Amherst: University of Massachusetts Press.

Osofsky, G. (1963) *Harlem: The Making of a Ghetto: Negro New York, 1890–1930*, New York: Harper & Row.

Palmer, B.D. (1990) *Descent into Discourse: The Reification of Language and the Writing of Social History*, Philadelphia: Temple University Press.

Pounds, N.J.G. (1989) *Hearth and Home: A History of Material Culture*, Bloomington: Indiana University Press.

Rapoport, A. (1969) *House Form and Culture*, Englewood Cliffs: Prentice-Hall.

Rubenstein, J.M. (1989) *The Cultural Landscape: An Introduction to Human Geography*, 2nd edn, Columbus: Merrill.

Spain, D. (1992) Gendered Spaces, Chapel Hill: University of North Carolina Press.

Steinitz, M. (1989) 'Rethinking geographical approaches to the common house: the evidence from eighteenth-century Massachusetts', in T. Carter and B.L. Herman (eds) *Perspectives in Vernacular Architecture, III*, Columbia: University of Missouri Press, 16–26.

Trinder, B. (1982) *The Making of the Industrial Landscape*, London: Dent.

Upton, D. (1985) 'White and black landscapes in eighteenth-century Virginia', *Places, A Quarterly Journal of Environmental Design* 2: 59–72.

Wagner, P.L. and Mikesell, M.W. (eds) (1962) *Readings in Cultural Geography*, Chicago: University of Chicago Press.

Walker, H.J. and Haag, W.G. (eds) (1974) *Man and Cultural Heritage: Papers in Honor of Fred B. Kniffen*, Geoscience and Man Vol. V, Baton Rouge: Louisiana State University, School of Geoscience.

Warf, B. (1988) 'Regional transformation, everyday life and Pacific Northwest lumber production', *Annals of the Association of American Geographers* 78: 326–46.

Watson, S. (1988) 'Women and housing or feminist housing analysis', *Housing Studies* 1: 1–10.

Wells, C. (ed.) (1982) *Perspectives in Vernacular Architecture*, Annapolis: Vernacular Architecture Forum.

—— (ed.) (1986) *Perspectives in Vernacular Architecture, II*, Columbia: University of Missouri Press.

Zunz, O. (1985) *Reliving the Past: The Worlds of Social History*, Chapel Hill: University of North Carolina Press.

7

PUBLIC HOUSING IN SINGLE-INDUSTRY TOWNS
Changing landscapes of paternalism
Don Mitchell

The Columbia Steel Company (a subsidiary of US Steel) ran an advertisement in the 2 August 1943 issue of the Pittsburg (California) *Post-Dispatch* announcing the need for skilled labour in their mills. The advertisement centred around the availability of 'New homes for workers in the Columbia Park Project'. Columbia Park was the first federally financed, publicly produced housing in the Pittsburg region, the first visible evidence of the struggles over housing for poor workers and the underclasses that had culminated in the landmark 1937 Wagner–Steagall Housing Act – an act which finally made public housing a national priority. When Columbia Park opened two weeks later it was obvious that the project was intended for skilled workers at Columbia Steel (with some space reserved for the nearby Pittsburg Army Cantonment) and was not intended to house the poor. The *Post-Dispatch* reported on 17 September that priority in housing assignments at Columbia Park was given to workers *recruited* by Columbia Steel, who were just arriving in town. In essence, then, Columbia Park was company housing in so far as hiring practices determined housing eligibility.

At Johnstown, Pennsylvania, the first public housing was also directed to workers in the steel industry. A local public housing agency, the Johnstown Housing Authority, was constituted in 1941 and the first housing opened on a hill above the Gautier Department of the Cambria Steel Company (a subsidiary of Bethlehem Steel) in 1943. Here, too, the housing was directed, in the first instance, at defence workers for whom 'there is no ceiling on salaries' (TSAC 1942). In Johnstown, unlike Pittsburg, the first public projects were to be permanent and 'would again become a low-cost housing project' (ibid.).

Perhaps it is not surprising that during the Second World War public housing would be reserved for war workers. What is significant, however, is the combination of timing and *language* that made public housing palatable to both of these single-industry towns. In both communities, housing shortages and poor housing quality were described as 'chronic' for the years immediately prior to the war (OHWS–NRPB 1942; WPA 1941). The 1937 Housing Act vested in the local community the responsibility for planning and construction of low-cost, low-rent publicly owned housing. The role of the federal government was, in large, limited to the provision of loans to the local authority and to guaranteeing returns on loans of private capital (Ebenstein 1940). In both Pennsylvania and California, state legislation that enabled the incorporation of local housing authorities was passed the same year as the federal Act. In Johnstown, however, a local housing authority was not established until 1941 – well after the incipient war economy had spurred increased production at Cambria. Similarly, there is no record of public housing being considered for the Pittsburg region until early 1942.

With the onset of war came a rhetoric of public housing as patriotic duty: public housing was 'owed' to defence workers for their role in furthering the aims of the country. Thus, public housing was no longer designed for those who could not afford to live in private market housing of a decent standard. Rather, publicly built housing was for *deserving* workers. It is within this rhetorical milieu that the provision of public housing in Pittsburg and Johnstown must be placed. And it is from within this rhetoric that the contours of the production of the landscapes of order and control within the industrial city of the post-war era begin to emerge. To structure the lived spaces of employees is also to structure the terms upon which the relations of production within a particular community are contested. The provision of public housing *as company housing* may be viewed as a continuation, though by altered means, of company domination over the external lives of both its employees and the other residents of the community. The landscapes of public housing within the company town stand as evidence of the struggles and contests over subjugation and domination, over order and insubordination, that came to a head with the closing of the Depression.

The new 'company housing' in Pittsburg and Johnstown represented a reworking of the nature of the 'social contract' that had evolved between the companies and their workers in single-industry towns – a social contract that may be understood to be the result of a *contested* emplacement of various philosophies of corporate paternalism. From its beginnings in the mills of Pawtuckett in the 1790s, American corporate paternalism – of which worker housing was just one part – came to represent over the next century something like a contract between a company and its workers (Zahavi 1988: 2). This paternalism combined a corporate benevolence and a concern for the health and well-being of the work-force with a wide-ranging set of corporate philosophies and policies of control over both the internal and external conditions of the lives of the company workers and their families. Though the social contract implied control over the lives of workers, corporate paternalism on the eve of the twentieth century also implied company *responsibility* for the welfare of employees and their families. As Garner puts it: 'Despite the resentment it may have fostered among workers, resulting from attempts at social engineering, paternalism was considered by many nineteenth-century businessmen to be a moral responsibility, protecting society while furthering business' (1984: 53). And it was in the language of morality that the claims of paternalistic prerogative were pushed by industrialists.

With the advent of the welfare state in the first half of the twentieth century, company responsibility for the health, welfare and housing of workers was moved more squarely to the realm of the 'public', thereby reworking the terms of the social contract that had developed in earlier decades. To understand the significance of this shift it is important to understand the moral and philosophical underpinnings of corporate paternalism, and the vestiges of paternalism that remain in the incipient welfare state. Such an understanding will carry with it a need for a rethinking of the role of corporate hegemonic ideology[1] in the lives of workers that is more complete, and more flexible, than is evident in most works on the corporate landscape.

PATERNALISM AS HEGEMONIC IDEOLOGY

Paternalism[2] within the nineteenth-century company town may be viewed as an incomplete hegemonic ideology, as a struggle to define the language of legitimation of corporate capitalism in a rapidly industrializing world. Increasingly, justification of a 'factory system (that) appeared just as "unnatural" to the bourgeoisie as it did to those who had to live out their daily lives under its regimen' (Harvey 1978: 26) relied on an idealized and romanticized vision of feudal society that was seen to be rapidly disappearing. The language of paternalism was itself based on a set of perceived social relations of production, and on the social contracts within which these relations were wrapped, that evolved under a wholly different set of economic and social conditions.

To view paternalism as a dominant hegemonic ideology is to adopt a view that sees hegemony as a *construction* of an ideology through which relations of power are expressed – and it is to understand that the construction is rarely complete. Contrary to the growing acceptance of a view of cultural hegemony that Mills (1988) (following Williams 1980) described as 'the "spontaneous philosophy" of a society, something which so saturates social consciousness that it defines the very substance and limits of common sense', I understand hegemonic ideologies (which both express and rework manifestations of 'culture') as the result of the ongoing, uneven struggles over definitions, legitimacies and relationships between and within opposing groups. The 'evidence' of the outcomes of these struggles – i.e. the cultural, and in this case landscape, manifestations – in turn, works to structure the very terms of the continuing contests (see Scott 1985: ch. 5). Thus, hegemony, whether called cultural or ideological, is more a result of struggle or 'negotiation' than it is a thing that arises 'spontaneously'. With this in mind, hegemony should not be defined in a 'structuralist' sense whereby 'ideology always pre-exists and pre-empts any authentic criticism' (Willis 1977: 175; quoted in Scott 1985) by those being dominated (Williams's spontaneous philosophy?). Rather we should understand that 'Social agents are not passive bearers of ideology, but active appropriators, who reproduce existing structures only through struggle, contestation, and a partial penetration of those structures' (ibid.). Consequently, the set of social relations implied under the banner of corporate paternalism is subject to ongoing reinterpretation by *individuals* (individual workers and individual capitalists), who, either individually or banded together as a class, take the bundle of policies, actions and pronouncements that are corporate paternalism and rework them according to their own understandings.

A philosophy of corporate paternalism, then, is actively constructed through the process of domination by, and penetration of, an ideology of benevolence that attempts to 'euphemize' the domination of workers in company towns under the very real concern for their mental and physical welfare. There are two important concepts here that are manifest in the landscape of company towns and that need to be examined in turn. The first is the idea of the 'penetration' of a dominant ideology.

Scott notes that the traditional concept of hegemony 'ignores the extent to which most subordinate classes are able, on the basis of their daily material experience, to penetrate and demystify the prevailing ideology' (1985: 317). In other words, dominant ideology is rarely simply accepted at face value. Thus, in single-industry towns, workers continually interpret and reinterpret the policies and moralistic pronouncements of their employers and the local business elite. In Johnstown, the sustained housing programmes of Cambria Steel, through which the company sought to 'buy' the loyalty and docility of its workers, and through which it attempted to control especially the employees' 'outside' existence in all its facets ('Helping the workingmen to help themselves' 1912), were often interpreted (as will be shown later) as a means of liberation (albeit a rather limited liberation) from the dictates of company control. To the purchaser (especially if an immigrant), the purchase of a house, even if it meant being impossibly mortgaged to the company, implied the protection that being a home owner afforded under state and federal law. To own, rather than to rent, suggested a further step towards security for foreign or 'Americanized' workers (Morawska 1985). Moreover, the public pronouncements of Cambria Steel, published in both local newspapers and trade journals such as the *Iron Trade Review*, provided workers with a reference point against which the actions of the company could be judged.[3] The very language of the company, as it attempted to set the permissible bounds of behaviour for its work-force, was in turn used by the workers to judge the company by its own standards. The hegemony of the company was penetrated by the workers as a means of holding the company to account.[4]

The second concept is that of 'euphemization' of domination – an idea that will take on increasing prominence as the story of public housing in Johnstown and Pittsburg unfolds. Euphemistic expressions are particularly valuable in setting the terms of power relations, in setting the bounds of discourse. Again following Scott:

Here we are concerned with appearances, with the mask that the exercise of economic or political power typically wears. Things in this domain are rarely as they seem; we should expect disguises, *whether they are self-conscious or not....* The executive who fires some of his work force is likely to say that he 'had to let them go'. This description of his action not only implies that he had no choice in the matter but that those 'let go' were being done a favour.... Those who are beneficiaries of this magnanimous act are likely to take a different view; their metaphors are usually more colorful: 'I got the sack,' 'I was axed'.

(1985: 204–5; emphasis added)

Scott's example is instructive because it points to the fact that it is not necessary for the reproducers of the lexicography of domination to *intend* domination of subordinate classes. It also illustrates the ways in which such a lexicography is pierced by those who come in contact with it.

Moreover, the language of domination – through its use of euphemism to mask the naked reality of power relations – binds company to worker just as much as it binds worker to company. A number of commentators and researchers (for example, Watts 1988) have demonstrated the extent to which dominant groups are tied to those that they subordinate through the language that they use to win domination. Within the company town of the first part of the twentieth century, the language of dependency became one of the most important avenues of domination and subjugation, and an important point of euphemization and penetration: by defining workers as dependent, the company established itself as responsible not only for, but also to, those dependants.[5] Companies such as Cambria and Columbia found that they had tied themselves to the maintenance of their dependants – a maintenance that grew increasingly expensive in the inter-war years. Thus, the extreme reaction of the Johnstown workers during the 'Little Steel' strike of 1937 may be seen, in part, as a reaction to Cambria's abrogation of its self-professed responsibilities during the Depression (McPherson 1972; Mitchell 1989).

What emerges from this discussion is the centrality of the *language* of paternalism as dominant hegemonic ideology as it is understood by both companies and workers. It is a language that ties worker and company in a web of mutual obligations that are the result of a struggle over the definitions of paternalism: a 'class bargain' is struck. To understand how such a language translates into the landscape of corporate paternalism, first in the company town, and then (in altered form) in public housing, it is first necessary to trace the ideological lineage of worker housing and housing reform.

THE IDEOLOGY OF HOUSING WORKERS

My interest is in the roots of social contracts – in the interplay of corporate control over, and corporate responsibility for, the lives of workers as they change through time and as they become evident in the landscapes of public housing. Therefore, it is important to trace the trajectory of paternalist philosophy as it is manifest in both corporate ideology *and* reformist thought. The latter was of particular influence in the form of paternalistic company policies in the early part of the twentieth century (Creese 1966) and then in the struggle for a national housing programme, though not necessarily in the form of early public housing itself (Bauman 1987; Parsons 1984).

One of the most important publicly *professed* ideological currents that has shaped the debate over the extent of company involvement in the lives of workers is the idea of worker housing as *moral* obligation. In fact, such an ideology was the backbone of public housing policies in the 1940s. Historically, the philosophy of company-provided worker housing has been infused with three inter-related streams within Western thought. There is, first, a strong and constant belief in environmental determinism that is evident in the works of paternalistic companies and the writings of housing reformers alike: good housing makes good workers and better citizens. Second, there is a belief that rational (and rationalized) landscapes are necessary for social order (see Foucault 1973, 1979). Finally, the first two streams are infused with a romantic feudalism and a longing for the sylvan settings that industrialization was destroying – expressed, for example, in the medieval English design elements of workers' housing in the company-built industrial village of Port Sunlight (see Leverhulme 1918).

Early industrialists adopted paternalistic policies towards their workers as a 'ploy' to attract labour to work in their mills and factories (Garner 1984: 53–4). In North America, such policies were rationalized at the beginning of the nineteenth century by referring to a Jeffersonian pastoralism coupled with a Calvinist Protestantism and a sincere sense of *noblesse oblige*. In Britain, the ideal harkened to a pastoral and feudal past where the vision was one of an ordered landscape blending the proper amounts of town and country in order to 'protect and uplift' the work-force – a work-force that often consisted of young girls, for whom early industrialists were charged with almost parental responsibility.

One means of fulfilling parental obligation was through the provision of housing. And it was through the provision of housing that reference to an idealized feudalism could be made. For, as the *Halifax Guardian* noted in 1842, the feudal social order, which was rapidly disappearing, was the proper social order:

> There was so much of a home character in their little half farmstead, half clothing-shop; the master and his men and domestic apprentices were so much associated in friendly, almost family, intercourse, that the destruction of such a system cannot but be productive of evil, succeeded as it has been by the gloomy factory system.
>
> (quoted in Creese 1966: 18).

Thus, the goal of early paternalists was to reproduce conditions whereby 'master and men' could be associated in such 'friendly, almost family, intercourse'.

Pastoralism was expressed through the belief that proper housing in a proper setting meant a contented and productive work-force that understood its role in relation to its place in the company. In this, environment was paramount. Creese notes that the early English paternalists were following 'the main stream of British thought from Pugin to Ruskin in believing that economic, social and moral uplift would be futile unless accompanied by a drastic change in the quality of the physical environment' (1966: 148). Such an environmental determinism – a determinism that is decidedly anti-urban – may be traced through the literature and debates over worker housing to the present.[6] With environment as the determining factor, paternalists sought to create a rational landscape in which social order could be manifest. It is here that the interplay between responsibility for, and control over, the work-force is most readily apparent. As Reynolds puts it: 'A principle of environmental order – the provision of decent housing – and a principle of moral reform – the "elevation of the working classes" – had been brought together in a programme of social reform in which paternalism could be effectively focused' (1983: 261). In the nineteenth-century paternalistic landscape, social control was inseparable from philanthropic goals. The idea was to provide decent housing to ensure decent workers – and to 'persuade working men that the interests of Capital and Labour were identical' (ibid.: 258).

The ideological basis of early corporate capitalism was reinforced by the newly emerging social sciences in the last two decades of the nineteenth century. When Lever claimed in 1898 that children in an urban environment, without access to the sight of nature, 'grow up depraved, and become a danger to the State; wealth-destroyers instead of wealth-producers' (quoted in Creese 1966: 114), he could just as easily have drawn upon the work of (for instance) political economist Carroll Wright, who held in 1882 that 'material prosperity depended on the health of ... workers and their surroundings' (Garner 1984: 57). Economists were not alone in the late 1800s in promoting good environments for good labour – social reformers, just as equally convinced of the determinacy of the environment as they were of their duty to help industrialism grow and prosper, repeatedly warned of the danger of a disordered environment.

Reform writing and its expression in the provision of workers' housing did not truly flourish, however, until the turn of the twentieth century. This writing drew heavily upon the successes and failures of paternalistic company towns in England and the United States even as it echoed much of the same rhetoric as these early paternalists. Environmental determinism was expanded to cover a range of housing conditions. Not only was access to the countryside important, but the size and arrangement of tenements and rooming houses also played an important role in the determination of a productive work-force. Typical of the importance placed on the quality of a worker's home to the maintenance of the industrial order was housing reformer and sanitary engineer Knowles's postulation that 'Labor unrest is not due entirely to the lack of sufficient pay, but in many cases to the psychological effects of the laborer's family upon himself due to poor living conditions' (1920: 14).

The rhetoric of housing reformers, geared as it was to the stirrings of public sentiment towards 'responsible capitalism', reinforced popular conceptions of the proper role of housing in the lives of labour. It would be for the betterment of business that good housing would be built – and it should be built, according to the reform line of the first two decades of the twentieth century, by private enterprise, particularly by single industries that could exercise complete control over the planning process so as to ensure a rational, ordered landscape (see Buder 1967: ch. 1).

The type of housing suited for workers was precisely prescribed by reformers:

> It should be recognized from the outset that the normal method of housing the working population in our American cities is in small houses, each house occupied by a separate family, often with a small bit of land, with privacy for all, and with a secure sense of individuality and opportunity for real domestic life. Under no other method can we expect American institutions to be maintained. It is useless to expect a conservative point of view in the workingman, if his home is some huge building in which dwell from twenty to thirty other families, and this home is his home only from month to month.
>
> (Veiller 1910: 6–7)

If there is any doubt as to which American institutions are to be upheld by the provision of decent housing, Knowles spells it out very clearly for us. To him, the clearest benefits of 'industrial housing' include a decrease in labour turnover, an increase in the health, cheerfulness, 'virility' and efficiency of the work-force, worker loyalty and a 'regulated payroll' (1920: 14–16). The importance of a regulated payroll is obvious to Knowles, who claims that 'a modern industrial town, planned in proper relation with the plant, permits a conscious control over the selection of classes of employees' (ibid.: 15). Moreover, control may be exercised over 'the percentage of skilled to unskilled workers; the percentage of foreign to native workers, the number of women workers and minors can be regulated to produce maximum efficiency, by building the town to suit the plant' (ibid.).

The language now is a language of control as well as a language of responsibility. And perhaps the most significant aspect of twentieth-century corporate paternalism, as it allied itself with the goals of the housing reform movement, was the increased ability to 'regulate the payroll' – an ability that employers struggled to maintain throughout the economic restructurings of the 1930s. The desire for such control remained an important backbone of paternalism within the public housing policies of the 1940s.

THE PENETRATION OF HOUSING IDEOLOGY

Paternalistic housing ideologies within single-industry towns were not merely imprinted upon the landscape, but were both sanctioned and penetrated by the workers whom they affected. As Zahavi (1988) has shown in the case of the

Endicott Johnson Corporation, the 'Square Deal' served to entangle both employee and employer in a web of mutual obligations, from which neither could easily extract themselves. In Johnstown, the establishment by Cambria Steel of a range of 'public goods' such as a free library and night school, and a (very) limited programme of compensation for dependants of workers killed or maimed on the job, represented not only a benevolence, but also a policy of control – a control of which the workers were not ignorant. Advancement within the mills required attendance at either citizenship or 'Americanization' classes where it was insisted that hard work and docility were integral to gaining acceptance within the American community. Workers, however, knew that such classes, like the purchasing of a company house, were the price to be paid for 'a further step on an uneven, slow, and difficult road toward a slightly more comfortable existence' (Morawska 1985: 145). To the Cambria worker, therefore, owning a home, even if it was built and financed by the company, stood as a symbol of achievement and independence – including a degree of independence from the company. The housing policy of Cambria Steel – like the library and schools – represented both a hegemony over the lives of workers and a means towards the independence of those same workers.

In Johnstown, then, the hegemony of the company's paternalism was penetrated by the workers and reinterpreted (at least to some extent) along new lines. The constant redefinition of corporate ideologies was consolidated in the social contracts that supported the social relations of production. During the depressions of the 1920s and 1930s, and the industrial restructuring that followed them, these social contracts and the responsibilities that they implied became a liability to paternalistic corporations – a liability that was increasingly turned over to the state.

JOHNSTOWN AND PITTSBURG: PUBLIC HOUSING AS WORKER HOUSING

In Pittsburg and Johnstown, the new social and economic realities of the post-Depression recovery within the steel industry made overtly paternalistic policies uneconomical. In other words, the welfare capitalism of the 1910s and 1920s had been penetrated by workers to a degree that such responsibilities were more a burden than a benefit. But with the reality of an increasingly legitimate unionism to contend with, companies were unwilling to jettison all control over the internal and external conditions of their employees' lives. It was in the interest of companies in single-industry towns, then, to 'euphemize' their domination through the local state. But just as the ideology of corporate paternalism was not merely imprinted over the lives of workers in the last century, the new euphemized paternalism was likewise contested. One outcome of these contests was the early public housing projects in single-industry towns like Pittsburg and Johnstown where the hidden landscapes of paternalism may be discovered.

The Wagner–Steagall Housing Act of 1937 vested the local state with the power and the responsibility for the provision of public housing. The original intent of the Act was to

> remedy the unsafe and insanitary housing conditions and the acute shortage of decent, safe, and sanitary dwellings for families of low income, in rural and urban communities, that are injurious to the health, safety and morals of the citizens of the Nation.
>
> (US Housing Act 1937: section 1)[7]

Due to a 1935 Supreme Court decision which declared that the exercise of eminent domain and the provision of low-cost housing were vested with the several states of the union, or their subdivisions, and not with the federal government, the Wagner– Steagall Act established the US Housing Authority (USHA) as a grant-giving institution. At the same time, owing to a federal distrust of state governments, the power to clear slums and build low-rent housing was not given to the states but rather to local authorities which were to be enabled by state legislation (Ebenstein 1940: 38–40).

As noted above, local housing authorities were enabled in Pennsylvania in 1937. The Johnstown Housing Authority (JHA), however, was not incorporated until 1941. By that time, the emphasis of the federal housing programme had shifted from provision of housing for low-income families to the provision of housing for low-income *workers*. As a writer in Johnstown, echoing a USHA publication (USHA #16–8857), put it:

> Slums mean bad health. Bad health means inefficiency. No matter how bright and airy working conditions may be, workers who live in ill-ventilated, insanitary homes cannot perform the best possible work. Unhealthy living conditions mean more accidents on the job, more absences from the job on account of sickness. The rehousing of slum families in decent homes, on the other hand, means healthier workers and greater industrial efficiency.
>
> (Hornbeck 1941)

Public housing was to be in the service of industry, and *for* industrial workers, a definition that made a public housing programme palatable both to the Bethlehem Steel Company – which previous to the incorporation of the Johnstown Housing Authority (JHA) was the largest residential landholder in the city (Cambria County Assessment Records 1939; Morawska 1985: 162) – and to the local and national steelworkers' union whose members would be eligible for public housing thus defined.

Public housing was further defined as housing in the service of industry when Congress passed the Lanham Act in October 1940. This act authorized the Federal Works Agency to spend $150 million for the provision of defence worker housing, as part of the general war mobilization. By May 1942 President Roosevelt had been forced to increase Lanham Act funding by $600 million

(Bauman 1987: 59–66). Though Bauman claims that the war 'obliterated the distinction between war workers and non-war workers' (ibid.: 66), the evidence from Johnstown and Pittsburg seems to indicate otherwise. The Lanham Act provided the means whereby Cambria and Columbia Steel could build a new 'company town', a town that would function in much the same way that company towns have traditionally functioned. It was a company town where responsibility for the well-being of the residents was vested with the state and removed from the company. But company influence and company control were never far off.

In Pittsburg, Columbia Steel was actively involved in shaping war housing policy through its agent, James C. Wood. Wood was the Pittsburg Commissioner of the Housing Authority of Contra Costa County (HACCC) which was responsible for setting housing policy, as well as for the actual development of public housing in the Pittsburg area. In other capacities Wood 'represent(ed) Columbia Steel' in meetings with local US Congressmen (*Pittsburg Post-Dispatch*, 23 April 1943). Locally, HACCC established close ties with the Pittsburg Chamber of Commerce (dominated by Columbia agents) and the County Development Board, upon which two Chamber of Commerce Commissioners sat. Together, these agencies established policy for public housing during the war years. There is no record of representation in any of these agencies of social service workers or activists.

In Johnstown there was at least nominal input by those involved in social work. But, none the less, the early public housing was, in essence, Cambria housing. Of the 147 decipherable[8] records of tenant application and review by the Tenant Selection Advisory Committee (TSAC) from 1943, 131 are from Bethlehem employees. Of the 119 that were approved by TSAC, 106 were Bethlehem employees. Only three applicants can be positively identified as unemployed or single mothers. All three were approved by the committee (TSAC n.d.).

Scott Keyes writing for USHA and the JHA around 1942 outlined a 'six-year housing program' that included slum clearance and rehousing as its major goals. Citing the 'Johnstown housing survey of low rent housing needs' (WPA 1941), Keyes outlined three neighbourhoods in need of immediate attention: Prospect, Cambria City–Minersville and Bedford–Conemaugh Boro. Public housing projects were to be built in Prospect and Bedford–Conemaugh, as well as in Oakhurst, the last of which would be used to 'drain' the Cambria City–Minersville area so that slum areas could be cleared (Figure 7.1). In all three 'slum' neighbourhoods, Bethlehem Steel was a major residential landowner. In Prospect, the median income of tenants living in substandard housing occupied by a single family (for whom, presumably, public housing was to be built) was $1,348. Yet the median income of those approved for public housing (for whom records remain) was $1,783. Similarly, in the Bedford–Conemaugh area, median income under the same conditions as those stated for Prospect was $1,282 for whites, and $978 for blacks. The median income for those approved by TSAC was $1,806.[9]

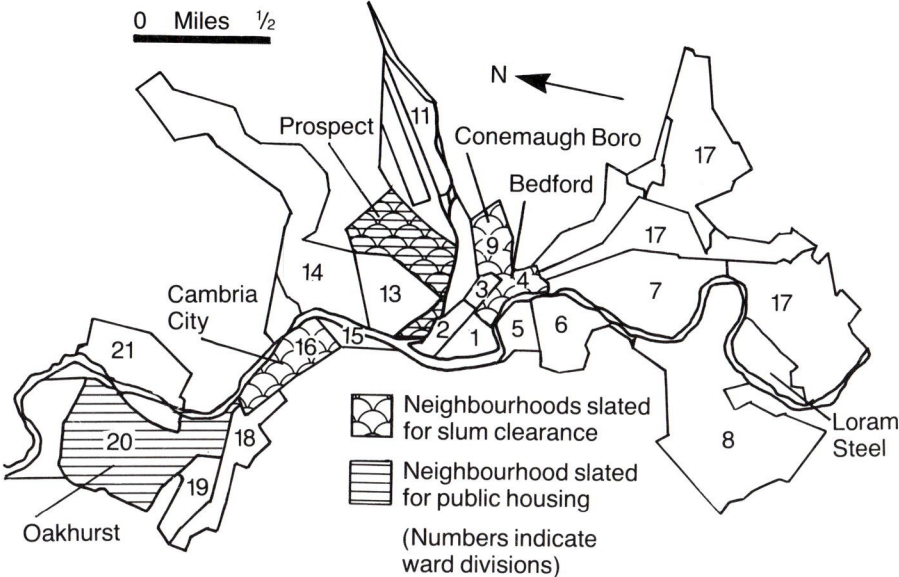

Figure 7.1 Map of the Johnstown Housing Authority Plans of 1943. Prospect and Oakhurst neighbourhoods were to get the earliest public housing. Conemaugh, Bedford, Cambria City and Prospect were slated for slum clearance.

Sources: Drawn from data in Keyes (n.d.); WPA (1941); JHA (1944)

Thus, within these neighbourhoods, public housing was not being provided for the lowest-income groups, nor was it being provided for the lowest-paid workers. Rather, to ensure model tenants, a stringent application review process was instigated which included 'an investigation or home visit to determine citizenship, housing needs, wages [,] employment in defense work, family composition'; moreover, an annual reinvestigation was planned for each resident of the new project (TSAC n.d.).

In Pittsburg, the Chamber of Commerce established itself first as a clearing house for information concerning the Columbia Park Project, and later as the processor of tenant applications for residency in the project. In Pittsburg, as in Johnstown, the intent was clear. The Columbia Park Project was designed for workers – and primarily for workers at the Columbia Steel Mills (*Pittsburg Post-Dispatch*, 23 April 1943). Housing in Pittsburg served a dual function in the eyes of its planners: the obvious function of easing a critical housing shortage, and the equally important function of social control. With the planned expansion of the Columbia Mill by 1,225 workers over three months in 1942, a sound public housing programme would aid in assuring what one researcher called 'social protection' in the Pittsburg area (OHWS–NRPB 1942: 3), with HACCC acting as a *de facto* planning agency.

Social protection and social control were to be achieved by replicating the existing social structure of the town. Thus one remnant of corporate paternalism is evident in the locational decisions and segregation policies of JHA. In Johnstown, as Keyes notes: 'the location of the many plants of the steel mills has been dictated by the topography, a fact which to no small extent dictates the shape of the housing program' (Keyes n.d.). The location of the housing projects was to be subservient to the mills – a familiar theme in corporate paternalism. Racial segregation – a segregation that reflected occupational segregation – was actively pursued by JHA. Projects were to be built with a specific racial quota in mind. Indeed, Keyes recommended that JHA take an active role in increasing the level of segregation in Johnstown, suggesting that blacks be housed in Prospect, in an effort to induce whites to move to other parts of town (ibid.). The Oakhurst project was to be entirely white. Developed in a context of exceptional racism (Sherman 1963), JHA's segregationist policies had a profound effect on the subsequent racial make-up of contemporary Johnstown. In 1980 the Prospect–Minersville neighbourhood was 42.2 per cent black, while the Oakhurst neighbourhood, which has the highest percentage of public housing units to total housing in the city, was only 7.3 per cent black. The overall percentage of blacks in Johnstown in 1980 was 7.7.

Control over the external conditions of Bethlehem employees, then, was to be maintained through a replication of existing social conditions, with blacks moved to a fringe neighbourhood overlooking the Gautier Department of Cambria Steel. Since residence in a JHA project was, *de facto*, contingent upon employment in a Cambria mill, Bethlehem Steel had lost little of the control over employee behaviour it possessed when an employee lived in a company house or was mortgaged to the company or its agents. But such control appears one step removed: the domination of the company in the single-industry town has been euphemized within the local state – in this case within JHA.

The benevolence of the corporation was shifted to the housing authority, and in the language of the authority a second remnant of paternalistic ideology may be found. Perhaps the most prevalent theme in the language of the authority is the idea of public housing granted as a gift. Or, turned around a bit, public housing is provided to extraordinary workers during extraordinary times. When things return to normal after the war, such gifts will no longer be as freely given. In a 'Statement of policy' dated 18 July 1944, the Johnstown Housing Authority made it clear that they were 'unanimous in the opinion that private enterprise should and will have the largest part of the tremendous task of providing housing in the post-war period' (JHA 1944). Thus the wartime role of JHA was to be curtailed, or altered, to reflect the more normal times to which the city was returning. The role of JHA would be limited 'to actual slum clearance' (ibid.). Likewise, in Pittsburg, where public housing was largely 'temporary', public housing programmes were not to be an important part of post-war planning.

But public housing had been legitimized in terms of housing workers, and that role became, to a degree, cemented in the newly emerging social contract

Plate 7.1 Garden apartments originally designed for white residents in the Oakhurst section of Johnstown, Pennsylvania

Source: Photo by D. Mitchell

between management and labour. It was now the role of the local state to provide those welfare services formerly provided by the paternalistic employer. One piece of evidence of the endurance of this contract in the post-war period is that many of the 'demountable' houses in Pittsburg still stand.

The philosophical basis of paternalistic ideology is preserved in the public housing project as well. An abiding faith in the fact that good houses, well planned, make good citizens, and an equally abiding faith in rationalism coupled with the 'suburban pastoralism' of the model company town guided the public housing planning process in Johnstown, if not in the hastily constructed project in Pittsburg. Keyes suggested that the planned Oakhurst project extension should be located 'on the fringe of the city' where 'the type of the development should be in keeping with the surroundings, i.e. . . . detached houses if possible, with a good bit of yard space, and possibly garden space' (Keyes n.d.) (Plate 7.1). Oakhurst was planned as white public housing, and thus (presumably) the prospective tenants could be expected both to afford such amenities and to benefit from them. In the black projects at Prospect, where tenants would be drawn from employees of a lower social and economic position, that lower position was to be reinforced by the built design, with austere apartment blocks planned rather than

Plate 7.2 Apartment blocks originally designed for black residents in the Prospect
neighbourhood of Johnstown

Source: Photo by D. Mitchell

garden communities (ibid.) (Plate 7.2). Thus, the social relations of the workplace were to be reinforced through a carefully planned and rationalized public
residential landscape.

CONCLUSION: THE CHANGING LANDSCAPE OF PATERNALISM

Aspects of the emerging social contract of the post-war period can be divined in
the landscapes of early public housing. Through influence over the direction of
policy within the local state, companies were able to ensure that control over the
external social conditions within their towns was largely retained. In both
Johnstown and Pittsburg, early public housing grew out of an ideology of
corporate paternalism that defined the nature of the product. Rationalism
coupled with pastoralism and a sincere sense of *noblesse oblige* was perhaps most
profoundly evident in company towns such as these. Though this ideology was
hegemonic in the sense that it was used to define the internal and external living
conditions of workers in company towns, the hegemony was never complete.
Rather, the hegemonic ideology of paternalistic corporations was penetrated by

the working classes, and redefined in their own terms. This struggle over meaning inextricably bound corporations to responsibility for the health and welfare of welfare state. Thus, they were able to maintain a certain degree of control over employees, a bond which grew increasingly diseconomic during the economic restructurings of the 1930s when companies in single-industry towns seized the opportunity to euphemize their control over employee lives in the budding the external conditions of 'their' towns, while hiding behind the myth of disengagement from corporate paternalism. The ideology of corporate paternalism was still manifest in the policies and practices of the local state – ready to be reinterpreted, reworked and contested by those who must live out their lives within this new hegemonic order.

NOTES

1   I use the term 'hegemonic ideology' as a sort of shorthand to describe the emplacement by dominant classes of 'their own vision of a just social order, not only on the behaviour of subordinate classes, but on their consciousness as well' (Scott 1985: 304–5). The imposition of a dominant ideology on subordinate classes is not unproblematic, but is in fact contested by the subordinate classes both in thought and in deed. Thus, to some extent at least, a hegemonic ideology must be sanctioned by those upon which it is imposed – and that sanction is not given freely. A struggle over the emplacement of ideology is integral to the construction of a 'social contract' – or as Clark (1986) calls it, a 'class bargain' – that defines the parameters of subsequent class struggle.

2   Here I am concerned with paternalism as a general corporate philosophy and am well aware of the fact that corporate paternalism as it relates to company towns may appear in many guises, from the raw control of company over worker in the Appalachian mining patch, to the welfare capitalism manifest in the 'Square Deal' of the Endicott-Johnson Corporation in Johnson City, New York (see Buder 1967; Creese 1966; Garner 1984; Haraven 1982; Reynolds 1983; Zahavi 1988; and for Johnstown, Morawska 1985).

3   A similar argument is made in Zahavi's (1988) case study of the shoeworkers' response to the paternalism of the Endicott-Johnson Company in New York.

4   For further elaboration on the penetration of hegemonic idiologies, see Lears's (1985) review, and the excellent case study of power relations among Thai villagers by Scott (1985).

5   I develop this theme more fully in Mitchell (1989).

6   For a sampling of this language, see Howard (1902); Knowles (1920); Veiller (1910); and Wood (1940).

7   It is open to speculation as to whether the 'that' in this declaration refers to substandard housing conditions or to low-income families.

8   It is obvious that these records were not kept for the benefit of the researcher looking at them forty-five years after the fact. Thus my interpretation of some of the scribblings may be slightly in error. What is obvious on almost all, however, is the abbreviation 'Beth' in the space marked 'Employer'. Equally clear is the yearly income of the applicants, which will be examined below.

9   There are only seven records that can be positively identified as coming from Cambria City–Minersville. Of those applicants, the median was $1,759 which seems to be

roughly equivalent to the values for the other areas. The median income of the Cambria City–Minersville area is $1,372.

ACKNOWLEDGEMENTS

Thanks to Anne Mosher, Rich McCluskey, Maxx Dilley, Deryck Holdsworth and David Ley for help on various drafts of this essay.

REFERENCES

Bauman, J. (1987) *Public Housing, Race, and Renewal: Urban Planning in Philadelphia, 1920–1974*, Philadelphia: Temple University Press.

Buder, S. (1967) *Pullman: An Experiment in Industrial Order and Community Planning, 1880–1930*, New York: Oxford University Press.

Clark, G. (1986) 'The crises of the midwest auto industry', in A. Scott and M. Storper (eds) *Production, Work, and Territory: The Geographical Anatomy of Industrial Capitalism*, Boston: Allen & Unwin, 127–48.

Creese, W. (1966) *The Search for Environment: The Garden City, Before and After*, New Haven: Yale University Press.

Ebenstein, W. (1940) *The Law of Public Housing*, Madison: University of Wisconsin Press.

Foucault, M. (1970) *The Order of Things*, New York: Vintage.

—— (1979) *Discipline and Punish: The Birth of the Prison*, trans. A. Sheridan, New York: Vintage.

Garner, J.S. (1984) *The Model Company Town: Urban Design Through Private Enterprise in Nineteenth-Century New England*, Amherst: University of Massachusetts Press.

Haraven, T. (1982) *Family Time and Industrial Time in a New England Industrial Community*, Cambridge: Cambridge University Press.

Harvey, D. (1978) 'Labor, capital, and class struggle around the built environment in advanced capitalist societies', in K.R. Cox (ed.) *Urbanization and Conflict in Market Societies*, Chicago: Maaroufa, 9–37.

'Helping the workingmen to help themselves' (1912) *Iron Age*, 6 June: 1213–19.

Hornbeck, F. (1941) *Survey of the Negro Population of Metropolitan Johnstown, Pennsylvania*, Johnstown: Johnstown Tribune and Johnstown Democrat (mimeo in Pennsylvania Room, Johnstown Public Library).

Howard, E. (1902) *Garden Cities of To-morrow*, London: S. Sonnenschein.

JHA (1944) 'A statement of policy', USWA Collection, District 13, Box 1, Labor Archives, Pattee Library, Pennsylvania State University.

Keyes, S. (n.d.) 'Rehousing Johnstown: an analysis of the need for slum clearance and a six-year housing program', USWA Collection, District 13, Box 1, Labor Archives, Pattee Library, Pennsylvania State University.

Knowles, M. (1920) *Industrial Housing*, New York: McGraw-Hill.

Lears, T.J.J. (1985) 'The concept of cultural hegemony: problems and possibilities', *American Historical Review* 90: 567–93.

Leverhulme, W.H. (Lord) (1918) *The Six-Hour Day and Other Industrial Questions*, London: Allen & Unwin.

McPherson, D. (1972) 'The "Little Steel" strike of 1937 in Johnstown, Pennsylvania', *Pennsylvania History* 39: 219–38.

Mills, C.A. (1988) '"Life on the upslope": the postmodern landscape of gentrification', *Environment and Planning D: Society and Space* 6: 169–89.

Mitchell, D. (1989) 'A history of homelessness – a geography of control: the production of order and marginality in Johnstown, Pennsylvania', unpublished master's thesis, Department of Geography, Pennsylvania State University.

Morawska, E. (1985) *For Bread with Butter: Life-Worlds of East Central Europeans in Johnstown, Pennsylvania, 1890–1940* Cambridge: Cambridge University Press.

OHWS–NRPB (1942) Composite report on the Pittsburg–Port Chicago area, Northern Contra Costa County, State of California, prepared by the Office of Health and Welfare Services, Region No. XII, and the Natural Resource Planning Board, Region No. 8, Bancroft Library, University of California, Berkeley.

Parsons D. (1984) 'Organized labor and the housing question: public housing, suburbanization, and urban renewal', *Environment and Planning D: Society and Space* 1: 75–86.

Reynolds, J. (1983) *The Great Paternalist: Titus Salt and the Growth of Nineteenth-Century Bradford*, New York: St Martin's Press.

Scott, J. (1985) *Weapons of the Weak: Everyday Forms of Peasant Resistance*, New Haven: Yale University Press.

Sherman, R. (1963) 'Johnstown v. the Negro: southern migrants and the exodus of 1923', *Pennsylvania History* 30: 454–64.

TSAC (1942) Minutes of the Tenant Selection Advisory Committee, 20 November 1942, USWA Collection, District 13, Box 1, Labor Archives, Pattee Library, Pennsylvania State University.

—— (n.d.) Tenant selection records, Johnstown Housing Authority, Tenant Selection Advisory Committee, USWA Collection, District 13, Box 1, Labor Archives, Pattee Library, Pennsylvania State University.

Veiller, L. (1910) *Housing Reform: A Hand-Book for Practical Use in American Cities*, New York: Charity Publications Committee.

Watts, M. (1988) 'Struggles over land, struggles over meaning: some thoughts on naming, peasant resistance and the politics of place', in R. Golledge, H. Couclelis and P. Gould (eds) *A Ground for Common Search*, Santa Barbara: Santa Barbara Geographical Press. 31–50.

Williams, R. (1980) *Problems in Materialism and Culture*, London: Verso/New Left Books.

Willis, P. (1977) *Learning to Labour*, Farnborough: Saxon House.

Wood, E. (1940) *Introduction to Housing: Facts and Principles*, Washington, DC: Federal Works Agency, US Housing Authority.

WPA (1941) Johnstown survey of low-rent housing needs, USWA Collection, District 13, Box 1, Labor Archives, Pattee Library, Pennsylvania State University.

Zahavi, G. (1988) *Workers, Managers, and Welfare Capitalism: The Shoeworkers and Tanners of Endicott Johnson, 1890–1950*, Urbana: University of Illinois Press.

8

CO-OPERATIVE HOUSING AS A MORAL LANDSCAPE

Re-examining 'the postmodern city'

David Ley

The folk house has provided a compelling interest to North American cultural geography and to an earlier anthropo-geography in Europe. The source of this fascination is clear enough, for the folk house presents a synthesis of pre-modern geography, a tight regional integration of local materials and local traditions in the context of a local physical environment. The connections between culture and environment are writ large in the visible texture of built form. Indeed the folk house is such a primary indicator of the structure of pre-modern geographies that it has frequently been employed to demarcate regional boundaries, or in trans-atlantic migration, to provide a marker tracing the movement of peoples in the colonization process.

But however refined and painstaking has been the description of culture areas, and the diffusion of culture traits including folk housing, in at least three areas the literature of cultural geography has remained remarkably silent. First, the men, women and children who were the builders of the structures do not appear in the accounts. In part of course this is a problem of historical reconstruction, of retrieving the sources which would illuminate the routines and meanings of ordinary lives. But the omission is not related to data alone, for recent work in social history and in some historical geography has found ways of reconstructing the meaning of ordinary places (e.g. Pred 1990). Indeed, even as unpromising a landscape as temporary buildings dating from the earliest period of settlement of the American colonies has been studied in an account which progresses from detailed archaeological excavations, permitting a description of built form, to an imaginative interpretation of the cultural milieu in which the structure was produced (Carson *et al.* 1981). This is not a choice which many cultural geographers have opted for in the past. Rather human agents have been regarded as the carriers of a culture without concern for the rules and rhythms of their everyday lives. The focus has instead been on the visible artefacts, after the fashion of material culture studies: 'Traditionally content to collect and to describe objects, many material culturalists still resist ... the imperative to deduce, wherever possible, the culture behind the material' (Schlereth 1983: 114).

There has been a second and related silence. The social relations implicated in the creation of spatial form have not been included in the research agenda. Society has been held constant: the contingencies of family relations, social structure and political authority have been largely unseen and unspoken for. The description and classification of house forms have been an end in themselves which has not required a questioning posture towards social processes. Instead an often undifferentiated view of a people is held, for example of an ethnic or racial group, who function as carriers of a folk tradition.

A third silence follows on from the second. The literature of cultural geography has little to say about the landscapes which succeeded folk society, about the geographies that were created with the passing of the world we have lost. In his seminal interpretation of *The Making of the English Landscape*, W.G. Hoskins effectively ended his account with the onset of the modernizing impulses of the Industrial Revolution: the twentieth century is 'a distasteful subject, but it must be faced for a few moments' (Hoskins 1955: 231; Meinig 1979). The pre-modern paradigm of local traditions and a local physical geography conspiring together to form a distinctive regional landscape was being superseded by the vigorous place-transforming energies of industrial capitalism. The mass production of housing units for a diversified migrant population in the burgeoning industrial cities largely eroded the effects of local vernacular styles. Housing solutions such as the working-class tenement, later the middle-class bungalow (King 1984) and later still the high-rise public and private apartment buildings in the austere proportions of the modern style (Frampton 1985) all generated a repetitive landscape with a continental and eventually a global reach.

In recent years new directions in cultural geography, fortified by a broader literature in social and cultural theory, have challenged these omissions. The appearance of the domestic landscape, for example, has been cast in a wider interpretative net ranging from philosophical reflections on the meaning of the modern suburb (Relph 1976) to more focused research examining the role of interior and exterior house design in constructing contemporary social identities (Pratt 1981). Outside the advanced industrial nations, Duncan and Duncan (1980) have noted the neighbourhood choice and architectural selection made by two distinctive elites in India who wish to use the residential landscape to advance either a traditional or a modern Western identity. In these and other works, the issue of the authorship of place looms large (Samuels 1979; Western 1986). The relationship between housing and identity also has clear gender dimensions which are the subject of a growing literature (e.g. Hayden 1981). Attention to the social structure may usefully be pursued through an appreciation of landscape which sees built form as an outcome of changing social and economic processes. Evincing some frustration with the 'dead-end questions' of a purely morphological approach, Holdsworth (1990: 8) approaches the interpretation of shelter for the gang labour of coal-miners, loggers, fishermen and others as a nexus 'for noting relationships between masters and men, capital and labour, and the creation of places' in the historic shift from mercantile to industrial economic systems.

Inequality in the market-place and before the power of the state, then, becomes an important interpretative theme, a state of affairs which resonates with the presence of conflict and resistance (Jackson 1989) or, equally problematic, its absence. The domestic landscape, for example, can be made to reinforce identities, whether the subordination of minority groups (Anderson 1988) or the upward mobility of ambitious young professionals (Mills 1988). Landscape is, as a result, inherently ideological, or in Stephen Daniels's phrase, duplicitous, a setting variously enabling and oppressive (Daniels 1989). For the landscape as an accomplished social fact not only is an expression of dominant values, but also tends to reproduce them as part of the 'natural' order. It thus evokes political as well as cultural relationships, and a current literature has begun a critical reading of the landscape as text, aiming to denaturalize its facticity and disclose the cultural politics which it reinforces but obscures (Duncan 1990; Ley 1987; Schorske 1981).

Nowhere has there been more enthusiastic landscape criticism in recent years than has been directed by some authors at the form of 'the postmodern city'. This generic entity has been widely criticized for concealing, beneath a user-friendly façade, the cultural politics of the new right. To be sure, this argument carries some weight, though not as much weight as its sponsors imagine. Elsewhere we have addressed this round of criticism at a conceptual level (Ley and Mills forthcoming). In this essay I wish to develop an argument, through an empirical study of the cultural politics of co-operative housing in Vancouver, that post-modern landscapes, like others, also need to be seen as duplicitous, or better, ambivalent, not simply showcases for a new bourgeoisie but also capable of supporting humane, indeed moral, public values.

MODERN DOMESTIC LANDSCAPES AND BEYOND

In its interpretation of modern urban landscapes recent cultural geography has overcome the earlier reluctance to leave the folk societies of the pre-modern world. But modern societies follow different organizing principles and as such require a separate theoretical strategy in interpretation. I have argued before that in the rapidly growing, culturally diverse and severely inequitable industrial cities of the late nineteenth century a cardinal social, political and moral problem was the problem of the masses, and linked to it was the problem of establishing a new social order in the cacophony of the industrial city. Reflecting a broadly shared sentiment, Gustave Le Bon's *Psychologie des Foules* (1895) charged that 'The substitution of the unconscious action of crowds for the conscious action of individuals is one of the principal characteristics of the present age.' It was also a portentous development, for, Le Bon reasoned, 'The crowd state and the domination of crowds is equivalent to the barbarian state, or a return to it' (quoted in Brantlinger 1983: 168).

More generally the ontology of personhood in modernist theory is an ontology of the mass: 'the twentieth century is the century of the masses ...

science, technology, mass locomotion, mass production and consumption, mass communication' (Pevsner 1968: 7). In a mass society, new answers were needed for urgent social problems. The old answers, old housing solutions, for example, would not do. The traditional house form, opined Le Corbusier (1927: 31), was 'an old and hostile environment', altogether inappropriate for the spirit of a new age. To modernists like him the alternatives in the cities were as sharply drawn as architecture or revolution. Domestic architecture should be the architecture of mass production: 'we must create the mass production spirit' (ibid.: 12). The furious onslaught upon nineteenth-century bourgeois culture, including middle-class homes – viewed by Le Corbusier (ibid.: 91) as 'the intolerable witnesses to a dead spirit', by the young Frank Lloyd Wright (1975: 235) as 'mere notion stores, bazaars or junk-shops' – was uttered with all the confidence of those certain that history was on their side.

The spirit of mass production harnessed the apparently progressive forces of science and technology into a machine-age aesthetic of urban design and planning, an aesthetic that dominated such influential leaders of the modern movement as Le Corbusier and the Bauhaus (Galison 1990; Ley 1989). Le Corbusier in particular was infatuated with the crisp, clean lines of industrial design. He named one standardized house design Maison Citrohan, a pun upon the name of the car manufacturer. Modern people, he wrote, '*can* be proud of having a house as serviceable as a typewriter' (1927: 241), for after all, what was a house other than a machine for living in, or what for that matter was a street other than a factory for producing traffic? As self-discovered claimants of the historical moment when the machine stood revealed as the incarnation of the world spirit, the modernists wrote with a heightened degree of moral conviction. 'A modern building,' observed Gropius of the Bauhaus, 'must be true to itself, logically transparent and virginal of lies or trivialities, as befits a direct affirmation of our contemporary world of mechanization and rapid transit' (Brolin 1976: 51).

The privileged status of science and technology in the machine-age aesthetic, together with an ontology of the masses, held two significant consequences for the built environment. First, it legitimated the modernists as a prophetic status group, heralding in the new age, and distanced them as a speech community from a larger public. It expedited the cult of the expert, the professional, and an increasingly centralized and corporate form of urban administration that was such a key feature of the Progressive era in the early decades of this century. It aided the rise of powerful civic bureaucrats like Robert Moses with extraordinary powers of command and control in the urban environment. The mechanical metaphor enjoyed a broad currency among intellectuals as a normative and not merely a descriptive tool. 'There is a city mentality which is clearly differentiated from the rural mind,' wrote the Chicago sociologist Louis Wirth (1925: 219). 'The city man thinks in mechanistic terms, in rational terms, while the rustic thinks in naturalistic magical terms.' And phrased this way, to deny the superiority of rational, mechanistic thought was to paddle against a flood tide.

A second consequence led from the first, as the mechanical metaphor left the factory and colonized the realm of culture. This was a domain that had been emptied of its nineteenth-century content in the relentless attacks by modernists like Adolf Loos on middle-class society, its taste and its landscapes (Frampton 1985; Janik and Toulmin 1973; Schorske 1981). But so too, in migration to the cities, folk culture had been severely eroded, as the Chicago sociologists sought to demonstrate in their detailed monographs on social disorganization. The disappearance of the old provided an opening for the new, as culture was reinterpreted in the rational discourse of the machine. Functionalism became the cultural *meaning* of products, standardization the method, universalism the goal. This was the higher good, the public interest, the rational response to the uncertainties following the passing of folk culture. It was right, from this perspective, that the city should be run like a business in a calculating, efficient manner, right that a civic bureaucracy headed by a city manager should become the control centre, right that the tainted fiefdoms of local politics, the wards, should be replaced by the more corporate design of the electoral at-large system. Such developments were not only scientifically correct, but they were also morally fitting. In a curious contradiction of their scientific epistemology, the modernists frequently bared their humanity in urging their values in peremptory style. Adjectives such as 'honest', 'pure', 'sincere' and 'truthful' are favourite terms in describing their own work. Giedion (1967), a student of Le Corbusier and an enthusiastic disseminator of modern planning and design theory, has a section on 'The demand for morality in architecture' in his famous book, *Space, Time and Architecture* (also Watkin 1977). A later postmodern sensibility would see, in this ubiquitous human disposition towards valuing an example of science as personal knowledge, a condition which undercuts the totalizing claims to objectivity of the modernists.

The orientation towards a scientific epistemology and an ontology of the masses among the modernists had predictable consequences for the urban landscape. Fuelled by development capital and state policy, the high-rise, freeway city with standardized suburban tract homes was a logical outcome. It was a landscape which systematically sought to remove the content of everyday culture: historic memory, regional distinctiveness, the personalization of space, all were treated sceptically. The individual was disappearing as a category before the engagement with the mass. 'The individual,' claimed the architect Mies van der Rohe in 1924, 'is losing significance; his destiny is no longer what interests us' (Watkin 1977: 38). The reactionary content of such a world-view became all too evident – on occasion tragically so – over the following forty years, and by the 1960s a variety of social movements were arrayed against it. A so-called rational epistemology, a mass ontology and centralized politics were challenged in a range of contexts: the environment, civil rights, the Third World, the arms race, appropriate technology and the city. The clamour against the military–industrial complex, multinational corporations and inaccessible bureaucracy all revealed the alienations of modernity and modernization. The ideal of an abstract and uniform public interest was exploded by vocal minorities excluded from decision-making and penalized by their omission: ethnic and racial minorities, women and, in the cities, neighbourhoods demolished by the likes of Robert Moses's urban 'meat axe' (Berman 1982: 194).

In urban development Jane Jacobs (1961) was an early, eloquent spokeswoman for what in hindsight has been called a postmodernism of resistance, and, significantly, she reversed the philosophical presuppositions of the modernists. Her epistemology was not professional and elitist, but populist, extolling the virtues of local knowledge; her administrative model was decentralized and participatory; her ontology acutely aware of diversity and difference. No less than the modernists, postmodern ideology was morally charged, though its quest for 'sensitive urban place-making' (Jencks 1981: 82) was very different. For some the agenda went even deeper to a 'Post-modern renewal of our imaginative and spiritual relationship to the world of nature' (Fuller 1987: 8). While the gods might take many forms, resistance to the elimination by fiat of the spiritual realm revealed that postmodernists aspired to an integrated and not a reductionist model of humankind, defending 'an adamant refusal to give up the imperatives of the spirit at a time when all systems of spiritual expression have been cast into doubt' (Jencks and Chaitkin 1982: 217). Pronouncements such as these contained the clear hope that the building of a home should be not less than a moral pursuit.

A NEW START IN CANADIAN SOCIAL HOUSING

During the 1960s discussion and mobilization developed around new forms of non-market housing in Canada, as elsewhere. They were prompted by a reaction against high-rise public housing in particular, housing for the masses which expressed all too well the agenda of the modern movement. The new critical discourse resonated with such phrases as 'ghettos for the poor', 'cell-like slabs' and 'centres of stigma'. What was at stake, however, was not merely the appearance of the structure, important though this was, but also the centralized process of construction and management that accompanied it. Community development workers urged greater participation and control by the residents of social housing, to lessen dependency and stigma. Building upon several grassroots experiments, labour, co-operative and church groups founded in 1968 the Cooperative Housing Foundation of Canada (Selby and Wilson 1988). The timing for this initiative could not have been better, for the new federal Liberal government under Pierre Trudeau established a Task Force on Housing and Urban Development the same year. Mindful of the extensive criticism of urban renewal and large-scale public housing, and enabled by the social experimentation of the early Trudeau years, the Hellyer Task Force produced a sternly critical report of both the design and the process of housing construction for the poor.

The big housing projects, in the view of the Task Force, have become ghettos of the poor. They do have too many problem families without adequate social services and too many children without adequate recreational facilities. There is a serious lack of privacy and an equally serious lack of pride.... There is a social stigma attached to life in a public housing project which touches its inhabitants in many aspects of their daily lives. If it leads to bitterness and alienation among parents, it creates puzzlement and resentment among their children ... the problems of living in public housing were akin to those experienced by some Indians who are on reserves.

(Government of Canada 1969: 53–4; in Selby and Wilson 1988: 9–10)

The critique was similar to, and as severe as, that issued against modernism by Jane Jacobs at the beginning of the decade. It provided an opportunity as well as establishing an objective to construct a new moral landscape, during a period of intense re-examination, redirection and innovation in social housing programmes. At the urban level a similar transformation of ideology occurred, particularly in the cities of Toronto and Vancouver where reform councils swept to power in 1972. In each city, there was 'a clear philosophical rejection of "modernist" approaches to urban design and architecture' (Hulchanski 1984: 41). Plans for the high-rise, freeway city, inherited from the idealism of inter-war modernists, were set aside (Jacobs 1971; Ley 1980, 1987). The centralized model of rational control was also invaded by greater public accessibility to decision-making (Hasson and Ley forthcoming). An ontology of the masses and an unproblematic public interest were seriously questioned; a more multi-vocal epistemology permitted a new if yet incomplete political interpenetration between system and life-world.

This historical conjunction of events enabled experimental programmes in social housing, including co-operatives, to be launched. In 1970 the federal government made available a loan fund of $200 million to support innovative approaches to affordable housing, and from it the Cooperative Housing Foundation was successful in bidding for $30 million to start up five co-op projects and gain credibility as a housing producer (Selby and Wilson 1988). Its achievements led to the incorporation of a co-operative housing programme in the wide-ranging amendments to the National Housing Act in 1973. In the next four years, 10,000 housing units were developed, and by 1988 co-operative housing units in Canada exceeded 52,000; through most of the 1980s co-operatives accounted for 4–5 per cent of annual housing starts. Moreover, because co-op projects are concentrated in the major urban centres with the highest level of affordability problems, their significance exceeds their share of the housing market (Hulchanski 1988).

Housing co-operatives offer a unique form of tenure. In Canada, membership in a non-profit housing co-operative frequently involves the contribution of share capital but not an equity investment. When a member leaves the co-operative that capital is returned with or without the addition of interest payments. This tenure form separates co-ops from condominiums with their potential for capital gains. What the co-op member receives instead is guaranteed and continuing affordable housing at below market rates and joint democratic control and management of the co-operative as a legal entity. There is considerable consistency in the enunciated objectives of co-ops. For example, an information package distributed by a housing society which has sponsored over forty co-operatives lists the six principles of co-operation established by the international Cooperative Alliance in 1966 (Inner City Housing Society 1988): open and voluntary membership; one member, one vote; limited or no return on invested capital; non-profit status; member education; and co-operation among co-operatives. These moral principles are translated into lists of rights, obligations and benefits.

	Rights	*Obligations*	*Benefits*
1	Security of tenure	To pay monthly housing charges	Control over rent
2	Goals determined by members	To participate in management	Sharing responsibilities
3	Housing charges set by members and cover costs only	To look after your own unit	Manageable-sized home in good condition
4	Planning done by elected and accountable co-op members	To protect and maintain common area for all to benefit	Caring atmosphere with security from a neighbourhood watch committee
5	Members set conditions of occupancy		Input into designing your home and ensuring high quality of construction
6	Information on planning and management accessible to and understood by all members		

CONSTRUCTING AN ALTERNATIVE RESIDENTIAL LANDSCAPE

From what has been said already, it is evident that an interpretation of co-operative housing must go beyond the issue of morphology, important and, as we shall see, in some ways distinctive though the built form may be. The alternative dimension of co-operatives extends also to non-hierarchical practices; to both the principles of ongoing management and also to the anticipated partnership between future residents and design professionals in reaching a design solution. Our study of co-operatives in Vancouver thus included the social process through which the built environment was constructed, and particularly the extent to which democratic ideals found their way into the adopted design. To illuminate these issues, site visits were made to 22 of the 95 housing co-operatives completed in the City of Vancouver between 1973 and 1986. This sample included the seven earliest co-ops, built before 1977, with the remainder all built in 1985, the year of highest production. In addition ten of the architects of these projects were interviewed; their firms were highly experienced in social housing and together had designed well over 100 co-ops in Greater Vancouver and beyond. Further interviews were conducted with several social housing resource groups, non-profit societies which are significant facilitators in the development of a co-operative, as well as officials of Canada Housing and Mortgage Corporation (CMHC), the federal agency which funds the co-op housing programme.

While there was no unanimity among architects, their involvement in co-operative housing in general involved an endorsement of co-operative principles, and for some, an explicit left-liberal political ideology. One architect, for example, received his first co-op commission from a housing resource group whose director he had earlier worked with in a grass-roots association that had resisted gentrification and promoted affordable housing. A second architect saw co-ops as an important vehicle offering not only shelter, but also consciousness-building among women, as female-headed households are a significant resident group (cf. Wekerle 1988). A third, an immigrant, who has specialized in co-ops for new Canadians, including refugees, spoke of an ideological commitment to social housing.

As established in the 1970s, the democratic model set out by co-operatives involved not only resident participation in management, but also involvement in the design of the structure. Housing resource groups aided the incorporation of a co-op, the selection of an architect and the procurement of government grants. While the design process is far more complicated with co-ops than with market units, architects evaluated positively their interaction with users. The personalization of this exchange was a welcome departure from the contractual nature of private housing, and its detachment from eventual occupants. The exchange implied also, of course, resident participation in design. That participation was variable and has certainly lessened in the late 1980s, but nevertheless was always mentioned to a lesser or greater extent as distinctive of co-ops. The most experienced architect, whose firm had completed over thirty projects, observed: 'Co-op housing is unique in itself. You can start from scratch with the co-op group. You meet with the group, you discuss site constraints. You blend in what they want to accomplish within the budget.' Yet the co-op group also faces significant constraints. Financial and some design parameters were laid out by

CMHC, including a so-called modesty clause in building appearance. Design conditions were imposed by the protocols of the City of Vancouver's planners, including their preference for 'neighbourliness' in the building's massing and façade, thereby ensuring its geographical compatibility with adjacent housing. In instances when the co-op was assembled by the resource group and had limited expertise and a poorly developed set of objectives, consultation was perfunctory. But where co-op groups were grass-roots entities they were often vociferous and full participants in the design process.

What design traditions were effective in developing co-op housing? Architects themselves proved to be surprisingly unselfconscious in nominating schools that had influenced them. Only one of the ten architects interviewed identified with postmodernism explicitly, while a second was clearly an anti-modernist, a third identified pro-modernist influences, and most of the others mentioned the importance of vernacular and traditional associations. Almost all were surprisingly inarticulate on this matter, and could not (or would not) even identify influences from individual architects. Some replied in terms only of their own intuition or artistry as inspiration for their buildings; in such personalized responses one can readily understand the hostility of being typecast within a general style category.

Explicitly or implicitly architects agreed that co-op housing did provide a text expressing social ideals (Plate 8.1). The dominant theme they identified was community and enhancing the co-operative vision in design:

> The built environment is a response to the social structure of the co-op. It is focused around communal space and the recreation room. Internal circulation all leads to this common space. This is unique to co-ops.

> Everything is designed around a courtyard which is the social heart of the project.

> Courtyards are important communal space. In co-ops you promote it. In market housing you're selling them their own private world, they don't want to mix.

> I tried to make it like Chilean housing. As a group Chileans like communal living. Every unit faces the courtyard. The courtyard is very intimate like Chilean housing projects.

As suggested by this last architect, with some grass-roots groups there are opportunities to promote a collective memory, or with a woman's co-op, a shared political consciousness. The courtyard design also permits surveillance, the 'eyes on the street' popularized by Jane Jacobs, as an explicit benefit of a consciously shared life-world:

> CMHC thought that the inner courtyard windows didn't allow enough privacy. They were confused by the women's need for surveillance.

Plate 8.1 Courtyard at the Sam Greer Co-op: note the personalization and domestication of this communal space, with planters and chair set up as one's own private patio

The kitchens look into the courtyard so the kids' play is protected. Bedrooms are on the exterior and more private.

Like Jacobs's Hudson Street, the co-op is characterized by a social mix, with rents adjusted according to income. Before 1986, 35–40 per cent of members received subsidies to their monthly charges, although all members benefited from low-interest mortgages and, commonly, mark-downs on land prices. Besides income mix, there was also some variation in age and family status. Mix has been a central, and sometimes controversial, objective of co-operatives, intended to promote 'healthy communities' in contrast to the tendency of public housing to be 'ghettos for the poor'. 'When Kanata was constructed co-ops were just starting, Columbia Housing [a housing resource group] had a mandate to mix people and break down the public housing image of social housing.'

Given the deliberate communal identity of co-ops, a difficult philosophical problem is whether to emphasize difference between individual units within the overall design. Most architects sought to escape standardization and introduce distinctiveness and variety to each unit, or at least groups of units. 'There's some individuality – each unit is a bit different'; 'We tried for individuality, with a series of [linked] individual houses facing the street'; 'There are 35 unit types out of 36 units. Yes, there's a lot of uniqueness.' But other architects identified a contradiction in enhancing individuality within a co-operative protocol. 'I first tried to create a sense of privacy, a zone for personal expression and identity. But

Plate 8.2 Façade articulation of Wit's End Co-op: diversity of shape, colour (manilla and orange) and protruding vertical planes, plus division into three buildings, break up the mass of this 52-apartment structure into smaller more personable units

the communalism is more important now. I learned that co-ops don't stress the individual'; or, again, 'They're part of a group. We emphasize group identity rather than individual identity.' The disagreement among designers about how far to carry the personalization of individual living space is instructive, for it shows the strength of the reaction against the impersonal standardization of public housing, with its ontology of the faceless and unvarying masses. In co-operatives, instead, we see an ontology of difference, a personalization of each project, if not of each unit. 'The co-op takes on a group character. The group has its own personality'; 'Because user groups have involvement in design, each co-op has its own personality'; 'I've done 30 co-ops and they've been 30 totally different clients.'

Two further features of modern design in public housing, massing and scale, have been addressed. 'My objective was to integrate with the neighbourhood by reflecting single family homes, to break up the [public housing] stereotype.' Undifferentiated massing was attacked by the diversity of design elements and materials, colour and the postmodern strategy of façade articulation (Plate 8.2), with recesses, projections, gables and, in one instance, a double wall providing 'protection' from a major thoroughfare (Plate 8.3). Equally important was the issue of scale. Although located in the inner city, all buildings were ground-oriented family developments, either row or stacked town houses or low-rise

Plate 8.3 Complexity and contradiction at Charleston Co-op: the architect wanted this 60-unit structure to be 'memorable', 'a good neighbour', 'a source of pride'; the gables and clapboard (in powder blue) are taken from nearby older houses, the postmodern programme (in cream and russet) from new town houses in this gentrified district; notice the roof gardens and false chimneys (e.g. extreme left) in this complex project, part of the iconography of 'home'

apartments. Favourably citing Schumacher, one architect described his structure as 'a domestic scale of architecture, low scale in a dense neighbourhood'. The separation of automobiles from pedestrian circulation contributed further to the human scale of the built environment.

Indeed in many ways, these co-operatives represent the literal domestication of social housing (Plate 8.4). Encouraged by the strong design guidelines of a neighbourhood-oriented city planning department, compatibility with the local area was an important requirement:

All details were taken from the neighbourhood: traditional windows, 3 × 3 wood rails, and the scale is like adjacent [single family] homes.

The shapes had to relate to smaller single family residences in the area and the form of the street. The City has a strong policy on this. Scale and articulation were important. Because the area has a French flavour, we used château style.

The symbolic adoption of château style by this architect carried further a

Plate 8.4 The domestication of social housing: a strongly vernacular iconography of the picket fence and the trellis leading to a landscaped front garden at Sitka Co-op; note the contextualism of design as the picket fence communes with its neighbour in an old house across the street, while the trellis repeats the lines of the gable

programme of historic and geographic contextualization employed by most. He acknowledged his predilection for 'contextual movements' in architecture, sympathetic to the geographic and historic setting, and a number of architects reflected on their use of a traditional vernacular style (Plate 8.5). Indeed the only architect who identified his building explicitly as postmodern declared that this style had become the true vernacular in the adjacent neighbourhood!

Decoration and symbolism were limited, in part because of cost constraints and CMHC's modesty guidelines for social housing. But colour was employed to animate the setting, and in one instance to provide historic symbolism, where the orange and red roofs of row housing were intended to reflect upon the orange railway boxcars that had once occupied the formerly industrial site. With the exception of feminist imagery in one co-op, other symbolism was resolutely domestic, with such elements as porches, gables, chimneys and picket fences providing the textual discourse of 'home'. At least one building had chimneys, but no fireplaces (Plate 8.3), in order to consolidate the domestic symbolism. Not merely the symbolism but the whole design programme was calculated to reinforce the supportive meanings of home. The intent of this design for difference was to inculcate a sense of pride among residents and remove the sense of stigma identified by the Hellyer Task Force in public housing. Of this the architects were keenly aware:

> It's important for co-ops to have pride. I wanted to create a grand building, a building with an image in the neighbourhood. I wanted to create a good image for co-ops.

> Co-op housing expresses social concerns: a sense of community, of individuality within the community. It's designed to be distinctive. They express themselves as a group and the pride shows. It's not a social housing image – they don't want project stigma.

The same message had been well learned by the City. The concept of 'neighbour-liness' enunciated by the Director of Planning (who had been appointed by a reform council in 1973) was similarly intended to remove the image of a spoiled identity for the residents of social housing.

The landscape of Vancouver co-operatives, with a design predicated upon personalization and resident participation, was not merely innovative, but more fundamentally it was oppositional to the alienations of modernism: 'The modern movement has alienated people. Now I think you'll see decades of very interesting architecture, with amazing changes in approach to the texture of multi-family housing.' The same conclusion has been drawn by a recent review of Vancouver's social housing. Acknowledging its 'sense of urbanity ... awareness of streetscape, a celebration of the collective, a sense of exuberance', the author notes also 'a rejection of post-war Modernist housing theories as epitomized in the high-rise tower and slab-block estates' (Berelowitz 1988: 34). 'It's undeniable,' he claims, 'that over the past decade some of the best urban housing in Canada has come out of the social housing sector in Vancouver' (ibid.) But there is also a

Plate 8.5 Contextualism at Grace McKinnis Co-op: this building is near Sitka and integrates a similar local vernacular; note the picket fence and the gable end, dormer additions and clapboard siding which reproduce neighbourhood elements illustrated in the old houses in Plate 8.4

Plate 8.6 Election posters at Marpole Terrace: like other co-ops, this building offers visible endorsement for social democratic candidates; architecturally and politically this is a landscape of resistant postmodernism

second opposition, and one which has been more troubling to conservative politicians (Chouinard 1989). In both design and lifestyle co-ops suggest departures from the market system and its supportive ideologies. There was an unexpected consensus among the architects as they compared the design of market and co-op housing. The former was described as 'more generic', as 'very formula-oriented', 'very standardized'. Co-op housing was more challenging with the participation of both co-op members and a set of urban managers, but the outcome not only was more satisfying professionally, but also led to a more substantial product. Three architects contrasted the 'liveability' of co-ops with the 'marketability' of condominiums:

> Yes, design requirements are different. I deal with liveability in co-ops but market units are designed for first impressions by potential buyers.

> There's a difference in philosophy. For market housing, what is most marketable? For co-ops, what is most liveable?

One architect who strongly endorsed co-ops commented on the high standards expected by co-op members, their 'litigative' tendencies and the direct interaction between designer and user, all of which tended to enhance quality. In contrast market housing was becoming 'just a pretty face'. Its generic character, the isolation between architect and user and the cost-cutting tendencies of the

144

developer did not guarantee as substantial a product.

The co-operative philosophy is also an antidote to the individualism of the market-place. Co-ops are commonly a haven for social democratic politics (Plate 8.6). Particularly in the early years co-op members often comprised politically articulate and critical grass-roots groups. Consider the Sam Greer Co-op constructed in 1976 (Plate 8.1). The site was bought by subscribing members of a community group with an Alinsky-style ideology of community development, strongly resistant to the waves of gentrification which were destroying affordable housing in their district. They outbid a condominium developer for the site, by posing as developers themselves in purchasing the land from its previous owner. The older units were not destroyed but added to, in an infill project. In the words of its founder, the co-op was 'a political statement'. The name too is symbolic, for Sam Greer had been a pioneer in the district whose land had been expropriated in a massive City-engineered land grant to persuade the Canadian Pacific Railway to push its west coast terminus into the Vancouver town site. Greer had refused to part with his property and was eventually imprisoned. In its bold association with this precedent, the Sam Greer Co-op was a landscape of opposition, a postmodernism of resistance.

CONCLUSION

In the midst of a highly partisan discourse over architectural postmodernism, which frequently avoids careful definition of its concepts and engages in a façadism offering the thinnest of landscape descriptions, this interpretation is intentionally oriented towards the empirical. I have defined an oppositional postmodernism in the built environment in terms of an ontology of difference, a multi-vocal epistemology and a politics of participation. These principles are refined in the built environment into a set of themes which we see realized to some degree in the co-operative housing movement in Canada. Born from alternative and oppositional visions in the late 1960s, this movement has sought to dislodge the alienating ethos of public housing and elevate an empowering vision of social housing.

But if empirical, the argument has sought not to be empiricist. The definition of postmodernism is theoretical and represents a level of conceptualization beyond that presented by the actors themselves; while most architects employed postmodern idioms such as contextualism, the vernacular, design articulation and symbolism, only one alluded to postmodern style directly. But the language of architectural style is incomplete for the theoretical project defined here. Façades are important, but to see landscape only as built form is to repeat the descriptive and morphological bias of an earlier cultural geography which suspended questions of theory and process. So too, to see contemporary culture only as a generic postmodern condition is to perpetuate the old error of the 1920s, when culture was treated as superorganic, a totalizing and uncontested entity.

For to see landscape as process is to recognize also the contingent geographical and historical contexts which enable particular landscape forms to emerge. I have identified the innovative cultural politics of the late 1960s which permitted the appearance of co-operative housing as part of an oppositional design for difference. But the co-operative housing movement has always been peripheral to the dominant forces in the land market, and to a degree subversive of them (Chouinard 1989). In 1986 the federal Conservative government significantly redefined the co-op programme, thereby encouraging a privatization of the design process in developer-led turnkey projects, with an attendant reduced role for the co-op both in the design process and in the recruitment of members. The initiative for co-op development began to pass from grass-roots organizations and housing resource groups to private developers who would launch a co-operative on land they owned which might be unsuitable for market housing. At the same time the subsidized proportion of the units has declined, and in British Columbia half of subsidized tenants are now selected from government housing lists. There is some evidence that this fiat from above has been aided by a waning idealism from below. The early co-ops had a strong philosophical base sustained by social activists. For example, De Cosmos Village, a co-op predating the 1973 Housing Act revisions, sustained its diversity through a system of internal rent subsidies. In contrast there seems to be a stronger orientation today among some co-op members to individual housing benefits, and the co-op serves only as a way station to eventual home ownership. Indeed in the words of one housing resource group, the structure of new co-ops is now one of market housing with a social housing component, unlike the non-market ideals of the earlier period. In the opinion of the founder of the Sam Greer Co-op: 'Somewhere along the way co-ops have been co-opted.' This has been noted by both architects and the resource groups:

> In the early years residents weren't looking for a good housing deal but were interested in living in a co-op. They understood more about co-operatives. Now in the last few I've designed you are designing for the board of co-op directors and looking after their self-interest as they start manoeuvring to maximize their own suites. Before everyone was more socialist, more democratic.

> It used to be a movement with a cause. To house people, mixing incomes in a manner which was beneficial to all. It's really market housing now and you're competing openly.

> Co-ops have really become a form of rental housing for the pre-home-ownership group. Politicians are afraid of the co-op movement. It stinks now.

Whether or not this assessment is correct it is a reminder that the imminence and massive presence of the built environment should not distract us into seeing it as natural. Like all cultural forms, the urban landscape is socially constructed and contingent, the product of an ethic that evokes both morality and politics. For oppositional movements with an alternative ethic that contingency is explored and tested in the public realm through political mobilization and struggle.

ACKNOWLEDGEMENTS

I am very grateful for the generous advice and assistance of Shelagh Lindsey and the skilled field research of Julie Podmore in the completion of this project. The photographs in Figures 8.1–8.5 were taken by Julie Podmore.

REFERENCES

Anderson, K. (1988) 'Cultural hegemony and the race-definition process in Chinatown, Vancouver: 1880–1980', *Environment and Planning D: Society and Space* 6: 127–49.

Berelowitz, L. (1988) 'The liveable city: social housing in Vancouver', *Canadian Architect*, February: 34–7.

Berman, M. (1982) *All That Is Solid Melts into Air: The Experience of Modernity*, New York: Simon & Schuster.

Brantlinger, P. (1983) *Bread and Circuses: Theories of Mass Culture as Social Decay*, Ithaca: Cornell University Press.

Brolin, B. (1976) *The Failure of Modern Architecture*, London: Studio Vista.

Carson, C., Barka, N., Kelso, W., Stone, G. and Upton, D. (1981) 'Impermanent architecture in the southern American colonies', *Winterthur Portfolio* 16: 135–96.

Chouinard, V. (1989) 'Social reproduction and housing alternatives: cooperative housing in post-war Canada', in M. Dear and J. Wolch (eds) *The Power of Geography*, London: Unwin Hyman, 222–37.

Daniels, S. (1989) 'The duplicity of landscape', in R. Peet and N. Thrift (eds) *New Models in Geography*, vol. 2, London: Unwin Hyman, 196–220.

Duncan, J. (1990) *The City as Text: The Politics of Landscape Interpretation in the Kandyan Kingdom*, Cambridge: Cambridge University Press.

Duncan, J. and Duncan, N. (1980) 'Residential landscapes and social worlds', in D. Sopher (ed.) *An Exploration of India*, Ithaca: Cornell University Press, 271–86.

Frampton, K. (1985) *Modern Architecture: A Critical History*, London: Thames & Hudson.

Fuller, P. (1987) 'Towards a new nature for the Gothic', *Art and Design* 3 (3–4): 5–10.

Galison, P. (1990) 'Aufbau/Bauhaus: logical positivism and architectural modernism', *Critical Inquiry* 16: 709–52.

Giedion, S. (1967) *Space, Time and Architecture*, 5th edn, Cambridge, Mass.: Harvard University Press.

Government of Canada (1969) *Report of the Federal Task Force on Housing and Urban Development*, Ottawa: Queen's Printer.

Hasson, S. and Ley, D. (forthcoming) *Neighbourhood Organizations and the Welfare State*.

Hayden, D. (1981) *The Grand Domestic Revolution: A History of Feminist Design for American Homes, Neighborhoods and Cities*, Cambridge, Mass.: MIT Press.

Holdsworth, D. (1990) 'The landscape and the archive: texts for the analysis of the built environment', paper presented to the Symposium on Cultural Landscape Interpretation, Berkeley, March.

Hoskins, W.G. (1955) *The Making of the English Landscape*, London: Hodder & Stoughton.

Hulchanski, D. (1984) *St Lawrence and False Creek*, School of Community and Regional Planning, Papers No. 10, Vancouver: University of British Columbia.

—— (1988) 'The evolution of property rights and housing tenure in postwar Canada', *Urban Law and Policy* 9: 135–56.

Inner City Housing Society (1988) Non-profit housing co-operative information package, Vancouver.

Jackson, P. (1989) *Maps of Meaning: An Introduction to Cultural Geography*, London: Unwin Hyman.

Jacobs, J. (1961) *The Death and Life of Great American Cities*, New York: Random House.

—— (1971) *City Limits*, Ottawa: National Film Board.

Janik, A. and Toulmin, S. (1973) *Wittgenstein's Vienna*, New York: Simon & Schuster.

Jencks, C. (1981) *The Language of Post-Modern Architecture*, New York: Rizzolli.

Jencks, C. and Chaitkin, W. (1982) *Architecture Today*, New York: Abrams.

King, A.D. (1984) *The Bungalow: The Production of a Global Culture*, London: Routledge & Kegan Paul.

Le Corbusier (1927) *Towards a New Architecture*, London: John Radker.

Ley, D. (1980) 'Liberal ideology and the post-industrial city', *Annals of the Association of American Geographers* 70: 238–58.

—— (1987) 'Styles of the times: liberal and neoconservative landscapes in inner Vancouver, 1968–1986', *Journal of Historical Geography* 13: 40–56.

—— (1989) 'Modernism, postmodernism and the struggle for place', in J. Agnew and J. Duncan (eds) *The Power of Place*, London: Unwin Hyman, 44–65.

Ley, D. and Mills, C. (forthcoming) 'Can there be a postmodernism of resistance in the urban landscape?', in P. Knox (ed.) *The Restless Urban Landscape*, Englewood Cliffs: Prentice-Hall.

Meinig, D. (1979) 'Reading the landscape: an appreciation of W.G. Hoskins and J.B. Jackson', in D. Meinig (ed.) *The Interpretation of Ordinary Landscapes*, New York: Oxford University Press, 195–244.

Mills, C. (1988) '"Life on the upslope": the postmodern landscape of gentrification', *Environment and Planning D: Society and Space* 6: 169–90.

Pevsner, N. (1968) *The Sources of Modern Architecture and Design*, New York: Praeger.

Pratt, G. (1981) 'The house as an expression of social worlds', in J. Duncan (ed.) *Housing and Identity*, London: Croom Helm, 135–80.

Pred, A. (1990) *Lost Words and Lost Worlds: Modernity and the Language of Everyday Life in Late Nineteenth-Century Stockholm*, Cambridge: Cambridge University Press.

Relph, E. (1976) *Place and Placelessness*, London: Pion.

Samuels, M. (1979) 'The biography of landscape', in D. Meinig (ed.) *The Interpretation of Ordinary Landscapes*, New York: Oxford University Press, 51–88.

Schlereth, T. (1983) 'Material culture studies and social history research', *Journal of Social History* 16(4): 111–43.

Schorske, C. (1981) *Fin-de-Siècle Vienna: Politics and Culture*, Cambridge: Cambridge University Press.

Selby, J. and Wilson, A. (1988) *Canada's Housing Cooperatives*, Research Paper No. 3, Ottawa: Cooperative Housing Foundation of Canada.

Watkin, D. (1977) *Morality and Architecture*, Oxford: Clarendon.

Wekerle, G. (1988) *Women's Housing Projects in Eight Canadian Cities*, Ottawa: Canada Housing and Mortgage Corporation.

Western, J. (1986) 'The authorship of places', *Syracuse Scholar* 7: 5–17.

Wirth, L. (1925) 'A bibliography of the urban community', in R. Park, E. Burgess and R. McKenzie (eds) *The City*, Chicago: University of Chicago Press, 161–228.

Wright, F.L. (1975) *In the Cause of Architecture*, New York: Architectural Record Books.

9

MYTHS AND MEANINGS OF GENTRIFICATION

Caroline Mills

Cultural geography today draws nourishment from two root-stocks. One, firmly anchored in geographical soil, is the Berkeley tradition which maps out the fine texture of material artefacts in the landscape and traces their association with ways of life. The other stem, rooted in cultural history and literary theory, alerts geographers to the politics of landscape and the instability of meanings assigned to the world of material objects; as Matless (1989) suggests, here 'text' rather than 'texture' is the dominant theme, grasped through the languages of post-structuralism and postmodernism.

It is the quality of postmodernism which has proved to be particularly intriguing to students of contemporary landscapes. Some geographers have dealt with postmodernism at the most sweeping of scales, playing it off against other master concepts such as capital accumulation and urbanization. This essay picks out some finer details; operating at the interface of the two strands introduced above, it offers a grounded portrait of the meaning of inner-city gentrification.

GENTRIFICATION AND IDENTITY

The literature on gentrification encompasses three broad approaches to explanation, couched in terms of the spheres of consumption, production and reproduction. Changing consumption and changing lifestyle are the bread and butter of popular accounts in the Sunday magazine supplements, as well as some of the planning-oriented literature (for example, Allen 1980; Laska and Spain 1980). But this form of explanation neglects the forces by which gentrification is produced. Production-side explanations may take a managerialist turn, pointing to the influence of government and financial institutions at a local or intra-national scale (for instance, Hamnett and Williams 1980; Williams 1978). A structural Marxist interpretation defines a more general production-side theory of gentrification: formalized in Smith's rent-gap theory (Smith 1979, 1982) this focuses on the switching of investment back into inner-city areas and the recommodification of derelict environments. In turn, the third model of explanation critiques the emphasis on production of 'gentrifiable' locations; instead, with reference to the sphere of reproduction, this 'production of gentrifiers' approach interprets urban change in terms of the restructuring of class and gender relations consequent upon changes in the organization of work and the domestic sphere (Beauregard 1986; Rose 1984).

Gentrification attracts attention, both popular and scholarly, in part because of its visible form. The details will vary according to local conditions: in one place there will be 'whitepainting'; in another, 'brownstones' are renovated; elsewhere, warehouses are converted to residential use, or old buildings demolished and replaced as high-rent apartments or condominiums. But, whatever the details, commentators agree that there has been a widespread transformation of landscapes in many of the inner cities of the advanced industrial nations. In terms of geographical research, however, this most visibly conspicuous aspect of gentrification has been downplayed as the surface manifestation of more fundamental economic or social changes.

The 'new cultural geography' seeks to recover an essentially *geographical* dimension: the intimate relationship between people and their environmental setting. Landscape is a way of seeing the world, a codification of social order which 'reminds us of our position in the scheme of nature' (Cosgrove 1989: 122). Places encapsulate and communicate identity. At the finest scale, a house may be an expression of individual selfhood (Cooper 1974). By exploring alternative symbolic qualities of the house, however, geographers have demonstrated the error of essentialism associated with an appeal to environmental archetypes (Duncan 1981; Pratt 1981). Each society's 'moral order' is reflected in its particular spatial order and in the language and imagery by which that spatial order is represented. Conversely, the social is spatially constituted, and people make sense of their social identity in terms of their environment. Their place of residence offers a map of their place in society: we produce not housing but 'dwellings of definite sorts, as a peasant's hut or a nobleman's castle ... [in] a continuous process of social life in which men [*sic*] reciprocally define objects in terms of themselves and themselves in terms of objects' (Sahlins 1976: 169). To 'place' someone, to 'know one's place': this language of social existence is unmistakably geographical. Cultural geography thus calls for a decoding of landscape imagery, a reading of the environmental 'maps of meaning' (Jackson 1989) which reveal and reproduce – and sometimes resist – social order.

What statements of social order do gentrifiers write into, and read from, gentrified landscapes? Much of the literature on gentrification does not concern itself with meaning. The rent-gap theory explains gentrification with reference to cycles in the circulation of capital; its implications for consumers are couched in images of that popular pariah the 'yuppie' (Smith 1987), depicted as a rather one-dimensional figure intent on economic advancement. As a cultural icon of a 'lifestyle', however, this character of (post)modern myth also deserves a 'cultural' treatment. The 'production of gentrifiers' approach tempers the rude depiction of the shallow yuppie lifestyle by characterizing gentrification as a strategy by which actors conduct a workable style of life within a changing field of opportunities and constraints in waged work and in the domestic sphere. Through our unpacking of the 'chaotic concepts' of 'gentrification' and 'lifestyle', and our sharpening up of the theoretical categories for a realist analysis, the active construction of domestic and public life by people involved in gentrification is highlighted (Rose 1984). However, in attempting this process of abstraction to reveal 'real' processes underlying urban restructuring, we necessarily deny the language by which participants make sense of their own lives.

By denying that language we cannot fully understand human agency, for, as the 'new cultural geography' insists: 'Culture is not a residual category, the surface variation left unaccounted for by more powerful economic analyses; it is the very medium through which social change is experienced, contested and constituted' (Cosgrove and Jackson 1987: 95). In everyday language, ways of living – strategies for coping – are objectified as 'lifestyle'. And, through language, places are given meaning. What kinds of meanings are attached to the inner city? American cities are popularly depicted as 'urban jungles': the New York skyline of King Kong and Superman (Cosgrove 1989) represents a social world poised on the edge of perdition, awaiting deliverance by the disciples of civilization. Smith (1986) has drawn attention to the vocabulary of 'wilderness', 'frontier' and the gentrifier as 'urban pioneer' through which Americans have construed the inner city; like the imagery of homesteading in the American west, this implies that existing inhabitants are barely social beings – indeed, merely part of the natural environment – and thus provides the moral justification for certain kinds of urban redevelopment. On the other hand, imagined as a world apart, 'inner-city problems' are also conveniently explained away through a set of 'deeply engrained discourses about physical, social and cultural conditions in "the city"', as Burgess (1990: 148) suggests in her discussion of press reports on urban riots in Britain.

The city, and the inner city, are thus made the subject of myths. Myths are, of course, contestable. Ley (1980, 1987) shows how, during a period of political liberalism in Vancouver, Canada, the old 'rational ideology' of the efficient, business-oriented city was replaced by a fresh discourse of 'liveability'. Re-interpreted in terms of a doctrine of urban amenity, the city was to be a 'people place' which valued human experience over money-making. In response to these principles, small-scale private gentrification flourished and the public sector initiated sensitive redevelopment of redundant industrial land to pleasant landscapes of social-mix housing and public park. One might, then, interpret gentrification as the victory over a hegemonic urban imagery by a new symbolism coupled to an emergent cultural manifesto. However, new visions may be co-opted in the reforging of hegemonic discourse by the machinery of dominant culture. For example, once digested and regurgitated as 'spectacle for the masses', urban liveability in Vancouver provided populist neo-conservatives with a mandate for urban corporatism and a new paradigm of the city as a free-enterprise money machine, thus furnishing the ideological basis for large-scale inner-city redevelopment to market housing by the private sector. In sum, as Crilley (1990: 235) points out for the case of Docklands-style redevelopments, the symbolic and aesthetic dimensions are 'essential accompaniments to the more material and economic transformations'.

In investigating a detailed example from one Vancouver neighbourhood, this essay explores how the material experience of gentrification is recast as myth. The process of myth-making involves the codification of symbolic landscapes by developers, planners and architects; the representation of socio-spatial order through the discourse of advertising; and the incorporation or contestation of meaning in the construction of their social identities by the gentrifiers themselves.

'WE'RE SELLING A LIFESTYLE'[1]

Fairview Slopes is a small inner-city neighbourhood on a steep hill with dramatic views over the business centre of Vancouver. Before the mid-1970s the landscape was dominated by modest wood-frame houses, plus some light industrial activities. By the late 1960s, the physical fabric of much of the rooming-house accommodation had deteriorated, and it attracted growing numbers of residents with a 'counter-cultural' lifestyle. The area was perceived as a quiet backwater: 'The Slopes tend to border on a rural lifestyle caught up in the urban animal of Vancouver' (Elligott and Zacharias 1973: 11). Yet only a few years later, Fairview was lauded as 'one of the few neighbourhoods in Vancouver with an urban feel' (Gruft 1983: 320). Today the population is dominated by upwardly mobile singles or dual-career couples, plus some older adult empty-nester households.

This social change was accompanied, in a limited way, by the classic form of gentrification: housing renovation. But contingent conditions – the poor physical quality of some of the property, plus the rise in land values sparked off by public redevelopment of the adjacent False Creek shoreline (Ley 1980) – encouraged redevelopment instead. Business preferences for developing high-rise residential and commercial uses were headed off by the liberal planning regime of the 1970s which instituted zoning regulations for medium-density town houses, condominiums and apartments. By operating a kind of 'bonus' system through which projects could earn higher densities, these regulations inspired innovative design solutions to tricky site conditions, in particular mimicking the architectural codes of the vernacular style, and maintaining an intimate scale and texture. By the early 1980s, Fairview had taken on a distinctly postmodern demeanour (Plate 9.1). While some projects make the most cynical use of façade postmodernism to disguise cramped and mediocre apartments, the neighbourhood as a whole has attracted possibly the highest density of architectural awards in the country.[2]

In common with other examples of gentrification which have taken the form of redevelopment and reconstruction (see, for example, on London Docklands, Burgess and Wood 1988; and Crilley 1990), Fairview's transformation is underwritten by an explicit marketing text, a strategy of 'place advertisement' which, accentuated by the compelling products of postmodern architectural 'imagineering', defines a commodity laden with mythical content.

Plate 9.1 Postmodern Fairview: La Galleria I, developer André Molnar, 1985

Source: Photo by C. Mills

In Fairview Slopes, both the direct advertising message and the motifs of landscape form are received and retransmitted (and sometimes rejected) as cultural signals by the gentrifiers. As one advertisement claims, 'We don't just sell you a townhome, we offer you an exciting new lifestyle.' And what does this lifestyle entail? It is introduced as 'designer living', 'an extraordinary lifestyle' and 'an upwardly mobile lifestyle'. Most tellingly, the lifestyle for sale is defined by a place – the city. Advertisements lure the buyer by the best in 'city living', 'urban living', the 'urban lifestyle', 'downtown style living' and even 'inner city living': all imagery which confirms the geographical constitution of social life.

The 'texture' of Fairview Slopes is accessible by means of a variety of 'texts' – presented in advertising, expressed in interviews[3] and made concrete in the landscape itself – which, together, define a 'textual community' of social actors (Duncan and Duncan 1988) who share a particular reading of the gentrified inner city. Before exploring that reading, however, the next section evaluates the treatment of cultural issues in the approach to gentrification which focuses on the role of producers in defining 'gentrifiable' locations, locations which offer new opportunities for capital accumulation.

'A DISTINCTIVE AREA FOR A LIVING INVESTMENT'[4]

According to Gregory (1987: 246), the dominant representation of postmodernism in the geographical literature takes a stance which is 'all too modern'. In this modern rendition, urban restructuring and landscape change are indicators of a new strategy of flexible capital accumulation through consumption-side growth (Smith 1987), coupled to the 'economics with mirrors' practised by image-conscious politicians and city yuppies (Harvey 1989). This is manifest in an array of conspicuous consumption items – shopping and leisure facilities, gentrification itself – which offers 'a more specialized and "discriminating" edge to it compared to the mass consumption of the 1960s. Above all, the city has to appear as an innovative, exciting, creative, and safe place to live, play, and consume' (Harvey 1987: 12).

In Fairview Slopes, indeed, developers have pioneered new kinds of commodities with such a 'discriminating' edge, shaping an environment which has attracted attention in local 'lifestyle' magazines as well as the national architecture press. 'Award yourself' was the advertising slogan for one award-winning project. Developers like André Molnar (responsible for a number of projects in the area) are local celebrities: he describes his work as 'the Pierre Cardin approach'. As well as profiting from home buyers who desire a distinctive kind of property, developers have established some creative means of acquiring return before selling starts: units were leased to foreign executives during the 1986 World Exposition in Vancouver; one project featured as a film set in a production of *Airwolf*, the television series starring a helicopter; André Molnar also donated units for 'Suite Dreams', a Red Cross fund-raising event where interior designers displayed their skills. In sum, Fairview Slopes is a complex and ingenious commodity.

Drawing on Debord, Harvey has argued that 'Urban life, under a regime of flexible accumulation, has ... increasingly come to present itself as an "immense accumulation of spectacles"' (Harvey 1987: 276). The themes of display and spectacle are epitomized in the launching of 'open houses' for new projects. On Fairview Slopes, browsers are offered free chocolates, or a draw for a weekend's skiing; one project opened with a hot-air balloon moored in the courtyard and a colour-coordinated pop group. To avoid any sense that these consumers are part of the common masses, the spectacle is personalized. André Molnar says his customers can have 'confidence in a person who's well known in the community'; they can contact him with any problems once they have bought one of his condominiums.[5] According to the marketing material another developer, Matthew Briscoe, fits 'his own "buyers profile"', and designs homes 'perfectly adapted to an upwardly mobile lifestyle'. By means of display and personalization, therefore, congeniality and joy are made tools in commodification.

Not only the developers but also the customers are interested in exchange value. One architect claimed of the marketing strategy, 'the most important thing they're trying to put over is that it's a good investment for people'. A resident said:

> I expect in the next ten years the value to appreciate substantially, having looked at similar areas of Chicago and Washington, DC, areas like this double, triple, quadruple in value in very short periods of time because of the concentration of people who want to be close to all these [central city facilities].

During periods of rapid construction activity in Fairview there has often been a brisk turnover of units immediately after a new project comes on the market: according to one architect, 'people who "pre-buy" these units can, if it's a well-marketed building, turn round after it's completely finished and make some profit' by selling them again. A large proportion of absentee owners, encouraged in the late 1970s by a federal programme to boost rental supply, has expanded further as a result of investment from the local Chinese Canadian community and from the overseas Chinese (developer André Molnar has pursued these customers through a marketing programme in Hong Kong).

Districts such as Fairview Slopes comprise a new arena for capital accumulation. The aesthetic of redevelopment – predominantly postmodern – has been described as the 'cultural clothing' of flexible accumulation (Harvey 1987: 279), the 'cultural logic' of late capitalism (Jameson 1984). In retrospect, indeed, these distinctive dwellings for the moderately affluent may be interpreted as capital's rational response to economic conditions in the 1970s. However, viewed as a contemplated act, gentrification appears as a more complex practice demanding a different kind of interpretation. The form of redevelopment is open to negotiation as actors with different skills and ambitions haggle over the outcome. Architects may not always agree with their developer clients, especially when they are motivated by social and aesthetic challenges and expect a pay-off in terms of professional status. One architect described himself as 'always fighting' his developer clients to put in special features that he believes will help to sell the units. Another architect, who made himself unpopular with some developers by insisting on the aesthetic integrity of his designs, was pleasantly surprised by the enthusiastic market response to one project which, by all accounts, was too expensive and too well appointed to compete in the expanding condominium market. If the details of gentrification cannot be taken for granted, neither can its ultimate consequences for capital accumulation. An important issue from the point of view of the rent-gap theory is how one might identify the potential ground rent of an inner-city area. Investors can make poor predictions: in Fairview, changing market conditions have meant that some projects have failed to sell; moreover, the high percentage of rented units means that the resale value of some owner-occupied units has suffered. Some projects are poorly maintained, and new buyers will not perceive them as attractive alternatives to the newest developments in other neighbourhoods.

A process of exploration precedes the establishment of a workable formula which will bring a satisfactory pay-off. The first moves to redevelop Fairview Slopes were made by small developer architects when the area was virtually red-lined. One architect described the problems of setting up his practice in what he claimed had been 'a house of ill-repute. . . . We had a terrible time with the Planning Department because we were a non-conforming use!' The first residential units he built were of an 'experimental kind'; and since they were located in a 'slum', the anticipated customers were 'non-establishment people'. In contrast, big developer André Molnar was slow to participate:

> I looked at Fairview Slopes in the 1976–77 era. . . . I liked the view, I liked the location, but I didn't like the atmosphere, plus I was terrified of pioneering the area because it was so wishy-washy as far as acceptability was concerned for the middle classes. . . . There was a lot of squatting and it was a terrible neighbourhood at that time – and I was frightened that my customers wouldn't find it acceptable. Some other more courageous developers were not frightened away by this and started to develop. . . . [Then once it was started] I got involved.

Compared to the bolder developer/architects before him, Molnar's hesitancy was highly rational, given the larger size and popular style of his practice.

From the point of view of the production-side approach, gentrification is a rational response to the opening of a rent gap; moreover, product differentiation (for instance, new aesthetics) is a rational response to changing market conditions. But the notion of rationality is not acultural: as Sahlins (1976: 215) points out, production is always a cultural intention, and 'Rational production for gain is in one and the same motion the production of symbols.' Indeed, not only does economic activity produce symbols; it is made possible by symbols:

'The accumulation of exchange-value is always the creation of use-value. The goods must sell, which is to say that they must have a preferred "utility", real or imagined – but always imaginable – for someone' (ibid.: 213). Not until social and cultural conditions were ripe did the gentrification commodity achieve the necessary 'imaginability' for production to proceed in Fairview Slopes. That imaginable utility is expressed by a real estate agent who said, of André Molnar's customers, 'His units *do* have such eye appeal. They can't resist. They say, "This is smaller than I wanted, smaller than I need, but I *want to buy*".' As Burgess and Wood (1988: 95) point out, 'Myth is more important than reality in selling places.' The following section describes how imaginability is exercised in the selling, but more particularly the buying, of Fairview Slopes.

'CITY LIVING AT ITS BEST'[6]

Rational production strategies are constituted by, and constitutive of, cultural categories. According to Sahlins (1976: 184–85):

> production is organized to exploit all possible social differentiation by a motivated differentiation of goods.... The product ... constitutes an objectification of a social category, and so helps to constitute the latter in society; as in turn, the differentiation of the category develops further social declensions of the goods system.

The advertisement is the point of translation where the structure of one sign system of social differentiation is applied to give structure to another system, that of consumer goods. Effective advertising provides 'points of coherence around which consumers can organize social experience into meaningful patterns' (Leiss 1983: 15). The practice of advertising is part of a 'magical' system (Jhally *et al.* 1985; Williams 1980) in which, through various rituals, meaning is drawn out of material goods which have been 'supercharged' with significance (McCracken 1986).

If advertisements are 'selling us ourselves' (Williamson 1978: 13), then for what textual community does gentrification offer a terrain of meaning? Harvey (1987: 274) refers us to Bourdieu's notion of 'symbolic capital', defined as 'the collection of luxury goods attesting the taste and distinction of the owner'. But this is imprecise. Is there a firm line between goods bought for conspicuous consumption and those necessary for human subsistence? Human creatures reproduce themselves as *social* beings as well as physical beings. The slogan to advertise one Fairview development – 'Can you afford not to be living at Emerald Court?' – evokes connotations of both pecuniary and positional necessity. New social constituencies articulate their identity with new cultural insignia, as both Moore (1982) and Jager (1986) have suggested in the context of gentrification. Features of the landscape are means for fixing social position; gentrification is a cultural tactic by which a new geographical space authenticates a new 'place' in the interstices of social structure.

What does this new space look like? Consider the potential of inverted signs, the vocabulary of 'inverted snobbery'. Gentrifiers do cultural 'work' with symbols of working-class culture. Classically, the gentrifier may take possession of a neglected historical neighbourhood, characterized for instance by the aesthetics of Victoriana (Jager 1986). Alternatively (perhaps when the 'do-it-yourself' culture becomes a little bit too widespread) a gentrifier might choose a new, postmodern dwelling that looks more like a humble cottage or even a factory than a conventional 'executive home'. Just in case anybody should mistake such a dwelling for a real working-class cottage, it will be ornamented with selected symbols of prestige such as a portico or a classical arch (Plate 9.2). Such postmodern 'double coding' (Jencks 1986) of stigma and status is very visible in the landscape of Fairview Slopes and, as I have argued elsewhere (Mills 1988, 1989), this corroborates the paradoxical status of a new class which Gouldner (1979) characterizes as both emancipatory and elitist. Advertising strategy reflects this ambiguity: one Fairview project, for instance, 'is designed to appeal to residents who prefer a standard of luxury and lifestyle amenity normally considered executive level, thereby further assuring a compatible attitude towards a lasting investment in real estate and real sensitivity to others'.

Plate 9.2 The triumphal arch to Willow Arbour Townhouses, an award-winning design by James Cheng

Source: Photo by C. Mills

What social identity for the Fairview Slopes gentrifiers is reproduced in the language of marketing? Some advertisements present the neighbourhood as the home of 'people on the way up', contributing to 'a statement about social achievement' and offering 'an address for those who like to set the standards'. But, though signals may be transmitted, they are not necessarily given a welcome reception; Fairview Slopes residents do not willingly concede to this characterization of mere social rank. Their status is rather more subtle. An examination of the language used by both 'producers' and 'consumers' reveals the predominant symbol to be thoroughly geographical: that of *urbanity*. Fairview is advertised as 'city living at its best', 'A new city "centre"', 'the most desirable metropolitan location in Vancouver', 'Vancouver's premium urban neighbourhood'. One development introduces 'truly distinctive city homes – homes that city people will treasure for the pleasure they add to an urban lifestyle – the pleasure which is 'all part of daily urban life'.

The notion of 'city people' echoes another familiar epithet. Although few residents were willing to claim the mischievous label for themselves, they do not hesitate from describing the reputation of Fairview Slopes as a 'yuppie' neighbourhood.[7] To geographers, the most striking trait of the yuppie must be – not their youth or their professional status – but their *geographical* constitution: they are young *urban* professionals. In interviews, residents volunteered similar self-definitions. Phil said, 'I think the people here are very "downtown people".' To Gill, Fairview Slopes 'seemed the ideal place for us city dweller types'. Erica adopted the same tone: 'We both considered ourselves city people. I think that we like the amenities that cities have to offer. I think we like that sense of taking advantage of what we consider to be urban things.' Even when her behaviour does not fit this self-depiction (she shops in an upmarket suburban mall), Erica employs the geographical counterpart in a self-deprecating quip: 'Yes, I'm terribly suburban, I go to Oakridge!'

Indeed, defining a social identity by a geographical referent depends on a code of oppositions and differences which thus define at the same time what a person is *not* (Williamson 1978). The sign only has value in relation to other signs. Fairview residents locate themselves – literally and metaphorically – in terms of a dualism of city space. One couple describe the lifestyle they prefer: 'We're not North Vancouver people ... we're not commuters.' More indirectly, Helen reports that her co-workers identify her directly with her place of residence, as 'being in the city, and "you're just a single swinger"'. Faced with the prospect of suburban life, gentrifiers react in horror. Gill complained, 'I can't bear the thought of moving into the suburbs'; and Gordon said:

> I hope and pray that if I ever *have* to move ... I'll still be able to live in the city. I do *not* want to live in the suburbs. ... Even if one was forced into some awful situation like the single family house ... one can still do that in the city.

Would Eric ever move to the suburbs? 'Never, never. If I'm going to be living in the city – I *like* cities, I like *busyness* – I want to be right here in the city. Being in a suburb is, like, forget it! It's not where I want to be.' Pat refers to the Vancouver suburb of Richmond: 'My sister, as much as it shames me to say it, has moved to Ditchmond'. To Andy, suburbia stands for a bland existence: 'I've lived in Toronto, I'm a believer in, if you're going to live in the city, *live in the city.* I'm not a suburban or a country person. I like living close to the urban area.' His wife, Alison, treats the suburb as a generic banality:

> I've lived in the Burnaby's of the world, and New Westminster, and I have lived in Surrey, and I don't think I ever enjoyed living in Surrey. I don't think I would enjoy living as much in New Westminster or Burnaby.... I love this area.

Alison would rather forget her suburban experiences in Burnaby.[8] But if the suburb is grasped generically, so is the inner city. The power of the urban metaphor is reinforced with patterns of urban life elsewhere. The architecture incorporates icons of the Georgian mews (some projects are named 'Mews'); brick row houses in Boston and Toronto also provide models for condominium designs. Some residents and marketers detect a correspondence between the Fairview landscape and urbane San Francisco: two projects are christened 'The San Franciscan' (I and II). Two interviewees picked out another resemblance: Neil claims 'it reminded me of a . . . very New York style of living', whereas Keith reports, 'We've become New Yorkers.'

For this textual community, city and suburb operate as moral orders, furnishing 'visions of a way of life' (Schwartz 1976: 326). Schwartz calls upon the novels of John Cheever to describe the image of suburbia:

> contemptible because of its spiritual bleakness, its 'shallow and despicable life-style', which transforms poetic dreams of pastoral bliss into the cheap prose of the 7:50 and the power mower . . . suburbia organizes a life without meaning, whose logical indictment is death itself.
>
> (ibid.: 337)

This is a remarkably extreme view. But to Fairview residents, suburbia is a powerfully negative signifier. Fairview itself is marketed with an anatomical metaphor of vitality. Here, located 'pulse beats away' from numerous amenities, one enjoys access to the 'pulse and vigor of a dynamic city'; here, 'in the heart of Vancouver's Fairview Slopes' which is 'pulsing with excitement', one occupies 'the pulse – the heart – the centre of the city'. Conversely, when I asked one resident about moving to the suburbs, he exclaimed, 'I think I would die!'

Notwithstanding the image of urbanity, however, the neighbourhood is valued for its quality of being *of* the city, yet somewhat *detached* – perhaps distinguished – from it. Developments are advertised as 'so close to downtown, yet removed from the hustle and bustle', 'a part of the city and yet apart from it'. One resident favourably describes Fairview as enjoying a 'position above the city, but in it'. Fairview is indeed topographically 'above the city', commanding spectacular views of the urban landscape below. And the language of visual observation saturates the advertising texts. As cultural geographers have shown, landscape is an ideological concept: 'it represents a way in which certain classes of people have signified themselves' (Cosgrove 1984: 15). Advertising blends the language of ownership and observation, presenting the cityscape as a quality to be commended and possessed: 'Southport has been specifically designed to capitalize on spectacular views'; at Seascape Vista, 'A panorama of the city is yours for the viewing' (echoes of appropriation?); The Courtyards offers 'Perspectives on the city's most desirable new neighbourhood.... A view towards value'.

Located 'Above the city. Above all expectation' (according to another advertisement), the neighbourhood is defined with respect to its external prospect. It is, of course, a selective view, and a distinctive one: the marketers of Alderview invite us to 'Come, share our views of life atop Fairview Slopes.' And at Southport you can acquire 'a new outlook on city living': 'Southport shows you just how good the best city living can be: magnificent views from every home; a location close to Vancouver's business core.... Let a new home at Southport refresh your outlook on urban living.' A real engagement with what is on the ground is confounded by distance and elevation, for, as Williams points out, 'the very idea of landscape implies separation and observation' (Williams 1973: 120). Vancouver's urban core becomes a silhouette, to be employed as a marketing logo (Plate 9.3). One architect contrasts this perspective with the style he adopts in designing for co-operative housing.[9]

> It's quite a hedonistic approach in that every suite is designed for the view, and they're designed to sort of turn their back on everyone else. You're in your little world, you're up there in your own box looking at the city and that's great. Whereas in a co-op you don't want everyone to turn their back on each other, you want group interaction.

The city is framed and offered up: as the builders of Emerald Court proclaim, 'We create exciting windows on city living.' The contemplated purchase of a product thus admits the possible purchase on a world.

At the same time as being spectators of the dazzling panorama, residents are invited to become players in the urban theatre. The marketers of Dover Pointe warn the buyer (with the black-cat logo for restricted films) of 'scenes suggestive of a lifestyle you've always wanted to live'. Fairview Slopes is presented as not merely a place to 'live' – that is, to reside – but also the stage upon which one may practise the art of living. The extended metaphor of performance invites customers of this and other developments to 'première openings' and open house 'sneak previews'.

A simple materialist reading of the texts examined above would suggest that, as spectator or role-player, the Fairview resident is merely following a script composed by the agents of capital; the cultural meaning is all part of the commodity composed by the Fairview Slopes 'producers'. However, the argument in this section has highlighted a more complex relationship between 'rational production' and cultural symbolism. The section below evaluates a

161

Plate 9.3 The urban overview offered at Dover Pointe

different materialist reading, one which focuses on the way in which gentrifiers actively incorporate gentrification in the reproduction of everyday life (the 'production of gentrifiers' approach). It will be argued that here, too, we must attend to the cultural dimension in order to understand the practice of gentrification.

SETTING AS STRATEGY, TOWN HOUSE AS TACTIC

In evaluating its status reputation, one renter admitted that Fairview would be a 'fairly strategic place to live'. As the 'production of gentrifiers' approach suggests, gentrification may be more literally a strategic act (Rose 1984). With the resources at their disposal, people articulate their daily 'paths' of reproductive practice with their longer-term goal-oriented 'projects'. The budgeting of time and space over habitual routines defines the limits of what is 'doable'; for some projects, gentrification may be an environmental solution to that tricky challenge.

In terms of Bell's three lifestyle types (Bell 1958), career-orientation and consumerism ('living the good life') feature as particularly important 'projects' for Fairview residents; few residents are raising families, though some expect to do so in the future and others are empty-nest households or divorcees. Many interviewees claimed that they sought a balance of the three lifestyle types, defining 'family' often in terms of an egalitarian relationship with a partner. One woman seeking this balance, for instance, claimed that 'the house liberates that ... because it looks after itself'. Similarly, a divorced man said, 'I like living in a town house because of the fact that you can just lock the door and go away, you don't have to worry about the garden and maintenance.' The enabling quality of environment may be captured by two images: the setting as strategy, and the town house as tactic.

Advertising expresses the strategic dimension of developments: one project is marketed as being 'strategically situated', and nearly all other campaigns emphasize accessibility. Asked to list the advantages of living here, one respondent answered, 'Location, number one'; another quipped, 'Basically location, location, location!' Rose (1984) speaks of gentrifiers' rejection of the time–space rhythms of suburbia (those conjured up by Schwartz, quoted above). Workplace location is one determinant of the rhythms negotiated each day, and a central location is desirable for dual-career households (Beauregard 1986) where both partners' jobs are treated as significant; for example, residents claimed, 'It's a very convenient place for us to live, for both of our businesses'; 'It's very central, it's good for both of us.... it seemed like a good compromise.' Many respondents mentioned the horrors of commuting:

It takes us exactly seven minutes from our garage to a car park downtown. It's much more convenient than West Van[couver] when you really don't know whether it's going to take you twenty minutes or two hours to get into the city.

A short journey-to-work is not the only strategic advantage: another is a short journey-to-play. Living close to work enables the freedom to enjoy certain leisure activities:

> I think there's a real time advantage. I get home from work in five minutes. I think that's a real advantage in my life particularly ... because I go out in the evenings quite a bit, so it means I can go home for an hour and a half and then go out in the evening.

Two residents mentioned the advantage of a short taxi ride which frees them from worry about drinking and driving. One woman who used to live near her workplace in the far suburbs has always socialized in the inner city; she found it 'a bind having to drive home whatever the hour of the evening ... after due consideration I decided I would rather do the driving on a morning on my way to work'.

Viewed from one perspective, therefore, gentrification is an 'environmental solution' to certain social constraints: it makes things 'doable'. What is legitimately 'doable', however, depends on what is 'thinkable'. I quoted earlier the couple who said, 'We're not North Vancouver people ... we're not commuters.' A single teacher in a suburban school evaluates the suburbs: 'I would feel very cut-off out there, and that's why I enjoy living away from the school, because I can be a different person when I come into town.' This expressive aspect – the confirmation of identity by place – marks the limits of a 'production of gentrifiers' approach that stops short of the treatment of cultural matters.

While the workplace is a prime institution that structures the pursuit of projects, the nature of the home itself can open and foreclose other options: 'I guess I didn't want a house, I wanted a town house, I didn't want the upkeep'; 'We like the convenience of condominium-type living because there is relatively little maintenance on the place'; 'I like paying 66 bucks a month "strata fee" and that's it; I don't want to be getting into the 200 dollars for getting the mower overhauled or raking the leaves.' Confirming Schwartz's characterization of suburbia, unprompted references to the drudgery of lawn-mowing were extremely frequent! With what would this interfere? Eric said:

> I don't have time for maintenance on houses and upkeep. I want as little time spent at home ... as possible, because I work a lot and when I want leisure I want leisure, I don't want to have to look after something like a house.

Two women agreed:

> Both my husband and I work hard during the week, so we don't want to spend a lot of time doing housework, cutting grass.... I really enjoy the lifestyle here ... it's trouble-free living.

> We moved from a neighbourhood where the house and the yard were more

than we could tackle in our waking hours.... this was a much more manageable ... way of living for us in terms of the kinds of household chores that there are, and us getting the leisure time and the work time that we want.

Town-house living offers particular advantages for such childless dual-career couples, especially those in the early stages of career-building who put in strenuous hours at work. And Fairview Slopes has attracted many single female buyers; indicators suggest that about a third of units in the projects built in the mid- to late 1980s sold to women alone, a quarter to men alone. Leila, a financial professional, explained why she wanted a town house: 'I'm single and I can't waste the time maintaining a home.... With a house you have ... maintenance, and you also have the problem of security because you're on your own.' Some marketing has been targeted at female buyers (Plate 9.4); the security systems in some buildings, and the aesthetics of interior and external decoration (described by a real estate agent as 'feminine'), apparently are designed with this in mind.

As the 'production of gentrifiers' thesis suggests, gentrification is integral to achieving certain projects. There remain hidden oppressions, as the social costs of new styles of life are displaced on to other groups; this essay makes no attempt to examine that dimension of gentrification. At the same time, the fortunes of gentrifiers are maintained often by very hard work as young adults, and by a judicious choice of housing. While there is clearly some validity in the view (discussed in an earlier section) that gentrification is a commodity produced as part of a new strategy for capital accumulation, there is also validity in the depiction of gentrification as a practice of human agents attempting to 'make their way' through a set of obstacles and opportunities.

However, neither of those interpretations is complete without considering the cultural medium through which economic and social processes operate and are experienced. They may be experienced, for instance, in terms of a language of 'lifestyle' which reifies the practical effort of making a living; as Helen says, 'I have adopted a lifestyle – adapted *to* a lifestyle – that fits my income. And I enjoy it, and I'm willing to work hard for it.' And such a discourse of social identity may also be geographically constituted. As one young man says of Fairview, 'It's great, suits my purposes perfectly.... it just fits my lifestyle perfectly – or maybe my lifestyle fits *it*, because it was here before me!'

CONCLUSION

In exploring the texture of gentrification in Fairview Slopes, I do not intend to imply that any one theoretical position could offer complete closure in the explanation of this phenomenon. By this, I reproduce 'local knowledge', expressed in conversations such as this one between a wife and husband:

Frances: 'I think it's going to ... become a place that people get quite committed to....'

165

Plate 9.4 Woman buyer, women seller: marketing the Maximillian

Source: Photo by C. Mills

Frederick: 'I think it's going to be a *very* desirable place when ... people realize how convenient it is, how *workable* it is. ...'

Frances: 'I think [the fact that] these homes are *here* is going to encourage people to change their lifestyle.... people seeing its virtues not just as a place you can live cheaply on your way to somewhere but as a place with a lifestyle of its own. And the fact that it's *here*, people will adapt their lifestyle to here.'

This conversation gives credence to a variety of accounts of gentrification: first, by asserting its desirability to consumers affiliated to a new lifestyle (sphere of consumption); second, by professing 'workability' in everyday life (sphere of reproduction); third, by implying that the place is a commodity offering symbolic capital into which one might rationally buy (sphere of production).

How are the new, positive meanings of the inner city established? Are they simply declarations of changing consumption preferences; do they emerge out of gentrifiers' experiences of constructing satisfactory lives; or are they manufactured as a bait to tempt consumption dollars? Exploring through ethnography rather than resolving by reduction, one highlights the constant process of negotiation which frames social practice and cultural meaning.

It is a fundamentally political process, for cultural objects are media for domination *and* for resistance. The history of postmodern architecture itself provides an example; its adversarial aesthetics once provided a new blueprint for the relationship between people and the built environment; later, postmodernism was co-opted by neo-conservative forces in a new landscape orthodoxy. Similarly, as producers pursue the growing market of independent women, feminism is packaged and processed in the commodification of sexual liberation. Furthermore, in another round of negotiation, contesting texts are erected against such hegemonic tendencies. Resistance may identify itself with emergent or excluded world-views: we have already encountered one such dissension in the Marxist decoding of postmodernism as flexible accumulation. Or resistance may be couched in the language of a residual culture (Williams 1980). Consider, for example, by way of a conclusion, this neo-conservative attack on Vancouver inner-city planning – specifically the doctrine of urban 'liveability' and the planning of 'people places' – by a local journalist:

I don't want to see [the area] turned into a people place. I want to see it turned into a *work* place.... Not full of boutiques.... Not smart shoppes in fake Palladian ... staffed by aspiring Beautiful People at $4 an hour.... Not places operated ... by slim young men of a kind I'm not, candidly, keen on.... I want to see, for a change in this ... flaccid city, something actually being *produced* – not consumed. And not cottage pottery or weaving either. Real stuff, made by real workers for real people and filling real needs ... even if it means a little smoke. I want to hear the noise of something being *built* – not the swish of yuppie commerce. Most of all, I want to see blue-collar workers ... paid $20 an hour, enough to buy detached houses on real lots and to raise decent families.... I often think fondly of Sigurdson's.... A great, green, ugly mill [adjoining Fairview Slopes] ... [it] gave honest work to generations of small businessmen and workers, who, in the days before Head Office was somewhere else, were bonded into a cast whose members ... understood one another.... This city is becoming a Narcissus, vain, precious, preening in the mirror-like pond that reflects its beautiful self but hides the shallow reality.

(Lautens 1987: B9)

In this polemical text, Lautens attacks the 'yuppies', postmodern architecture, gays, consumer culture, ecological concerns, flexible production; he promotes suburban living, 'traditional' family life, the family wage and paternalistic labour relations. Fairview Slopes would represent, to him, a symbol of a world turned upside-down. Lautens yearns to re-erect a lost master-narrative, to return people to their proper places, places to their proper people.

Moral orders are also spatial orders. Conservative, radical and liberal projects all invoke an 'authentic' organization of space to naturalize a mythical version of the way the world works. The project of gentrification is a discourse about landscape which offers, in turn, a discourse about society. A geography which takes culture seriously attests to this mutual reconfiguration of landscape texture and social text.

NOTES

1   Quotation from a conversation with a real estate agent in the study area.

2   Former residents were unable to sustain vigorous resistance to redevelopment. An active resident group sought to influence the new zoning regime in the mid-1970s, calling for a larger element of 'social' planning to maintain resident mix. By the mid-1980s, however, signs of protest had dissipated. This essay focuses on the experience of the gentrifiers, not on those already displaced.

3   Representatives of forty-five households living in new Fairview Slopes buildings were interviewed. Interviews were also carried out with selected developers and architects.

4   Quotation from a marketing pamphlet.

5   From an interview with Mr Molnar.

6   Slogan employed in the marketing of one town-house development.

7   The same image features in the only material evidence of conflict over gentrification which could be (literally) read in the Fairview landscape by the early 1980s: a graffito text reading: 'THIS USED TO BE A STREET OF HOMES, NOT YUPPIE. YOUNG URBAN PROF SHIT HEADS. $ STINKS'.

8   Compare the deprecation of suburbanites in *Talking Dirty*, a long-running Vancouver play set in an adjacent inner-city neighbourhood, in which one character is described as 'a refugee from Burnaby' (Snukal 1982).

9   Two co-operative developments have been built in this area, compared to over ninety new projects in total. See Ley (Chapter 8 in this volume) for a discussion of co-op aesthetics in Vancouver.

REFERENCES

Allen, I. (1980) 'The ideology of dense neighborhood redevelopment: cultural diversity and transcendent community experience', *Urban Affairs Quarterly* 15: 409–28.

Beauregard, R.A. (1986) 'The chaos and complexity of gentrification', in N. Smith and P. Williams (eds) *Gentrification of the City*, Boston: Allen & Unwin, 35–55.

Bell, W. (1958) 'Social choice, life styles and suburban residence', in W. Dobriner (ed.) *The Suburban Community*, New York: Putnam, 225–47.

Burgess, J. (1982) 'Selling places: environmental images for the executive', *Regional Studies* 16: 1–17.

——— (1990) 'The production and consumption of environmental meanings in the mass media: a research agenda for the 1990s', *Transactions of the Institute of British Geographers* (new series) 15: 139–61.

Burgess, J. and Wood, P. (1988) 'Decoding Docklands', in J. Eyles and D.M. Smith (eds) *Qualitative Methods in Human Geography*, Cambridge: Polity, 94–117.

Cooper, C. (1974) 'The house as symbol of the self', in J. Lang, C. Burnette, W. Moleski and D. Vachan (eds) *Designing for Human Behavior*, Stroudsberg, Pa: Dowden, Hutchinson & Ross, 130–46.

Cosgrove, D.E. (1984) *Social Formation and Symbolic Landscape*, London: Croom Helm.

——— (1989) 'Geography is everywhere: culture and symbolism in human landscapes', in D. Gregory and R. Walford (eds) *Horizons in Human Geography*, Basingstoke: Macmillan, 118–35.

Cosgrove, D.E. and Jackson, P. (1987) 'New directions in cultural geography', *Area* 19: 95–101.

Crilley, D. (1990) 'The disorder of John Short's new urban order', *Transactions of the Institute of British Geographers* (new series) 15: 232–8.

Duncan, J.S. (ed.) (1981) *Housing and Identity*, London: Croom Helm.

Duncan, J. and Duncan, N. (1988) '(Re)reading the landscape', *Environment and Planning D: Society and Space* 6: 117–26.

Elligott, R. and Zacharias, J. (1973) *Report: Fairview Slopes*, Vancouver: City Planning Department.

Gouldner, A.W. (1979) *The Future of Intellectuals and the Rise of the New Class*, New York: Continuum.

Gregory, D. (1987) Editorial, *Society and Space* 5: 245–8.

Gruft, A. (1983) 'Vancouver architecture: the last fifteen years', in *Vancouver: Art and Artists 1931–1983*, Vancouver: Vancouver Art Gallery, 318–31.

Hamnett, C.R. and Williams, P. (1980) 'Social change in London: a study of gentrification', *Urban Affairs Quarterly* 15: 469–87.

Harvey, D. (1987) 'Flexible accumulation through urbanization: reflections on "postmodernism" in the American city', *Antipode* 19: 260–86.

——— (1989) *The Condition of Postmodernity: An Inquiry into the Origins of Cultural Change*, Oxford: Blackwell.

Jackson, P. (1989) *Maps of Meaning: An Introduction to Cultural Geography*, London: Unwin Hyman.

Jager, M. (1986) 'Class definition and the aesthetics of gentrification: Victoriana in Melbourne', in N. Smith and P. Williams (eds) *Gentrification of the City*, Hemel Hempstead: Allen & Unwin, 78–91.

Jameson, F. (1984) 'Postmodernism, or the cultural logic of late capitalism', *New Left Review* 146: 53–92.

Jencks, C. (1986) *What Is Post-Modernism?* New York: St Martin's Press.

Jhally, S., Kline, S. and Leiss, W. (1985) 'Magic in the marketplace: an empirical test for commodity fetishism', *Canadian Journal of Political and Social Theory* 9: 1–22.

Laska, S.B. and Spain, S. (1980) *Back to the City: Issues in Neighborhood Renovation*, New York: Pergamon.

Lautens, T. (1987) 'Pray, save us from this awful phrase', *Vancouver Sun*, 10 October: B9.

Leiss, W. (1983) 'The icons of the market place', *Theory, Culture and Society* 1: 10–21.

Ley, D.F. (1980) 'Liberal ideology and the post-industrial city', *Annals of the Association of American Geographers* 70: 238–58.

—— (1987) 'Styles of the times: liberal and neo-conservative landscapes in inner Vancouver, 1968–1986', *Journal of Historical Geography* 13: 40–56.

—— (1993) Co-operative housing as a moral landscape, Chapter 8 in this volume.

McCracken, G. (1986) 'Culture and consumption: a theoretical account of the structure and movement of the cultural meaning of consumer goods', *Journal of Consumer Research* 13: 71–84.

Matless, D. (1989) 'Conference report: new directions in cultural geography', Institute of British Geographers, Social and Cultural Geography Study Group, Newsletter, 1–3.

Mills, C.A. (1988) ' "Life on the upslope": the postmodern landscape of gentrification', *Environment and Planning D: Society and Space* 6: 169–89.

—— (1989) 'Interpreting gentrification: postindustrial, postpatriarchal, postmodern? Vancouver', unpublished Ph.D thesis, Department of Geography, University of British Columbia.

Moore, P.W. (1982) 'Gentrification and the residential geography of the new class', mimeo, Toronto: Department of Geography, University of Toronto.

Pratt, G. (1981) 'The house as an expression of social worlds', in J. Duncan (ed.) *Housing and Identity*, London: Croom Helm, 135–80.

Rose, D. (1984) 'Rethinking gentrification: beyond the uneven development of Marxist urban theory', *Environment and Planning D: Society and Space* 2: 47–74.

Sahlins, M. (1976) *Culture and Practical Reason*, Chicago: University of Chicago Press.

Schwartz, B. (ed.) (1976) *The Changing Face of the Suburbs*, Chicago: University of Chicago Press.

Smith, N. (1979) 'Toward a theory of gentrification: a back to the city movement by capital not people', *Journal of the American Planning Association* 45: 538–48.

—— (1982) 'Gentrification and uneven development', *Economic Geography* 58: 139–55.

—— (1986) 'Gentrification, the frontier, and the restructuring of urban space', in N. Smith and P. Williams (eds) *Gentrification of the City*, Boston: Allen & Unwin, 15–34.

—— (1987) 'Of yuppies and housing: gentrification, social restructuring, and the urban dream', *Society and Space* 5: 151–72.

Snukal, S. (1982) *Talking Dirty*, Toronto: Playwrites Canada.

Williams, P. (1978) 'Building societies and the inner city', *Transactions of the Institute of British Geographers* (new series) 3: 23–34.

Williams, R. (1973) *The Country and the City*, London: Chatto & Windus.

—— (1980) *Problems in Materialism and Culture*, London: Verso New Left Books.

Williamson, J. (1978) *Decoding Advertisements: Ideology and Meaning in Advertising*, London: Marion Boyars.

Part III

ON REPRESENTING INSTITUTIONAL CULTURES

10

'THIS HEAVEN GIVES ME MIGRAINES'

The problems and promise of landscapes of leisure[1]

Stacy Warren

Landscapes of leisure and entertainment in North America seem to have taken on entirely new dimensions in recent years. What were once treated as separate, self-contained places within which one could escape from the rigours of daily life now are seen as not so much segregated sites but modes of representation that permeate virtually all landscapes and hence are inseparable from daily life. The popular culture that surrounds us acquires new significance as the ground upon which our cultural geographies are formulated, articulated, negotiated and lived; more and more academics are becoming intrigued with its problems and possibilities.

Underlying this relatively new concern for popular, leisure-oriented landscapes is a keen interest in the cultural conditions and practices associated with the spectacles of everyday life. The focus on culture, and particularly on popular culture, raises fascinating questions for cultural geographers, concerning not only what become acceptable targets of study but also how the interaction between culture and landscape is theorized. In this essay, after situating the discussion within the popular cultural landscape, I examine various approaches to culture, including cultural hegemony, mass culture vs popular culture and the cultural uses of fantasy, and discuss their ramifications for cultural geographers.

LANDSCAPES OF LEISURE

It is not surprising that the landscape in question has sparked lively debate, for it is one we encounter on a daily basis and one that forms an integral part of our dealings with the surrounding world. It is a landscape where the lines between leisure, entertainment and commodity become blurred. It is everywhere. The most audacious examples are the most famous. Disneyland (substitute Walt Disney World, or Tokyo's Disneyland, or Paris's Euro Disney) has raised many eyebrows, with its clever – some call it insidious – marketing of fantasy, consumerism and multinational capitalism (Eco 1986; Gottdiener 1982; Marin

1984; Real 1977; Stephanson 1988; Wallace 1985). The West Edmonton Mall
in Canada uses a virtually identical formula but perhaps more truthfully actually
calls itself a mall; the seeming incongruence of roller coasters and department
stores has not gone unnoticed (Hopkins 1990; Shields 1989). A World's Fair
such as Vancouver's Expo 86 offers a similar, if more impermanent, exuberant
display of leisure laced with consumerism (Ley and Olds 1988). The recently
revamped Canadian Museum of Civilization in Ottawa blends anthropology and
education with a 'Disney touch' of entertainment (Young 1987).

The examples that are less eye-catching are perhaps more instructive, for they
represent this landscape in its everyday articulation. The neighbourhood
McDonald's, for instance, may be on one level a place to grab a quick Big Mac
and fries, but its architectural themes and inclusion of entertaining elements,
from play equipment for children to homey fireplaces for adults, suggest that
leisure is also central to its popularity. Shopping malls of lesser ambition than the
West Edmonton Mall still rely on shades of fantasy to complement their retail
functions, most notably through the creation of an 'artificial' atmosphere of
traditional Main Street charm (Francaviglia 1974; Jacobs 1986; Kowinski 1985).
The new Disney stores springing up across North America may point towards a
new phase of entertaining consumption: instead of hiring store clerks, Disney has
hired 'cast members' trained to treat customers cheerfully as 'guests'. Stores are
meant to offer the 'magic' of a theme park experience (Stevenson 1990). Harvey
(1989) and Soja (1989) argue that this trend towards 'theme park experiences',
as epitomized by the Disney stores, in fact premeates much of the contemporary
urban fabric, from residential districts to revitalized downtowns. Sack further
notes that today's commercial landscape, much like the mass consumption that
fuels it, is 'at the foundation of modern life' (Sack 1988: 661).

Landscapes of leisure and entertainment are directly woven into North
American culture, and they are nearly inescapable. Most people, for whatever
reason, often end up at one or another of these places. Annual visitors to
destinations like Disneyland or the West Edmonton Mall alone number in the
millions, figures that no doubt do not even begin to approximate the total picture
when we include every visit to the local amusement area, shopping mall or fast-
food restaurant. Those who choose not to frequent such places would have a
much more difficult time avoiding exposure to the mass-mediated culture that
promotes their existence.

The popular culture represented by landscapes of leisure and entertainment is
the backcloth against which almost all our everyday cultural geographies are
lived. Though popular culture has not been a foreign concept to cultural
geographers, traditionally we have neither the empirical nor the theoretical tools
to study it adequately. A long-standing interest in the spatial distribution of
commonplace artefacts has permeated cultural geography since being popularized by Carl Sauer, yet this approach tends to privilege the folk over the mass-
mediated, and to discourage a wider arena of theoretical inquiry.[2] The turn
towards a more humanistic reading of the landscape as championed by the

journal *Landscape* and brought to full fruition by cultural geographers in the 1970s, while no longer ignoring 'mass' culture, often treats it, as Relph (1976) definitively phrased it, as 'placeless'.[3] It is only in recent years that cultural geographers have begun both to refocus their interest to the mass-mediated landscape and to seek new theoretical frameworks to understand it; culture theory, and in particular the concept of cultural hegemony, often has proved pivotal in these new perspectives.[4]

That cultural geographers for so long have neglected the rich possibilities of popular culture is not surprising; they are merely echoing sentiments widespread throughout the social sciences. The most sustained commentary on the everyday has resolutely dismissed it as dangerously mindless 'mass culture'. Taking popular culture seriously is a novel idea, and redeeming it from the hands of 'mass culture' theorists a formidable task. The task is made even more difficult because not only the 'popular' but 'culture' itself must be subject to theoretical scrutiny.[5]

THEORIES OF CULTURE

Any approach to popular culture is only as solid as the theoretical framework upon which it rests, and at the heart of that theoretical framework is the conceptualization of culture. The word 'culture', as Raymond Williams (1976: 76) points out, may well be one of the most complicated words in the English language. Constantly changing in definition, it has its own social history. In the twentieth century, two competing definitions of the word have reigned – a literary/moral one, emphasizing culture as an aesthetic or intellectual ideal, and an anthropological one, emphasizing culture as a way of life. In recent decades, a new variation of the term has been emerging. An amalgamation of the work of several groups of culture theorists, this latest approach grows from an anthropological basis and expands it in two related directions. First, it incorporates the notion of struggle to construct a view of culture as a fluid entity always being created, contested and recreated. Second, it attempts to situate the dynamics of cultural practices within the confines – and resources – of a mass-mediated world. These two threads – the broader realm of cultural practices in general, and the more specific arena of popular cultural practices in particular – can become the groundwork for studying the popular culture landscape, and as such constitute the major themes of this section.

CULTURAL PRACTICES

The social construction now emerging of 'culture' encompasses many intellectual currents, but perhaps can be best crystallized around the concept of cultural hegemony.[6] A brief examination of cultural hegemony's general parameters and their articulation in recent cultural writings can pinpoint the ideas and innovations most central to current cultural debates. Cultural hegemony, a concept far more complex than its cruder usages sometimes imply, refers to a set of ideas most directly associated with the Italian Marxist Antonio Gramsci concerning the moral, philosophical and political dimensions of state and civil society.[7] One of Gramsci's key interests lay in understanding how dominant social orders come to be dominant. He reasoned that in modern capitalist societies leadership is attained not through physical coercion but through cultural consent. From this seemingly simple notion, Gramsci developed remarkably nuanced insights into the workings of culture.[8] His insights can be divided roughly into three areas: culture's dynamic nature; its inclusion of many groups and subcultures; and the wide spectrum of cultural practices associated with it. These areas in turn mark three defining qualities of culture theory today.

First, Gramsci's concept of cultural hegemony hinges, above all, on its dynamic nature. It is, as Gramsci saw it, a 'moving equilibrium' – equilibrium, because some form of consent to leadership has been achieved; moving, because in a society composed of what he called a 'contradictory and discordant ensemble', consent will never last long. Most culture theorists today take pains to point out that hegemony is not a state of absolute or permanent social control. By definition, it cannot be. The groups of people involved, including ruling-class alliances and various subordinate groups, are always shifting (Clarke *et al.* 1981). The terrain upon which they struggle is also shifting, as cultural meanings are constantly forged, revised and rejected (Gitlin 1982; Hebdige 1979). Gitlin (1982: 206) points out that, given the inherently contradictory and changeable nature of Western society, 'the process of renegotiation is mandatory'.

Hegemony strikes a fragile balance, one that may be nearly impossible to freeze at a particular moment for dissection. Yet as a 'moving equilibrium' it provides a rich foundation for coming to grips with culture. Seeing hegemony as an active site draws attention to the transitory complex of experiences, relationships and activities that constitute cultural practices. It allows Hall (1981) to assert that culture is first and foremost an historical process, and hegemony is neither dominance nor resistance but the ground upon which the transformations are worked. Lears (1985) and Williams (1977) also argue that culture's fluidity lies in the coexistence of and tensions between dominant and resistant elements; both use the terms 'counter-hegemonies' and 'alternative hegemonies' to indicate the totality of what Williams (1977: 108) sums up as 'the whole social process'.

Second, Gramsci insisted upon acknowledging the existence of not one dominant group sharing a common goal but various ruling-class alliances often at odds among themselves, and not one subordinate group united in its resistance but a collection of oppositional groups. He hinted at a blending of interests and boundaries beyond the traditional Marxist divisions based on class, thus moving beyond class essentialism and opening the way for examinations based on gender, race and other cultural dimensions (Bennett 1986a; Bocock 1986).[9] Hegemonic plays for leadership result in mobile combinations of culture, as dominant groups incorporate opposing ideas and interests in an effort to win over the subordinate groups. Their success, of course, as Gramsci (cited in Clarke *et al.* 1981: 61) cautioned, is by no means guaranteed, as different groups, both dominant and subordinate, may react in different ways to the various 'relations of forces favourable or unfavourable to this or that tendency'.

In short, Gramsci spoke of the possibility of discussing *cultures* in place of *culture*, and of recognizing the distinct vantage-points from which these sub-cultures experience the totality of cultural practices. This multi-faceted perspective makes possible a more fluid examination of cultural phenomena, as witnessed by Clarke *et al.*'s (1981) investigation of both working-class sub-cultures and middle-class counter-cultures within the rubric of cultural hegemony, or Williams's (1977) well-known distinction between dominant, residual and emergent cultural forms.

Third, one of Gramsci's most appealing innovations in cultural thought is his careful consideration of hegemony at its most practical level – how consent is actually achieved. In societies where outright domination is frowned upon, consent is hard work. The hegemonic process by definition traces the give and take that inevitably lie behind any group's quest for consent. Negotiation and compromise are necessary on both sides, and, as Gramsci (cited in Lears 1985: 570) noted, ultimately consent often amounts to no more than a 'contradictory consciousness', a tenuous mix of approval and apathy, resistance and resignation.

Cultural expression and struggle become in this light increasingly sophisticated and difficult to pin down. The interplay between the two poles of willing approval and steadfast rejection takes many forms including incorporation, distortion, resistance and negotiation (Hall 1981: 236). Such ambiguity reinforces Gramsci's insistence that all cultural practices are historically and socially constructed. Hall (1981) and Bennett (1986a: xvi) stress that a cultural practice 'does not carry its politics with it'. Each practice must be evaluated and understood in its own context. Hebdige (1979) has applied this argument to style and appearance, demonstrating that even a seemingly non-commital object such as a safety pin can be transformed into a symbol of resistance when worn on the human body.

Finally, Gramsci's interest in the practical levels of hegemony, or in what he called the 'common-sense' level halfway between folklore and philosophy, reflects his belief that the 'simple' did matter (Bennett *et al.* 1981b; 203–4). A hegemonic order that forgot to remain in contact with the 'simple', he postulated, could never become 'life'.[10] This is perhaps one of Gramsci's most spectacular, if underrated, achievements, for it reveals his respect for the potentials of popular culture. Romano (1983: 42) speculates that were Gramsci writing today, he would likely pass up the academic press and write colloquially on subjects such as *Dynasty* and the *New York Times*, producing 'nuggets' of analysis 'that a worker could stay awake through'. Gramsci believed that social change could come from 'the simple' – 'all men are "philosophers"', as he said – and that the everyday, the popular, was the terrain upon which negotiation, compromise and resistance could be carried out (Bennett *et al.* 1981b: 200–1).

Gramsci's enthusiastic assessment of popular culture lies beneath the surface of much current cultural debate. Williams (1977: 110–11) and Clarke *et al.* (1981: 67) both argue that hegemony allows not just popular culture but specifically the realm of leisure and entertainment to emerge as a crucial domain of study. Yet within cultural studies there remains a distinct predilection not to take this realm to its full theoretical or empirical conclusions. Often the focus on 'the popular' is defined through the eyes of the working classes (see, for example, Hoggart 1958 or Willis 1977) or clearly marginalized subcultures (Hebdige 1979, for instance), thus side-stepping examination of middle-class, mass-mediated cultures (Hall 1981).

Studies that do engage directly these strands of the everyday reveal a depressingly consistent tendency to declare the 'popular' a lost cause. Interpretations of leisure-oriented landscapes are no exception. Most display a distrust of the culture of the masses articulated in the landscape, with 'postmodern' interpretations voicing perhaps the harshest criticisms of what they see as its artificial, superficial and generally inferior qualities (Eco 1986, Jameson 1984; Marin 1984; Stephanson 1988). This pessimism also echoes among geographers. Harvey (1989) and Soja (1989) are deeply disturbed by the increasing 'theme park' nature of urban areas. Shields (1989) and Hopkins (1990) express similar concern for the 'placelessness' of shopping malls. As Meyrowitz (1985: 125) sums up the argument, mass-mediated culture has 'destroy[ed] the specialness of place and time'.

To do justice to landscapes of leisure and entertainment, culture theory must be disentangled from its 'mass culture' overtones. This brings us to the second major thread of discussion, the arena of popular cultural practices. A brief examination of the long, time-honoured tradition of 'mass culture' will, first, demonstrate how ultimately futile the 'mass culture' vs 'popular culture' stand-off can be and, second, point us towards avenues for integrating 'real life' back into theory.

MASS CULTURE AND POPULAR CULTURAL PRACTICES

The delights of the masses have not gone unnoticed by all cultural observers. Many have focused on the trappings of an urban, industrial world, and have paid detailed attention to dime novels, movies, jazz recordings, seaside resorts, professional ball games and other mass-produced experiences. Almost unfailingly, interest in 'mass culture' has gone hand in hand with a certain cultural fatalism. The culture of the masses has for so long been treated as a prime cause of society's imminent downfall that it ceases to appear as a theoretically grounded construct and instead suggests a material reality. Only quite recently have culture theorists stepped back to consider the concept of 'mass culture' more critically and to chart out new terrain for dealing with the everyday (Brantlinger 1983; Fiske 1989a, 1989b; Frith 1988a, 1988b; Gruneau 1988; Jameson 1979; Laba 1986; Lazare 1987a; Real 1977; Ross 1989).

Brantlinger (1983) traces the pervasive and persistently negative 'mass culture' outlook back to the 'bread and circuses' commentary of ancient Greek and Roman civilization. He demonstrates that both the right and the left have embraced this disparaging view whole-heartedly; equating mass culture with social ruin came as an apparently natural reaction for writers as diverse as, for example, T.S. Eliot and Theodore Adorno. While the right tend to fear cultural decadence from below in the hands of 'the brutal empire of the masses', as Jose Ortega y Gasset (cited in Brantlinger 1983: 187) expressed it, and the left instead look above with a wary eye trained on the ruthless mass deceptions of the culture industry, all are united in their conviction that mass culture presents a clear threat to 'genuine', more 'authentic' forms of culture.

Early attempts to move beyond 'mass culture's' obvious pessimism centred on affirmative readings of everyday culture. Strongly associated with the Popular Culture Association and its *Journal of Popular Culture*, this approach is commonly referred to as a 'popular culture' perspective (Geist and Nachbar 1983). Its object is to rescue the everyday and the commercial from 'mass culture's' vile connotations. It attempts to dissolve the boundaries 'mass culture' places between the 'genuine' culture of the elite and the kitsch-infested culture of the masses, and it rejects the notion that the masses have been unwittingly manipulated (Lazare 1987b). The 'popular culture' approach offers instead a view of everyday culture as a valid and authentic expression of people's interests (Ross 1989). People consume popular culture because they want to, 'popular culture' proponents argue, and they want to because it reflects their values. As such, it presents a 'truthful picture' of what people do and think (Browne 1983: 17).

Popular culture thus defined provides a much-needed corrective to some of 'mass culture's' extravagances by reinserting agency and optimism into the cultural picture. However, in many ways 'popular culture' simply ends up turning 'mass culture' inside out — reversing, but not overcoming, its flaws. 'Popular culture' replaces unquestioning rejection with unquestioning acceptance, and hence maintains a monolithic view of culture. Everyone, it is still assumed, partakes of a single set of values, ideas and desires (Gruneau 1988; Lazare 1987b).

Ultimately, the two perspectives exist in polar opposition to one another and neither can provide the basis for study of a clearly mass-mediated, mass-marketed culture. Each perspective isolates certain meaningful cultural processes and dimensions — 'mass culture's' interest in how culture is shaped from the outside, 'popular culture's' assertion that the pleasures of culture are genuine — but their dichotomous placement fails to incorporate these processes as part of the same phenomenon. Neither considers culture as an active and continually contested process; neither can strike a balance between the twin poles of dominance and resistance within everyday life.

When dealing with a mass-marketed, corporate-sponsored popular culture, clearly elements of both arguments ring true. Culture industries do exist, and as

Maltby (1989: 8) states, they do 'steal our dreams and sell them back to us'. Yet their products would not achieve popularity if they did not hold genuine appeal (Fiske 1989a, 1989b). A theoretical framework for a culture that is both mass produced and popularly enjoyed must rest on the hegemonic play between the producers and the enjoyers. This entails, first, a re-evaluation of popular culture's defining characteristics and, second, a closer exploration of one of its most powerful dimensions, that of fantasy.

A defining characteristic of mass-mediated culture is that, due to its expression in virtually every corner of a consumer-oriented society, it 'inescapably shapes us all, for better or worse' (Lazare 1987b: 2). It is fruitless to attempt to evaluate cultural forms as either real or falsified, authentic or just an advertising gimmick, because popular culture *is*, to a considerable extent, a commodity; what is possible for us as consumers *is* constrained by the process of cultural production (Cohen 1989; Frith 1988a, 1988b). Since commodification is predominantly the form that cultural life takes today, Angus and Jhally (1989: 3) would add, the question to ask becomes what possibilities this opens up. Fiske (1989a, 1989b) refers to the products of mass-mediated culture as, simply, possibilities. They are the cultural resources from which people, through acceptance, negotiation, resistance and evasion, form their own popular cultures. Popular culture, though laden with commodities, is primarily about the circulation of meanings.

Stated in more theoretical terms, understanding mass-mediated culture becomes a task of understanding its hegemonic role. Ross (1989: 3) suggests popular culture is the realm where 'the struggle to win popular respect and consent for authority is endlessly being waged', where the desires and aspirations of ordinary people, elements of disrespect and opposition, and elements of explanation for the maintenance of respect all meet, intermingle and become transformed. Popular culture, Fiske further argues, bears both traces of power relations and signs of resistance to them; of particular interest to him is the popular vitality and creativity that make hegemonic incorporation such a constant necessity (Fiske 1989b: 20).

The resources of mass-mediated culture can ensure their popularity only by making themselves inviting terrain for this struggle (Fiske 1989a: 5). Thus their relevance to everyday life is central, and the pleasure they offer is genuine. Frith (1988a: 471) contends that the pleasure of the popular stems from its status as unofficial articulation of doubts, needs and desires. Popular culture is far from merely escapist; rather, it provides the ground upon which people can attempt to accommodate themselves to and indeed construct their everyday lives (Hall 1981; Jameson 1979; Ross 1989).

Few, however, would mistake popular culture for organized social struggle. The struggle popular culture offers occurs on a more subliminal level. It is acted out through the conduits of daily life in an atmosphere heavily imbued with a sense of fantasy and hence often dismissed as contentless. But the role of fantasy in popular culture is one of its most potent resources: as people take pleasure in recognition of and identification with elements that have resonance in their own lives, they not only partake in wishful and imaginary renditions of social conditions, they also actively construct new relationships and new constellations of meaning. 'Real life' is refracted through, not replaced by, popular culture's prism of fantasy.

THE CULTURAL USES OF FANTASY

We can begin to sketch in the role fantasy plays in popular culture as a forum where 'fundamental social anxieties and concerns, hopes and blind spots' can be expressed, and future reconciliations hinted at (Jameson 1979: 141). Under the guise of fantasy, popular culture can make statements about social needs and the best ways to meet them, yet never reveal how truly 'serious' its statements are. It easily can be dismissed, and often is, as offering no more than an 'optical illusion of social harmony', as Jameson (ibid.) ultimately concludes. What Jameson and others see as a weakness may also be interpreted as a strength: popular culture, perhaps unlike any other form of cultural practice, can confront head-on society's contradictions and conflicts in 'guerrilla' fashion (Fiske 1989b.) By speaking the language of fantasy, it can remain sublimely outside conventional structures of logic and always just beyond the reach of dominant hegemonic forces.

The most compelling examination of the uses of fantasy, and particularly of fantasy in the landscape, comes from two of the twentieth century's foremost, if unorthodox, Marxian cultural theorists, Walter Benjamin and Ernst Boch. Both look to the everyday – the shopping street, the arcade, the fun fair – to illuminate how within the commonplace, traces of fantasy can actually embody the Utopian seeds of society's transformation.

Benjamin (1986) and Bloch (1988) examine what Benjamin calls the 'residues of a dream world' that in a thousand configurations from permanent buildings to fleeting fashions permeate our everyday landscapes. Both see in the practices and artefacts of mass-mediated culture not only dominant ideology but also Utopian strands of resistance. Bloch (1988) situates the cultural uses of fantasy within the same dream-like conditions found in the fairy tale. The fairy tale, beneath its bizarre veneer, is a veiled dialogue about deep-seated hopes and fears where the small and the weak can be victorious through courage and ingenuity. Reading the fairy tale, the reader thinks 'about a great deal ... almost everything in their lives', and is spurred on to contemplate visions of the future (ibid.: 164).

The elements of fantasy embedded within popular culture function in much the same way as the irrational and absurd elements of the fairy tale: they are the conduits for examining from a 'safe' perspective the real social conditions that appear in fantastical guise. When articulated in the landscape, these 'residues of a dream world' hold the power to transform our imaginations and suggest possibilities for the future.

Benjamin (1986) investigates two nineteenth-century landscapes, the panorama and the arcade, and provides fascinating glimpses of the uses of fantasy within a wider web of capitalism, culture and commodity. The panorama was a short-lived object of entertainment, an exhibition no longer photograph but not yet moving picture. The makers of panoramas, with 'tireless exertions of technical skill', struggled to achieve what may have seemed to be the impossible: the incorporation of the passage of time into a static picture (ibid.: 149). In their imitations of natural landscapes, they wished to reproduce the time of day, the rising of the moon, the flow of water. The attraction that stemmed from this infusion of fantastic elements was not simply a matter of artistic flourish, Benjamin (ibid.: 150) argues, but a revolutionary expression of a new feeling about art, technology and life. Likewise, the arcade, an architectural form dating from the early nineteenth century, gained its Utopian strength from its somewhat fantastical existence as both solid building and glass-roofed infinity, as a functionally industrial setting out of which an elegantly luxurious shopping district blossomed. It was, says Benjamin (ibid.: 157) 'both house and stars'.

Bloch (1986) also discovers Utopian potential in the shopping street. Observing what he calls the obvious futility and deception of a commodity culture that flatteringly and corruptly arouses hope, he none the less also sees the shopping street as a street 'steeped in dreams' (ibid.:33). He contemplates by way of example a pair of lizard-skinned shoes in the shop window. A woman walking past, he surmises, might pause and think, first, of money, but also of what that money could be changed into, and what the shoes might then offer or symbolize. Thus she has a 'share of the wishful land'.

Bloch extends his examination of wishful landscapes to the fun fair. Though recognizing the dominant impulses it contains – 'a complete swindle', he calls it – he also finds in the fair intriguing possibilities. Almost every theme at the fair, he notes, would raise the irritation of the 'bourgeois conformist'. Taken as a whole, the fair represents a 'colorful rough fantasy' of unprecedented proportions (Bloch 1986: 363).

CONCLUSIONS

Neither Benjamin nor Bloch lived to see the latest phase of leisure-oriented landscapes; indeed, it is disputable whether Benjamin could perceive the West Edmonton Mall, like the arcade, to be 'both house and stars', or if Bloch could call EPCOT Center a 'colorful rough fantasy'. These places, judging from most commentary, seem far more superficial and blatantly commercial than their nineteenth-century counterparts. Yet as Benjamin and Bloch (1988: 5) would be the first to remind us, 'the content of the utopian changes according to the social situation'.

Our task as cultural geographers is to get beyond our distaste for current styles of expression and pierce the surface of cultural meanings and practices. The 'culture industries' behind the look of today's landscape, such as the McDonald's or Disney corporations, or the festival markets and malls of the Rouse Corporation, do create the 'fake' histories, artificial Main Streets and cleverly disguised palaces of consumption of which so many, in true 'mass culture' fashion, have complained. Their role in creating the landscape is indisputable, but analysis cannot stop there. Half of the hegemonic process is missing: we still must ask how people incorporate these places into the cultural practices of their everyday lives, and how these places form part of the 'moving equilibrium' of an always contested, always changing popular culture.

The element that is most often overlooked in analysis of the popular landscape is, surprisingly, the people. To move beyond a static image of culture, their presence is necessary. Those few studies that actually have leapt into the landscape to examine how people use it and why they think they use it hint at the vast potential of such an approach. Fiske (1989a), for instance, bravely followed teenage boys around shopping malls and discovered that most of their behaviour was aimed at subverting, not acquiescing to, the 'system' and its consumer ideology. Real (1977) similarly found, no doubt to Walt Disney's chagrin, that most visitors to Disneyland he interviewed were more inclined to compare their visit to a drug trip than to a nostalgic return to their home town or childhood. Ley and Olds (1988) report that, when surveyed, visitors to Expo 86 revealed that their attention was more on strengthening family ties and meeting new people than on the so-called messages of indoctrination associated with a World Fair.

Popular culture invites resistance, misinterpretation, evasion and distortion as well as acceptance; all of these cultural practices eventually result in transformation. Landscapes of leisure, as one of the sites upon which these transformations are worked, play a central role in people's everyday lives. Their popular and sometimes fantastic forms embody texts of domination and texts of resistance. By frequenting or generally paying attention to these landscapes, people find the resources to make sense of their lives. Geographers should also pay attention to landscapes of leisure to make sense of our popular cultures.

NOTES

1 Title suggested by the Gang of Four song and commentary on the built environment 'Natural's not in it'.
2 This tradition is exemplified by Sauer (1963 [1925]), Wagner and Mikesell (1962) and Zelinsky (1973). Rooney *et al.* (1982) in their atlas of popular culture provide an excellent collection of maps that typify this approach.
3 Lowenthal (1979), Meinig (1979), Relph (1976) and Tuan (1977) well represent the largely untheorized landscape-as-text approach of the 1970s.
4 Cultural geography's newly emerging engagement with hegemony and the popular is well illustrated by Anderson (1988), Burgess and Gold (1985), Duncan (1990) and Jackson (1989). Cosgrove and Jackson (1987), Daniels (1989) and Ley (1985) have been at the forefront of more general calls for a more nuanced cultural geography.
5 As Stuart Hall (1981: 227) once remarked, 'I have almost as many problems with "popular" as I have with "culture". When you put the two terms together, the difficulties can be pretty horrendous.'
6 Bennett (1986b), Gruneau (1988), Hall (1980) and Johnson (1979) provide various

accounts of the tug of war between structuralist and culturalist interpretations of culture that led to the current interest in cultural hegemony.

7   The interpretation of Gramsci I present here, 'cleansed' of its economic and political dimensions to the point of extreme over-simplification, is meant as a forum for examining the general cultural questions he raises.

8   Gramsci's most frequently referenced source on cultural hegemony is *Selections from the Prison Notebooks* (1971), written while he was imprisoned by Mussolini. Culture theorists and others have analysed, interpreted and argued extensively over Gramsci's use of cultural hegemony. See Bennett (1986a) Bennett *et al.* (1981a), Bocock (1986), Clarke *et al.* (1981), Hall (1980), Lears (1985), Romano (1983), Rosenthal (1988) and Williams (1977).

9   The orthodoxy of Gramsci's interpretation of Marxist categories such as class, and of Marxism in general, is a matter of debate. See, for instance, Bocock (1986), Romano (1983) and Rosenthal (1988).

10   Gramsci's concern was not idle: his project ultimately rested upon his belief that the hegemonic process could lead society past capitalism and into socialism.

REFERENCES

Anderson, K.J. (1988) 'Cultural hegemony and the race-definition process in Chinatown, Vancouver: 1880–1980', *Environment and Planning D: Society and Space* 6: 127–49.

Angus, I. and Jhally, S. (eds) (1989) *Cultural Politics in Contemporary America*, New York and London: Routledge.

Benjamin, W. (1986) *Reflections*, New York: Schocken.

Bennett, T. (1986a) 'Introduction: popular culture and the "turn to Gramsci"', in T. Bennett, C. Mercer and J. Woollacott (eds) *Popular Culture and Social Relations*, Milton Keynes and Philadelphia: Open University Press, xi–xix.

——— (1986b) 'The politics of "the popular" and popular culture', in T. Bennett, C. Mercer and J. Woollacott (eds) *Popular Culture and Social Relations*, Milton Keynes and Philadelphia: Open University Press, 6–21.

Bennett, T., Martin, G., Mercer, C. and Woollacott, J. (eds) (1981a) *Culture, Ideology and Social Process*, Milton Keynes: Open University Press.

——— , ——— , ——— and ——— (1981b) 'Antonio Gramsci', in T. Bennett, C. Martin, C. Mercer & J. Woollacott (eds) *Culture, Ideology and Social Process*, Milton Keynes: Open University Press, 191–218.

Bennett, T., Mercer, C. and Woollacott, J. (eds) (1986) *Popular Culture and Social Relations*, Milton Keynes and Philadelphia: Open University Press.

Bloch, E. (1986) *The Principle of Hope*, Cambridge, Mass.: MIT Press.

——— (1988) *The Utopian Function of Art and Literature*, Cambridge, Mass., and London: MIT Press.

Bocock, R. (1986) *Hegemony*, Chichester: Ellis Horwood.

Brantlinger, P. (1983) *Bread and Circuses: Theories of Mass Culture as Social Decay*, Ithaca and London: Cornell University Press.

Browne, R.B. (1983) 'Popular culture – new notes toward a definition', in C.D. Geist and J. Nachbar (eds) *The Popular Culture Reader*, 3rd edn, Bowling Green, Ohio: Bowling Green University Popular Press, 13–20.

Burgess, J. and Gold, J.R. (eds) (1985) *Geography, the Media, and Popular Culture*, London: Croom Helm.

Clarke, J., Hall, S., Jefferson, T. and Roberts, B. (1981) 'Sub-cultures, cultures and class', in T. Bennett, G. Martin, C. Mercer and J. Woollacott (eds) *Culture, Ideology and Social Process*, Milton Keynes: Open University Press, 53–79.

Cohen, E. (1989) 'The "hyperreal" vs the "really real": if European intellectuals stop making sense of American culture can we still dance?', *Cultural Studies* 3: 25–37.

Cosgrove, D. and Jackson, P. (1987) 'New directions in cultural geography', *Area* 19: 95–101.

Daniels, S. (1989) 'Marxism, culture and the duplicity of landscape', in R. Peet and N. Thrift (eds) *New Models in Geography*, Vol. 2, London: Unwin Hyman, 196–220.

Duncan, J.S. (1990) *The City as Text: The Politics of Landscape Interpretation in the Kandyan Kingdom*, Cambridge: Cambridge University Press.

Eco, U. (1986) *Travels in Hyperreality*, trans. S. Rendall, San Diego, New York and London: Harcourt Brace Jovanovich.

Fiske, J. (1989a) *Reading the Popular*, Boston: Unwin Hyman.

—— (1989b) *Understanding Popular Culture*, Boston: Unwin Hyman.

Francaviglia, R.V. (1974) 'Main Street revisited', *Places* 1: 7–11.

Frith, S. (1988a) 'Art ideology and pop practice', in C. Nelson and L. Grossberg (eds) *Marxism and the Interpretation of Culture*, Urbana and Chicago: University of Illinois Press, 461–76.

—— (1988b) *Music for Pleasure*, New York: Routledge.

Geist, C.D. and Nachbar, J. (eds) (1983) *The Popular Culture Reader* 3rd edn, Bowling Green, Ohio: Bowling Green University Popular Press.

Gitlin, T. (1982) 'Television's screens: hegemony in transition', in M. Apple (ed.) *Cultural and Economic Reproduction in Education*, New York: Routledge, 202–21.

Gottdiener, M. (1982) 'Disneyland: a Utopian urban space', *Urban Life* 2: 139–62.

Gramsci, A. (1971) *Selections from the Prison Notebooks*, London: Lawrence & Wishart.

Gruneau, R.S. (1988) 'Introduction: notes on popular cultures and political practices', in R.S. Gruneau (ed.) *Popular Cultures and Political Practices*, Toronto: Garamond Press, 11–32.

Hall, S. (1980) 'Cultural studies and the centre: some problematics and problems', in S. Hall, D. Hobson, A. Lowe and P. Willis (eds) *Culture, Media, Language,* London: Hutchinson, 15–47.

—— (1981) 'Notes on deconstructing "the popular"', in R. Samuel (ed.) *People's History and Socialist Theory*, London: Routledge & Kegan Paul, 227–40.

Harvey, D. (1989) *The Condition of Postmodernity: An Inquiry into the Origins of Cultural Change*, Oxford and Cambridge, Mass.: Blackwell.

Hebdige, D. (1979) *Subculture: The Meaning of Style*, London: Methuen.

Hoggart, R. (1958) *The Uses of Literacy*, Harmondsworth: Penguin/Chatto & Windus.

Hopkins, J.S.P. (1990) 'West Edmonton Mall: landscape of myths and elsewhereness', *The Canadian Geographer* 34: 2–17.

Jackson, P. (1989) *Maps of Meaning: An Introduction to Cultural Geography*, London: Unwin Hyman.

Jacobs, J. (1984) *The Mall: An Attempted Escape from Everyday Life*, Prospect Heights, Ill.: Waveland Press.

Jameson, F. (1979) 'Reification and Utopia in mass culture', *Social Text* 1: 130–48.

—— (1984) 'Postmodernism or the cultural logic of late capitalism', *New Left Review* 146: 53–92.

Johnson, R. (1979) 'Histories of culture/theories of ideology: notes on an impasse', in M. Barrett, P. Corrigan, A. Kuhn and J. Wolff (eds) *Ideology and Cultural Production*, London: Croom Helm, 49–77.

Kowinski, W. (1985) *The Malling of America*, New York: Morrow.

Laba, M. (1986) 'Making sense: expressiveness, stylization and the popular culture process', *Journal of Popular Culture* 19: 107–17.

Lazare, D. (ed.) (1987a) *American Media and Mass Culture*, Berkeley and Los Angeles: University of California Press.

—— (1987b) 'Introduction: entertainment as social control', in D. Lazare (ed.)

American Media and Mass Culture, Berkeley and Los Angeles: University of California Press, 1–26.

Lears, T.J.J. (1985) 'The concept of cultural hegemony: problems and possibilities', *American Historical Review* 90: 567–93.

Ley, D. (1985) 'Cultural/humanistic geography', *Progress in Human Geography* 9: 415–23.

Ley, D and Olds, K. (1988) 'Landscape as Spectacle: world's fairs and the culture of heroic consumption', *Environment and Planning D: Society and Space* 6: 191–212.

Lowenthal, D. (1979) 'The American scene', in D. Ward (ed.) *Geographic Perspectives on the American Past*, New York: Oxford University Press, 17–32.

Maltby, R. (ed.) (1989) *Passing Parade: A History of Popular Culture in the Twentieth Century*, Oxford and New York: Oxford University Press.

Marin, L. (1984) *Utopics: Spatial Play*, Atlantic Highlands, Humanities Press.

Meinig, D.W. (ed.) (1979) *The Interpretation of Ordinary Landscapes*, New York and Oxford: Oxford University Press.

Meyrowitz, J. (1985) *No Sense of Place*, New York and Oxford: Oxford University Press.

Real, M. (1977) *Mass-Mediated Culture*, Englewood Cliffs: Prentice-Hall.

Relph, E. (1976) *Place and Placelessness*, London: Pion.

Romano, C. (1983) 'But was he a Marxist?', *Village Voice*, 29 March: 41–2.

Rooney, J.F., Zelinsky, W. and Louder, D.R. (eds) (1982) *This Remarkable Continent: An Atlas of United States and Canadian Society and Culture*, College Station: Texas A&M University Press.

Rosenthal, J. (1988) 'Who practices hegemony? Class division and the subject of politics', *Cultural Critique* 9: 25–52.

Ross, A. (1989) *No Respect: Intellectuals and Popular Culture*, New York and London: Routledge.

Sack, R.D. (1988) 'The consumer's world: place as context', *Annals of the Association of American Geographers* 78: 642–64.

Sauer, C.O. (1963) [1925] 'The morphology of landscape', in J. Leighly (ed.) *Land and Life: Selections from the Writings of Carl Ortwin Sauer*, Berkeley and Los Angeles: University of California Press, 315–50.

Shields, R. (1989) 'Social spatialization and the built environment: the West Edmonton Mall', *Environment and Planning D: Society and Space* 7: 147–64.

Soja, E. (1989) *Postmodern Geographies*, London and New York: Verso.

Stephanson, A. (1988) 'Regarding postmodernism – a conversation with Fredric Jameson', in A. Ross (ed.) *Universal Abandon? The Politics of Postmodernism*, Minneapolis: University of Minnesota Press, 3–30.

Stevenson, R.W. (1990) 'Disney stores: magic in retail?', *New York Times*, 4 May: D1.

Tuan, Y.-F. (1977) *Space and Place: The Perspective of Experience*, Minneapolis: University of Minnesota Press.

Wagner, P.L. and Mikesell, M.W. (1962) *Readings in Cultural Geography*, Chicago: University of Chicago Press.

Wallace, M. (1985) 'Mickey Mouse history: portraying the past at Disney World', *Radical History Review* 32: 33–57.

Williams, R. (1976) *Keywords: A Vocabulary of Culture and Society*, New York and Oxford: Oxford University Press.

—— (1977) *Marxism and Literature*, Oxford: Oxford University Press.

Willis, P.E. (1977) *Learning to Labour*, Farnborough: Saxon House.

Young, K. (1987) 'Museum planners aided by Disney', *Calgary Herald*, 22 November: A13.

Zelinsky, W. (1973) *The Cultural Geography of the United States*, Englewood Cliffs: Prentice-Hall.

11

THE DEPARTMENT OF INDIAN AFFAIRS AND NORTHERN DEVELOPMENT
The culture-building process within an institution
Paul Kariya

INTRODUCTION

In June 1985, the Government of Canada passed Bill C-31 and repealed those sections of the Indian Act through which Indian women who married non-Indian men lost their Indian status. Furthermore provision was made for entitlement to reinstate Indian status and Indian band membership to those who had lost or been denied their status. In its preliminary estimates, the government announced that approximately 70,000 persons, primarily women and children, would be eligible to join the existing population of status Indians. In fact as at October 1989, 123,040 applications had been received, of which 65,433 applicants have been registered, bringing the total number of registered Indians in Canada at the end of September 1989 to 460,359.

Ironically, the Department of Indian Affairs and Northern Development (DIAND), the government agency so identified in Canadian history with the colonial problems of aboriginal people – not the least of which have been racism and discrimination – was now proclaiming that it was ending structural discrimination based on gender. But given the problems of being an Indian in Canada, why would those eligible for reinstatement want to apply to become Indians? An adequate response to this question requires an examination of the unique history and relationship which has existed between Indians and governments in Canada. There has been an ongoing contradiction in Canadian Indian policy between protection, wardship, assimilation and self-reliance. In a broader context, Western (1981) and Anderson (1988) point out that the existence of powerful government institutions defining people and places, and conferring or altering racial identity, is not new and has many parallels in the Western world.

In this essay, I examine the development and nature of the institution which has created, maintained and altered the definition of who is an Indian in Canada. If one accepts a definition of culture as the web of significance which people have created and are suspended in, then the task of this essay is to examine the 'webs' in the DIAND milieu and interpret how they have been created and maintained

(Geertz 1973: 5). Through an examination of the culture-building process and projection of these values upon the Indian world by DIAND, I argue that the social and economic definition of Indians has been a cultural abstraction. Furthermore, the unique and readily identifiable Indian reserve landscape in Canada exists as a constant reminder of this continued abstraction. I am concerned, then, to explore the cultural politics which link the Canadian state and aboriginal peoples, through an examination of the federal department mandated to address Indian concerns. Its frequently confused values are writ large on the cultural landscape of aboriginal Canadians.

BUREAUCRACY AND WORLD-VIEW: THE DEPARTMENT OF INDIAN AFFAIRS

While most social service agencies are established to provide a single service, the Indian and Inuit Affairs Program (IIAP) of DIAND with a 1989 budget of $2.1 billion and 4,250 employees is responsible for providing a wide spectrum of services to Indian people.[1] Instituted by the British imperial government in the mid-1700s, Nicholson (1984: 59) points out that 'Indian Affairs is the oldest continuously operating arm of government in Canada.' Frideres (1983: 227) refers to DIA as

> a 'total' institution in that it has a monopoly on the delivery of services to a captive clientele. Its organization is characterized by specialization, hierarchy, and regimentation, while its clients are uneducated, unspecialized, and varied. By limiting the choices available to its Native clients, the IIAP shapes and standardizes Native behaviour at minimal cost and risk to itself.

As well as the Indian Act, DIA's legislative scope encompasses 13 treaties and 37 other federal acts. Although referring to the US Bureau of Indian Affairs, Anderson's (1974: 250) description of Indian administration might also apply to the Canadian context. He writes:

> The bureaucratic regulation of reservation Indians is nearly monolithic. The Bureau of Indian Affairs is the closest thing to a total institution that any group of American citizens has ever experienced. The government's *Indian Affairs Manuals* fill 33 volumes, standing six feet high.

Interestingly, both Frideres and Anderson refer to the respective Indian Affairs bureaucracies as total institutions. This is a strong categorization, reserved by Goffman (1961), Foucault (1967, 1979) and others for mental health institutions, prisons and similar places of complete social control.

A theory of formal organization

Blau and Scott (1963: 1, 14) note that formal organizations are 'established for the explicit purpose of achieving certain goals,' and that they possess 'a formal status structure with clearly marked lines of communication and authority'. In terms of service delivery to a clientele, the bureaucratic approach is a particular strategy characterized by rationally organized action. Explicit is the separation between organizational and social lives, so that a person's needs (and a person) can be redefined within the role of client. Based on Weber's characteristics of an ideal-typical bureaucracy, superficially, one might conclude that DIAND is organized according to bureaucratic principles. Within DIAND there are specialist sub-organizations (i.e. Education, Economic Development, etc.); access to services are via formal criteria and standards; service delivery is decentralized; the organization is hierarchical in structure; functionaries as public servants receive limited rewards as office holders; performance standards exist; and career advancement opportunities are according to Public Service Commission regulations. Upon closer scrutiny, however, one finds many non-specialists occupying various positions (e.g. former schoolteachers working as economic development officers), and formal criteria and rules which are set aside to handle non-routine cases. DIAND is established on the principles of an ideal-typical bureaucracy in form and function, but in actuality it does not achieve the ideal. In this sense it could be labelled a symbolic bureaucracy (Jacobs 1969).

Even though positivist organization theorists distinguish between formal organizations and social organizations, the distinction is a tenuous one (Silverman 1970: 8). Post-Weberian theories of bureaucracy have shown that employees do not actually function as mechanized tools; clients approach social welfare agencies without narrowly articulated needs, while organizations are never solely the sum of formal arrangements of consciously co-ordinated activities (Monzelis 1969).

One assumes that social service agencies are generally established to provide services to a defined client group. With a formal mission or mandate, the type and quality of the service to be provided is defined in quantitative terms. Similarly, the eligibility of those who qualify as clients is demarcated rigorously. But, behind this opportunity approach to dispensing service and meeting needs, there exists what Galper (1975: 45) labels a contradiction. He notes that, at the same time as these agencies are trying to alleviate need and thereby liberate people, they also frustrate, deny and undermine the possibilities of human liberation (Handlemann and Leyton, 1978). Once in operation, a complex dynamic of client need, frustration, dependency and fatalism conflict with agency self-perpetuation, need redefinition and administration. Weber (1947: 225) explains that, ideally,

> The dominant norms are concepts of straightforward duty without regard to personal considerations. Everyone is subject to formal equality of treatment; that is, everyone in the same empirical situation. This is the spirit in which the ideal official conducts his office.

However, reality is very different, such that in many cases social service agencies embody a denial of their own best ideals (Henry 1973; Scheff 1966) DIAND could be characterized in this manner.

There exists a dialectic in bureaucratic relations between formal rules and social situations. Social life, action and organization occur in, and are a part of, formal organizations. As Silverman (1971: 23) explains:

> Any theory of organizations must specify the nature of their relationships with the wider society. Secondly, if similar processes may arise both in organizations and in other social institutions, then this suggests the inseparability of theories of organizations from theories of society and the need to pay attention to the theoretical orientations available from the study of the latter.

This reformulation focuses upon the dynamism and changing nature of organizations as ongoing and continuing processes. It does not matter if there are conflicting goals or apparent inconsistencies. What is important is that organization members behave or act as if the organization has consistent goals. As Thomas (1928: 572) has often been quoted: 'It is not important whether or not the interpretation is correct – if men define situations as real, they are real in their consequences.'

The contradictions of the Department of Indian Affairs

DIAND is an organization which could be characterized as caught within Galper's contradiction between the expressed goals of 'being helpful to Indians', while being perceived as contributing to the opposite. Rarely has an institution been as heavily criticized as DIAND, with the attack seemingly coming from all sides – Indians, Parliament, the media, academics and the public (Canada 1982; Ponting and Gibbons 1980; Ryan 1978). Despite having over two centuries to perfect its role, the ineffectiveness and possible inappropriateness of DIAND is illustrated by the appalling socio-economic level of living statistics which characterize Indian reserves today. In 1986, while only 17.1 per cent of the general Canadian population had less than a grade nine level of education, for the on-reserve Indian the figure was 44.7 per cent. While the employment rate for Canadians was 59.6 per cent in 1986, for the on-reserve Indian it was 28.2 per cent. Where 28.9 per cent of all on-reserve dwellings had more than one person per room (a census-based overcrowding index), for the rest of Canada the level was 1.7 per cent. While one might question the appropriateness of the goals which underlie the collection of these statistics, the point to be noted is that the improvement of the Indian standard of living against Western ideals has been the expressed objective of DIAND. It has failed in what it was supposed to do.

DIAND as a bureaucracy responsible for providing education, social welfare, community development, policing, housing, administration, economic development and trust services to status Indians has gone through numerous reorganizations, mandate reviews and policy changes over the years, and yet the basic fabric of its function remains – the administration of Indian affairs. Though patterned on a bureaucratic model, the functioning of DIAND is neither effective nor client-oriented. There are many internal and external factors which prevent this. Historically DIAND and, by extension, the federal government have not clarified policy objectives pertinent to Indians and, when they have attempted to do so, have not been able to eliminate conflicting objectives. As an example, while the current policy of self-government is commendable, its effectiveness and the sincerity of the government are undermined by an archaic legislative base, the Indian Act. Throughout the history of Indian administration in Canada, the ambivalence between protection, assimilation and self-reliance policy objectives has been evident (Tobias 1976).

Furthermore, contrary to the monolithic image presented in programme guidelines and service delivery manuals, the client group is not homogeneous, and the demands and requirements of Indian people vary widely. Juxtaposed to this heterogeneity is the fact that DIAND is responsible for a wide diversity of programmes. Add the ingredient of a client group struggling to lift the shackles of post-colonialism and to express a degree of self-reliance and one finds the paradox of a group unwilling to be served yet demanding service results from DIAND. Paton (1982: 23) aptly explains that for DIAND staff, 'as task uncertainty increases, the number of exceptions increases, until the hierarchy is overloaded'.

Adding to the ineffectiveness of the department is the ever increasing contradiction of paternalism. The government official working through government programmes cannot create Indian governments or Indian self-reliance. But if this is the case, is DIAND irrelevant, and how is it that it endures and functions? Part of the reason lies in the demands and expectations of many status Indians, particularly treaty Indians, that DIAND exist in perpetuity as a symbol of the agreements originally established between Indians and the Crown. From another perspective, DIAND exists as a buffer between the government decision-making machinery and a vocal client group. The departmental officials are really caught between the demands of the clients and the central government power brokers. Included in this latter group are the Inner Cabinet and senior officials from the Privy Council Office, Treasury Board Secretariat and Ministry of Finance. The contradiction of paternalism has come full circle where, as much as the Indians need DIAND, the federal government also needs DIAND as an intervening institution (Fudge 1983).

THE MAKING OF AN INSTITUTIONAL CULTURE

The transfer of responsibility for Indian Affairs from department to department is a revealing study of the place and meaning of Indians to policy-makers in Canadian history. This section provides an overview of the broad policy phases which the Canadian government has embraced in its relationship with Indian people. Officials of the government have sought through various policies to prepare the Indians for their place within an idealized Canadian society.

DIA within the federal bureaucracy

The Department of Indian Affairs and Northern Development is not a large federal ministry when compared to others. But since 1966, when it was combined with Northern Development, Indian Affairs has existed as a separate and independent department. During this time it has received publicity and attention far beyond its relative size and importance to the government bureaucracy. However, the relative autonomy of Indian Affairs has not always existed.

Over the last century, the administration of Indian concerns has changed hands no less than ten times between various federal ministries and departments. Among the incongruous departments Indian Affairs has been associated with have been Federal Supply and Services, Citizenship and Immigration, and Mines and Resources, denoting the shroud of colonialism which has shifted native people and their territory from sovereign independence to protectorates as mere third parties in intergovernmental or private enterprise–government negotiations. What these actions and movements express is an expediency towards how Indian administration has been conducted by government and related financial interests. Consider the placement of the Indian Affairs Branch within the Department of Interior in 1873. The primary mandate of this department was the disposal of Crown lands and the opening up and settlement of the western frontier. A definite conflict of interest existed where the advocate for the Indians was also the promoter and facilitator of increasing white settlement of Indian territories.

The meaning, importance and political strength of Indian people to the larger society have been reflected in the federal ministries which have claimed responsibility for Indian Affairs during the last century. It is no coincidence, therefore, that the Indian Affairs Branch has twice been subsumed by an immigration ministry. Another curious partnership was the tie with mines, minerals and natural resources. Most likely, the connection was one which linked Indians with the remote locations of natural resources.

As a further expression of the marginal role of Indians within Canada, they were not granted the federal franchise until 1960 when their administration was under the Department of Citizenship and Immigration. Only on the eve of Canada's centennial celebration, when the world's eyes turned to view the cultural mosaic of Canada, was Indian Affairs elevated to a semi-autonomous ministry as the Department of Indian Affairs and Northern Development (Sieciechowicz 1983). Viewed within an historical context, Indian administration has been subservient to other dominant interests. The Indian question has commonly been considered a problem to be solved within some broader departmental context. Whether the needs and priorities of government matched those of the Indian people has usually been incidental.

Approaches to the administration of Indian affairs

Rather than discrete phases in Indian policy development in Canada, there have been overlapping and concurrent approaches to Indian administration. While each may have occupied an ascendant position during a specific time period it is difficult to review the subject in a linear manner. The confusion in policy direction has existed because of the tension and ambivalence which have been inherent in the three basic objectives which have been a backdrop to Indian administration. The contradictions between protection, assimilation and self-government have never been resolved. Bouchard and La Rusic (1981: 51) write:

> If there is an ambiguity in the relationship with Indians in Indian Affairs, one will not find it at the level of overall policy. If one ignores the various labels it has been given over the years, overall policy has not changed: to civilize, to christianize, to assimilate, to integrate and to educate have embodied similar objectives. The consistency in policy is that of attempting to make Indians just like average Canadians (even if there is no such entity as an 'average' Canadian).

Indeed, the primary legislation governing Indians and Indian affairs, the Indian Act, has remained virtually unchanged for 100 years. It represents the embodiment of all three objectives noted earlier. However, each has been cast, interpreted and operationalized through programmes at different historic times, each reflecting the priorities and dominant attitudes of that day.

Protection

The British adopted early colonial policies of protecting the Indians from European encroachment upon their lands and from fraudulent trading practices. Boundary lines were established between Indian and European lands. Under the Royal Proclamation of 1763, Indian people of the 'new land' were viewed as sovereign powers by the British Crown – powers under royal protection (Cumming and Mickenberg 1972: 291–2). While Indians were considered to possess absolute title to land, and territory could not be occupied by settlers without royal permission and purchase from them, Indian territory began to disappear. Entrepreneurial corporations including the Hudson's Bay Company and the Northwest Company fostered a spirit which considered the new land and resources as theirs for the taking. Where they constituted the legal authority of the land, their factors saw no conflict of interest between respecting and interpreting British law and improving their economic circumstances. Despite this erosion, early legislation was passed to protect the Indian's land and use rights from violation by private financial interests and even government-endorsed settlement and expansion goals (Allan 1971: 17–20). As an example, consider two early pieces of legislation:

10 George IV (1829) Cap. 3 (Upper Canada). An Act the Better to Protect

the Mississauga Tribes, Living on the Indian Reserve of the River Credit, in their Exclusive Right of Fishing and Hunting therein [passed 29 March 1829].

13 & 14 Victoria (1850) Cap. 42 (Province of Canada). An act for the Better Protection of Lands and Property of the Indians in Lower Canada [passed 10 August 1850].

Concurrent with the need to protect the Indian from sovereignty encroachment was the humanitarian ambition of protection from the poorest examples of European civilization evident on the frontier. Drink, greed, prostitution, dishonesty and violence were manifest aberrations of white society which Indians were supposed to be shielded from. Again, consider an early piece of legislation which cast the Indian as a ward of the government:

20 Victoria (1857) Cap. 26 (Province of Canada). An Act to Encourage the Gradual Civilization of the Indian Tribes in this Province, and to Amend the Laws Respecting Indians [assented to 10 June 1857].

Laws prohibiting the sale of Indian land, Indian consumption of alcohol and prostitution of Indian women were enacted. Harper (1945: 132) notes that the reserve system was also a device instituted to isolate and protect the Indians while at the same time providing a means to 'civilize' them to partake of the 'white man's freedom to exploit the vast riches of a growing dominion'.

Though it is an unpleasant term to both the Indian and government officials, the notion of 'wardship', if not an explicit theme, has been an implicit one from the beginning of Indian administration in Canada. Protecting and defending the Indian who was thought to be unable to cope with European civilization and settlement led to the creation of a paternalistic administrative structure. The government through its Indian administration branch, it was posited, would control and supervise the Indians in their best interest (Miller *et al.* 1978: 100).

Today the Indian Act remains the primary legislative instrument to explain the legal basis of Indian administration in Canada. It is the rule book explaining who is an Indian, what benefits s/he has as such, the nature of reserve administration, the ownership and alienation of such lands, terms of development, and so on. A sample of the power of this act can be gleaned from the following examples:

No Indian is lawfully in possession of land in a reserve unless with the approval of the Minister possession of the land has been allocated to him by the council of the Band.

(Section 20, 1)

'Indian' means a person who pursuant to this Act is registered as an Indian or is entitled to be registered as an Indian.

(Section 2, 1, c)

This document makes it abundantly clear that the Indian and his or her rights are ultimately subject to the Minister of Indian Affairs whose decisions are made day to day by civil servants in the Department of Indian Affairs.

Even before the passage of the Indian Act in 1867, one finds a pattern of Indian administration developed based on administering land and people as protectorates and as wards.[2] A current-day outcome of this is that Indian reserve land is not owned by its occupants but by the Crown in right of the Indians. Hence, Indian reserves cannot be pledged directly as security for mortgages and loans. Part of the economic development problem for Indians has been an inability to raise capital and a key reason is that they have limited assets to pledge. For the most part, the principles of protection in subsequent Canadian Indian policy were fully established before confederation in 1867.

Assimilation

The second major government policy objective was 'civilization' and assimilation. Protection was a conditional step or process towards teaching the Indian to read and write, to be debt free and to lead a moral upstanding life. At the conclusion of this process, all legal distinctions between Indians and other Canadians were to be unnecessary. Once the Indian was ready, s/he could then function in society like other Canadians. Tobias (1976: 16) notes that this 'paradox of isolation and integration was to become and remain a characteristic of Canada's Indian policy'.

Where protection was the primary pre-confederation objective of British government policy, assimilation was the long-term post-confederation goal. In 1869 the goals of civilization and assimilation were added to the corpus of colonial legislation in Canada by the passage of an Act for the gradual enfranchisement of Indians (Statute of Canada 1869 32–3 Victoria, Chapter 42). Included in the statutes was the provision that the Governor in Council could shape an Indian local government with Euro-Canadian ideals. Indian councils could make by-laws, but they were subject to approval of the Superintendent-General (the Minister) of Indian Affairs. The notion of elected officers, a strong tenet of 'civilization', was a conditional responsibility, since the Governor in Council could remove any elected Indian considered unqualified or unfit (Tobias 1976: 17). While promoting assimilation, the new legislation incorporated the protective features of previous legislation. It was no different when the 'Act to amend and consolidate the laws respecting Indians' or the Indian Act was passed in 1876.

In 1920, the Deputy Superintendent-General of Indian Affairs speaking of proposed changes to the enfranchisement provisions of the Indian Act stated:

> I want to get rid of the Indian problem.... Our object is to continue until there is not a single Indian in Canada that has not been absorbed into the body politic and there is no Indian question, and no Indian Department.
>
> (Miller *et al.* 1978: 14)

The objective was to be in business until Indian customs, beliefs and languages were eradicated and Indians had become like other Canadians. Inducements were provided to encourage the process, which were created by, and more likely to appeal to, those already possessing a Euro-Canadian value system. As an example, the 'location ticket' was particularly aimed at more sedentary eastern and prairie Indians. Concomitant with Euro-Canadian values was the concept of real property and sedentary ties to it. Individual Indians could receive an assignment of land via a location ticket issued from the Superintendent-General. After a three-year probationary period, and evidence of occupation similar to the fashion that a Euro-Canadian might occupy land, the Indian could be enfranchised and given fee simple title to his land. The exception to this provision was in British Columbia, where the provincial government took the position that the land of enfranchised Indians reverted back to the province. What was considered an unenlightened position on the part of the province by federal authorities probably prevented more enfranchisements in British Columbia (Raunet 1984: 131).

Another means of acquiring a location ticket and enfranchisement without a probationary period was if an Indian earned a professional degree, and designation as a doctor, lawyer, minister or teacher (Tobias 1976: 18). It is clear that the government wanted to eliminate Indian political, economic, religious and educational systems. With enfranchisement, Indians would also be given their share of the former reserve lands and this would lead to the eventual disappearance of separate Indian lands.

Education of the Indian child as a 'civilizing' process has been a vital government concern. But early reserve schools were not well attended and were thereby rendered ineffective as a tool of assimilation. To combat this, in 1894 the Indian Act was amended to authorize the Governor in Council to establish school regulations which he thought necessary to empower him to commit children to residential and industrial schools founded by the government or religious orders.

Almost from the beginning of missionary contact, education next to religion has been considered an important priority in redirecting Indian culture. Harper (1945: 122; see also Ponting and Gibbons 1980) notes the history of collaboration between the federal government and the Church in the area of Indian education. The religious and government residential schools, which most adult Indians living today were sent to, were operated on the philosophy of assimilation through isolation. Children were free from the distractions of home life on the reserves. They were also protected from the distractions of the wider, particularly urban, society. Through the study of academic or vocational subjects and with religious education, the residential school graduate was expected to be able to step out into the mainstream of society. Unfortunately, the results were students basically ignorant of Indian society and unable to cope with modern Canadian society. The residential school when compared to the larger public school system also tended to equip the Indian student with a poorer quality of education. The small number of graduates were unqualified to advance to higher education and unable to compete in the labour market for jobs. The residential schools and the association between the federal government and the churches in administering Indian education began to be phased out in the late 1950s. Ponting and Gibbons (1980: 20) note, however, that education still remains a tool of assimilation.

Numerous explanations have been put forward explaining the failure of the colonial and Canadian governments' assimilation policies. They include the geographic isolation and dispersion of the Indian population; the linguistic and cultural diversity of aboriginal peoples; and social discrimination and enforced labour-force marginality practised by the dominant society and embodied in government policy. This latter reason in many ways encompasses all of the others.

Having established protective policies and separate reserves for Indians, it was impossible to introduce effective assimilation policies. Setting a people apart and legalistically and socially labelling them as Indians established a cycle of social reinforcement within the Indian communities and also within the dominant Euro-Canadian society. The people caught within the net of government policy objectives were no longer true Indians, nor could they be ordinary Canadians. By the very fact of special legislation and separate treatment, they had become people in limbo and assured of continuing in such a position with the creation of a government bureaucracy to enforce the various policies (Fisher 1976: 460–2). Possibly if the protectionist policies had been repealed once the assimilation policies had been enacted, then there might have been greater success. However, this was not the case and protection and assimilation were maintained as concurrent but ultimately conflicting goals.

Self-government

Contrary to Long *et al.* (1982: 190), who claim that the most recent Canadian policy has indicated an attempt to 'break away from the policy paradigm that has dominated the federal government's relationship since the beginning of confederation', one might note that Canadian Indian policy has from the earliest period included provision for some form of local self-management. Doerr (1974), Marule (1978) and Weaver (1981) would further argue that there has not been a policy paradigm shift at all, only a continuation of the same protection and assimilation objectives disguised with a new lexicon. As a means to accelerate the destruction of indigenous political systems, early Canadian policy encouraged the trappings of self-government under a British municipal model. In 1884, an Act was passed 'for conferring certain priviledges on the more advanced bands of Indians of Canada with the view of training them for exercise of Municipal Affairs' (SC 1984 47 Victoria, Chapter 28). The bill which became known as the Indian Advancement Act extended the powers of the band council beyond those of the Indian Act (Tobias 1976: 19). Councils could levy taxes on the real property of band members and were given expanded powers over police and public health-and-safety matters.

Harper (1945: 120) cites self-sufficiency as having been an important early policy objective of the Department of Indian Affairs. He quotes from departmental instructions sent to Indian agents in 1933: 'It may be stated as a first principle that it is the policy of the department to promote self-sufficiency among the Indians and not to provide gratuitous assistance to those Indians who can provide for themselves.'

Negating this type of policy were the contradictory policies pertaining to protection which provided for Indians as government wards. Furthermore, it is one thing to issue enlightened policy, but its implementation is another matter, particularly in the nether reaches of the country. Indian Agents whose jobs depended upon Indian dependency were not necessarily the best advocates of Indian self-sufficiency.

Prior to the 1884 Indian Advancement Act, the department despatched a circular to all Indian Superintendents and Agents asking them to report whether the bands under their supervision were 'sufficiently enlightened to justify the conclusion that . . . a simple form of municipal government among them would be attended with success'. The response from the field as recorded in the DIAND Annual Report (1880) states that, 'From the majority of its officers who have replied to the circular, the reports received lead to the conclusion that the Indian bands within their respective districts are not sufficiently advanced in intelligence for the change.'

The Indian Act makes provision for Indian participation in the local government of their reserves, but as Harper (1946: 313) notes, 'the Act is a non-democratic document, because it reflects so little faith in the Indians'. Almost every power or delegation of responsibility defined in the Act has a 'subject to' clause or a caveat of necessary ministerial approval. Section 83 is available to bands which have reached 'an advanced stage of development' to pass taxation by-laws, but even being permitted to operate under this section is a ministerial perogative. Similarly, Section 81 outlines the power of councils. Here, the Minister does not approve band council by-laws, but retains the power to reject them.

Indians unlike other Canadians are faced with a narrowly circumscribed form of local-level government through one federal department which administers the Indian Act. In its broad coverage its goals have been more oriented to social control than to self-government. A former Assistant Deputy Minister of the Indian Affairs Branch described it well when he stated (Doerr 1974: 40):

> The Indian Act is a Lands Act. It is a Municipal Act, an Education Act and a Societies Act. It is primarily social legislation, but it has a very broad scope: there are provisions about liquor, agriculture and mining as well as Indian lands, band membership and so forth. It has elements that are embodied in perhaps two dozen different acts of any of the provinces and overrides some federal legislation in some respects. . . . It has the force of the Criminal Code and the impact of a constitution on these people and communities that come within its purview.

However, implicit in the creation of an economic development programme in the 1950s and a retitling of the Community Affairs section to the Local Government Unit in the 1970s, DIAND has attempted to promote a measure of Indian self-sufficiency and self-government. Apart from the substantive issues involved, recent constitutional debates have left many Indian leaders and public officials questioning the meaning and utility of terms such as 'self-determination', 'self-government' and 'self-sufficiency'. The existing Indian Act provides a framework for local band management and self-administration of existing policies and programmes, which new legislation might take further to permit the creation of municipal-type governments. But this is not what most Indian leaders want, many of whom have placed on the bargaining table a demand for more in-dependent government status.

Although some definition of self-government or self-management has existed in Canadian Indian policy, its effectiveness and purpose have been questionable. The paradox of Canadian policy is that it has embraced a protection, assimilation and self-sufficiency stance in a cumulative manner. Successive policy phases have not been tested and discarded, but augmented. Indian policy today is rife with inconsistencies. Consider the fact that the department attempts to treat band councils as band governments or legal entities; however, their legal status remains in doubt.

The Indian Act in 1992 remains little changed from the Indian Acts of 1951 or 1924 and earlier. It is still the legislative backdrop to Indian policy in Canada. Bartlett (1980: 6) notes that, despite a century of continual revision and modification, the legislators, 'were always preoccupied with details and never contradicted the basic rationale of the Indian Act, which demanded civilization and responsibility from the Indian population while denying them control over the forces affecting their lives'. Berger (1981) notes that the Act has legally and constitutionally accorded Indians a special status but at the expense of true equality in the Canadian mosaic.

PROJECTION OF VALUES TO THE INDIAN WORLD

The message to Indian people from DIAND and the federal government on its Indian policy remains unclear. Even though formally announced policies and programmes refer clearly to support for Indian self-government, partnerships in economic development, and consultation on a changing formal relationship, recent actions indicate an ongoing dilemma of mixed signals. Despite the overall expression of positive change, the values projected to the Indian world by DIAND are in conflict. Two examples illustrate what Indian leaders have recently witnessed.

On 12 June 1989, the Minister of DIAND and the Minister of State for Small Business and Tourism announced the new Canadian Aboriginal Economic Development Strategy (CAEDS), which over the next five years would provide $873.7 million ($474.7 million through DIAND) to increase the number of aboriginal businesses, managers and entrepreneurs. More important than the funding, the Minister of DIAND stated, 'This will mean continuing the change in the role of Indian economic development in my Department – from one of direct program delivery to one of facilitating the capacity of Indians and Inuit to deliver more economic development programs' (Canada 1989: 10). Funding and programmes for the DIAND components were to be delivered by Indian bands and tribal councils. Moreover, except for some broad reporting requirements, programme design decisions would be left to local Indian authorities. Paramount was the objective of supporting Indian governments to deliver the services at the local level.

Despite an initial flurry of government activity, including press conferences and announcements, the Indian response to CAEDS has been mute. Many Indian leaders have expressed optimism about the new strategy but remain guarded in their assessment. Part of the scepticism may be linked to the fact that while the press announcements mentioned funding of $474.7 million over five years, this total included already approved funding. In terms of new funding, the annual increase was only to be in the order of $25 million per year over five years. Indian leaders have also been concerned about a phrase which is starting to be heard quite frequently from DIAND officials whenever a funding issue arises: 'You control the purse strings now; it is up to you to establish the priorities of what you want to fund.' This type of response may be acceptable where budgets are sufficient, but the issue of sufficiency, given needs, remains in doubt. Overall, however, Indian leaders have expressed support for CAEDS as the direction in which the Indian–government relationship should be progressing.

A more critical reason why CAEDS has received a cautious response from Indians could be related to another DIAND policy announcement which was made by the Minister prior to the CAEDS release. On 20 March 1989, the Minister announced changes to the Post-Secondary Students Assistance Programme. While it was introduced with a positive preamble, that the new policy would enhance the educational opportunity for Indian people and promote the development of self-government and economic self-reliance, almost all Indian leaders and other commentators reacted negatively to the proposals. To Indians this announcement was nothing more than a thinly disguised way of cutting funding and Indian entitlement. Possibly CAEDS was being tarred with the same cynicism by Indian leaders as was the Post-Secondary Students Assistance Program.

After announcing that 'for the government, the position is crystal clear – treaty references to education do not include post-secondary education' (*Edmonton Journal*, 22 March 1989), the Minister's release went on to outline some twenty changes to funding and entitlement categories under the programme for Indian post-secondary students. Major changes included a reduction of support from the existing 96 months for up to three degrees to 48 months with only consideration for additional postgraduate support. Categories for living allowances and child care support were to be reduced for certain single and married students. The choice of colleges was also not to be without geographical limits, particularly if programmes were offered at local institutions.

While the changes announced were perceived negatively, one should examine the whole set more carefully. Of the 20 proposed changes, 18 were targeted to result in increased benefits to students. As examples, incentive-based scholarships of $3,500 were provided for certain fields of study such as the pure sciences and business professions. Similarly, incentive awards of $1,500 were to be available for the top 5 per cent of the students receiving support. But the Indian response was sit-ins at government offices, demonstrations in many major urban centres and most dramatically a hunger strike by students. What went wrong? How did all the good news and the new era of consultation and co-operation get lost?

In analysing the background to the case, one finds that there is quite a significant difference between the values projected and those intended to be projected by DIAND. Why would the Minister attempt to communicate the changes to Indians as good news when indeed the indications were that any tinkering with the programme would be interpreted negatively by the clients? Indeed why would the communications packaging not have been more skilful? Some of the errors are probably attributable to myopia within DIAND ranks. If senior management is perceived as wanting a certain type of approach, such as cost reduction and financial controls, then certain options and negative concerns are automatically filtered out. During the generation of option papers and internal reviews, the positive, hoped-for scenario becomes stronger by inbred consensus. There is little incentive and reward in backing what management does not want to hear. Also as DIAND has changed from a service delivery field organization to a funding transfer agency based in Ottawa and major regional centres across the country, it has lost the information and intelligence contact it used to have with the Indian community. There are no longer district office staff to read the local-level political trends. Plainly, many DIAND officials believed that, as in previous eras, *they knew what was best* for the Indian or at worst that they could control any Indian reaction.

Moreover, in Ottawa, it is clear that departmental officials were caught within a larger political arena. Even though senior DIAND officials were concerned about escalating costs, much of this agenda was driven by officials at the central agencies (Treasury Board, Finance and Privy Council) who saw programmes in light of the government's broader fiscal framework. What they pointed out to DIAND managers was that in 1977 there were 3,500 Indian and Inuit post-secondary students receiving $9 million in government assistance, compared with 15,000 students receiving assistance at $130 million in 1988. The paradox is that only two decades ago grave concern was expressed over the lack of qualified university-level Indian applicants. Now, due to great success, government officials wanted to reduce the intake to reduce costs.

In defending the revisions, DIAND officials explained that full consultations had been undertaken with Indian leaders and that the objective was also to improve the programme and student success rates. It was further explained that, despite the large number of students entering the programme, the success rate to graduation remained poor. Having set the drop-out rate as the problem to be solved, to then focus upon the status of post-secondary education as not a treaty right is a surprising diversion. The message seemed to be that, as long as there were only nominal costs involved, programmes for Indians would be available.

Even from a communications perspective, the approach and release were mishandled. To lead with the treaty issue, with education and children perceived as targets, is to wave a red flag in the face of the Indian politicians. As the Commissioner for Human Rights noted, in contemporary Canada, an Indian youth has a far greater chance of going to prison than s/he does of going on to university. George Erasmus, the Grand Chief of the Assembly of First Nations, pointed out that the Minister was 'ill advised by departmental officials', who clearly never anticipated the reaction (*Globe and Mail*, 20 April 1989).

The values presented to the Indian world by DIAND remain ambiguous. Despite progress made in the areas of land claims, economic development and self-government, the underlying conflicting values of protection and assimilation remain. Within DIAND and the federal government, the contradictory values intimated by the new economic development programme and the Post-Secondary Assistance Programme, bear witness to this dilemma.

CONCLUSION

DIAND, as the government agency charged with Indian affairs, is in a state of contradiction because it has the responsibility to assist Indian people while in many cases it is perceived as the source of native problems. While recent policy changes and new statistical releases indicate some improvement in the social and economic level of living conditions of Indians and reserve communities, the gap between aboriginal people and mainstream Canadians remains very wide. A primary reason for this is that Indians have not been able to occupy a place they have been able to define for themselves in Canadian society. This has always been mediated for them by another, all too often DIAND. In this sense Indians have been, and continue to be, defined in someone else's image.

The bureaucracy of DIAND itself is as much a captive as are the Indian clients. The policy legacy in protection, assimilation and self-reliance is a reflection of the confusion which rests with senior officials and politicians in the government. Strong proactive steps to rectify aboriginal problems are seen as too unpalatable since they entail great financial cost and political effort. Similarly, strong negative actions (as perceived with post-secondary education) are emerging as costly political options in a liberal society. Given this, a stalemate has emerged, with incremental change the philosophy. It is in this sense that Indians remain a cultural abstraction in Canada.

NOTES

1   IIAP refers to the Indian and Inuit Affairs Program, which is the largest of the three programmes within the Department of Indian Affairs and Northern Development. The other three programmes are Northern Affairs, Transfers to Territorial Governments and Administration.

2   There are many legislative Acts concerned with Indian administration which pre-date this Act; however, it is generally cited as the first consolidated piece of legislation pertaining to Indian affairs and known as the 'Indian Act 39 Victoria (1876) Cap. 18 (Canada). An Act to Amend and Consolidate the Laws respecting Indians (The Indian Act) / assented to April 12, 1876'. Major revisions were incorporated in 1880, 1951, 1970 and 1986.

REFERENCES

Allan, R.S. (1971) *A History of the British Indian Development in North America (1755–1830)*, Ottawa: Department of Indian Affairs and Northern Development.

Anderson, C. (1974) *Toward a New Sociology*, Homewood: Dorsey Press.

Anderson, K. (1988) 'Cultural hegemony and the race-definition process in Chinatown, Vancouver: 1880–1980', *Environment and Planning D: Society and Space* 6: 127–49.

Bartlett, R. (1980) *Indian Act of Canada*, Saskatoon: Native Law Centre, University of Saskatchewan.

Berger, T. (1981) *Fragile Freedoms: Human Rights and Dissent in Canada*, Toronto: Clarke, Irwin.

Blau, P. and Scott, W.R. (1963) *Formal Organizations: A Comparative Approach*, London: Routledge & Kegan Paul.

Bouchard, S. and La Rusic, I.E. (1981) *The Shadow of Bureaucracy: Culture in Indian Affairs*, Montreal: Consulting Services in Social Services.

Canada (1851–7) *Statutes of the Province of Canada*. Toronto and Quebec: Stewart Derbyshire & George Desbarats.

—— (1880) *Annual Report, 1880*, Ottawa: Queen's Printer.

—— (1887) *Acts of the Legislatures of the Provinces now Comprised in the Dominion and of Canada*, Ottawa: Brown Chamberlain.

—— (1967–88) *Annual Reports, Department of Indian Affairs and Northern Development*, Ottawa: Queen's Printer/Dept of Supply and Services.

—— (1970) *The Indian Act, RS, c. 149, s.1*, Ottawa: Information Canada.

—— (1982) *Auditor General's Report to the House of Commons for the Fiscal Year Ended 31 March 1982*, Ottawa: Dept of Supply and Services.

—— (1985) *Statutes of Canada, 1984: RG–10 Vol. 3947, File 123, 764–3 Vol. 6908, File 470-2-3, Vol. 11*, Ottawa: National Archives.

—— (1989) Canadian Aboriginal Economic Development Strategy, *Transition* 2(7), 19 July: 10.

Cumming, P.A. and Mickenberg, N.H. (1972) *Native Rights in Canada*, Toronto: Indian-Eskimo Association of Canada and General Publishing.

Doerr, A.D. (1974) 'Indian policy', in S. Wilson (ed.) *Issues in Canadian Public Policy*, Toronto: Macmillan, 36–54.

Fisher, R. (1976) *Contact and Conflict: Indian–European Relations in British Columbia, 1774–1890*, Vancouver: University of British Columbia Press.

Foucault, M. (1967) *Madness and Civilization*, London: Tavistock.

—— (1979) *Discipline and Punishment: The Birth of the Prison*, trans. A. Sheridan, London: Penguin.

Frideres, J.S. (1983) *Native People in Canada: Contemporary Conflicts*, Scarborough: Prentice-Hall Canada.

Fudge, S.K. (1983) 'Too weak to win, too strong to lose: Indians and Indian policy in Canada', *B.C. Studies* 57: 137–45.

Galper, J. (1975) *The Politics of Social Services*, Englewood Cliffs: Prentice-Hall.

Geertz, C. (1973) *The Interpretation of Cultures*, New York: Basic Books.

Goffman, I. (1961) *Asylums*, Garden City, NY: Doubleday/Anchor.

Handelman, D. and Leyton, E. (1978) *Bureaucracy and World View: Studies in the Logic of Official Interpretation*, St John's: Memorial University.

Harper, A.G. (1945) 'Canada's Indian administration: basic concepts and objectives', *America Indigena* 4: 119–32.

—— (1946) 'Canada's administration', *America Indigena* 5: 297–314.

Henry, J. (1973) *On Sham, Vulnerability and Other Forms of Self-Destruction*, New York: Vintage.

Jacobs, J. (1969) 'Symbolic bureaucracy: a case study of a social welfare agency', *Social Forces* 47: 413–22.

Kariya, P. (1987) 'The Indian reserve as a negotiated reality', unpublished Ph.D thesis, Department of Geography, Clark University, Worcester, Mass.

Long, A.J., Little Bear, L. and Boldt, M. (1982) 'Federal Indian policy and Indian self-government in Canada', *Canadian Public Policy* 3: 189–99.

Marule, M.S. (1978) 'The Canadian government's termination policy: from 1969 to the present day', in I.A.L. Getty and D.B. Smith (eds) *One Century Later*, Vancouver: University of British Columbia Press, 103–16.

Miller, K.T. *et al.* (1978) *Historical Development of the Indian Act*, Ottawa: Department of Indian Affairs and Northern Development.

Mills, C.W. (1956) *The Power Elite*, New York: Oxford University Press.

Monzelis, N. (1969) *Organization and Bureaucracy*, Chicago: Aldine.

Nicholson, D. (1984) Indian government in federal policy: an insider's views', in L. Little Bear, M. Boldt and J.A. Long (eds) *Pathways to Self-Determination: Canadian Indians and the Canadian State*, Toronto: University of Toronto Press, 59–64.

Paton, R. (1982) *New Policies and Old Organizations: Can Indian Affairs Change?*, Ottawa: Carleton University.

Ponting, J.R. and Gibbons, R. (1980) *Out of Irrelevance: A Socio-Political Introduction to Indian Affairs of Canada*, Toronto: Butterworth.

Raunet, D. (1984) *Without Surrender, Without Consent*, Vancouver: Douglas & McIntyre.

Ryan, J. (1978) *Wall of Words*, Toronto: Peter Martin Associates.

Scheff, T. (1966) *Being Mentally Ill: A Sociological Theory*, Chicago: Aldine.

Sieciechowicz, K.Z. (1983) 'Activism to localism', paper presented at the Eleventh Meeting of the International Congress on Anthropology and Ethnography Studies, Vancouver.

Silverman, D. (1971) *The Theory of Organizations*, New York: Basic Books.

Thomas, W.I. (1928) *The Child in America*, New York: Knopf.

Tobias, J.L. (1976) 'Protection, civilization, assimilation: an outline history of Canada's Indian policy', *Western Canadian Journal of Anthropology* 6: 13–30.

Warwick, D.P. (1975) *A Theory of Public Bureaucracy*, Cambridge, Mass: Harvard University Press.

Weaver, S.M. (1981) *Making Canadian Indian Policy: The Hidden Agenda 1968–70*, Toronto: University of Toronto Press.

Weber, M. (1947) *The Theory of Social and Economic Organization*, trans. T. Parsons, Glencoe: Free Press.

Western J. (1981) *Outcast Cape Town*, Minneapolis: University of Minnesota Press.

12

MULTICULTURALISM: REPRESENTING A CANADIAN INSTITUTION

Audrey Kobayashi

> Although there are two official languages, there is no official culture, nor does any ethnic group take precedence over any other.

With this statement, then Prime Minister Pierre Elliott Trudeau established in Canada in 1971 an official policy of *multiculturalism*[1] (Canada 1971). Within a framework of official bilingualism, multiculturalism recognized the need to maintain the cultural heritage of all groups within a multicultural population, and the right of members of 'minority' groups to equality with members of the two 'charter' groups, of English and French ancestry. This chapter examines the *institution* of multiculturalism within the federal government in light of often contradictory processes that define and represent Canadian cultural life. Multiculturalism is a focal point for understanding, and theorizing, the relationship between culture and institution.[2]

Multiculturalism has occurred in three broadly defined stages:[3] first, the 'demographic' stage, a state of ethno-cultural diversity towards which no coherent official policy exists; second, the 'symbolic' stage, wherein it is official policy to recognize and promote multiculturalism, without a firm commitment to bring about its objectives; and third, the 'structural' stage, at which legislative reform provides the basis for social change. Whereas Canada has always been multicultural, shifting patterns of immigration made the nation more diverse ethnically, especially as a result of large-scale international migration after the Second World War. This fact was recognized by Trudeau with his symbolic implementation of multiculturalism policy in 1971, formulated in response to public and political debate surrounding the Royal Commission on Bilingualism and Biculturalism (1969; hereinafter the B&B Commission), which strengthened bilingualism, as a gesture towards the rights of French Canadians, and officially repudiated 'biculturalism', as a gesture towards the non-charter groups. Symbolic multiculturalism reached a crisis point a decade later, at which time, while awaiting an imminent announcement of the repatriation of the Constitution,[4] David Ley pondered the 'institutional integrity' of Canada as a plural state. He concluded (echoing Furnivall's assessment of European colonies in Southeast Asia)

205

that the country was 'a crowd not a community' (Furnivall 1956: 307–8; quoted in Ley 1984: 107) and that political divisions create plurality along a number of dimensions, including the 'multicultural'. Canadian unity, he suggested, takes the 'singularly unheroic' form of celebrating diversity. Multiculturalism policy is therefore limited to 'contributing to the consumption options of the middle class' (p. 106), while the plight of native peoples, the alienation of Quebec and the marginalization of other special interest groups represent a more ominous potential for challenging the national will. Ley's article was both a harbinger of the social difficulties that were to arise during the 1980s, Canada's constitutional decade, and a ringing indictment of the results of a decade of symbolic multiculturalism, interpreted not as a mechanism of equality rights but as an instrument to promote 'red boots' multiculturalism: folk dancing, cultural festivals and ethnic restaurants.

As I begin this essay, nearly a decade later still, the nation anticipates a meeting of the First Ministers on 23 June 1990, at which the *Constitution Amendment, 1987* (hereinafter referred to as the Meech Lake Accord) may or may not be ratified.[5] Public attention to this issue seems to be less indifferent than it was in 1981, probably as a result of recent debate and legal action surrounding the guarantee of equality rights introduced by the Constitution and the Canadian Charter of Rights and Freedoms in 1982. But the negotiation of the constitutional amendment invokes a sad sense of *déjà vu*. While the focus is on the status of Quebec and its inclusion as a 'full partner' in the national federation (Canada 1987a), the issues of 1981 – constitutional amendment processes, federal–provincial powers and differential provincial status, the rights of women, native Canadians and ethno-cultural and linguistic groups outside the French and English 'charter' groups – remain unresolved (but see postscript).

But the past decade has also seen strengthening of multiculturalism as an institution, culminating in the 1988 passage of the Canadian Multiculturalism Act and the ongoing restructuring of the Department of Multiculturalism.[6] Multiculturalism has a new social meaning and a new official character. Emphasis has shifted from heritage to equality, that is, from the first principle of 'celebration of diversity' to which Ley (1984: 106) restricted his argument to the second principle of equality rights. Although there is tension between the two notions, present policy is formulated with a view to strengthening both.

A CULTURAL GEOGRAPHY OF INSTITUTIONS

There has also developed over the past decade a stronger basis within cultural geography from which to address political culture. Cultural geography has become politicized (Cosgrove and Jackson 1987; Gregory and Ley 1988), shifting emphasis to the processes through which relations among cultural groups are negotiated. This shift has resulted in critical reinterpretation of social 'facts', as socially constructed and, moreover, institutionally mediated and contested (e.g. Anderson 1987; Smith 1989b).

A politicized notion of culture has two major analytical dimensions. The first is the material process by which a shared way of life is structured. Following the 'cultural studies' approach,[7] culture is at one pole a 'history of common life' (Williams 1961: 285) where 'social groups develop distinct patterns of life, and give *expressive form* to their social and material life-experience' (Clarke *et al.* 1976: 10), and at the other the collective expression of an historical epoch (Gramsci 1971). This double-edged project can be supported using Sartre's (1963, 1976) 'retrogressive-progressive' method which provides not only a means of connecting pre-given theory to emerging social conditions, but also the basis for understanding the terms according to which cultural representations are hauled forward in mediation between individual experience and social process (Kobayashi 1989). As Butters (1976: 71) expresses it:

> Sartre's strategy attempts a round-trip passage from a presumptive theory of the whole social order, to the level of the negotiated cultural nexus in which individuals make and live their experience, and back to the totality, carrying now some means to criticize the original account.

The geographical totality, organized as landscape, or 'solid appearance' (Inglis 1977: 489) of the dialectical relationships that structure and give form to cultural action, is neither monolithic nor static, but a multiplicity of expressions, overlapping and shifting along several dimensions. As a material process, culture is also place-specific. Jackson (1989) draws upon the metaphoric use of 'maps of meaning' (Clarke *et al.* 1976: 10) to show that culture is both 'socially constructed and *geographically expressed*' (p. 3).

A second point about culture, again drawing from Gramsci, is that it is 'always collective and always an arena of power' (Cocks 1989: 73). A more power *ful* cultural geography addresses the shifting, unequal relations among the geographical conditions that give rise to and result from political processes. Political processes, too, need to be understood at various levels, from the day-to-day negotiation of life within the family to the institutional settings wherein normative structures are set in motion, giving rise to the 'complex strategical situation' of society (Foucault 1978: 93).

Within a cultural studies perspective, institutions are identified as one of the 'sites of struggle' wherein cultural hegemony is 'produced and reproduced ideologically' (Mannette 1988: 177).[8] Clarke *et al.* (1976) use Althusser's (1971) notion of 'state apparatus' providing 'the stake, but also the site of class struggle'. They reshape the Althusserian notion of 'superstructure', however, from the concept of institutional power produced and *located* through the processes of capitalism, to one in which institutions are the 'terrain' across which power is negotiated between, to use Gramsci's (1971) terminology, 'the state and civil society'. In other words, sites of struggle are the domain of ideological representation. In a modern, Western context, the terrain is contoured by dominant forms of social relationship – including class divisions, patriarchy and racism – which rise up to empower those who maintain the institutions. The result is a lived condition of society, culturally constructed, in which hegemony sets the '*limits* within which ideas and conflicts move and are resolved' (Clarke *et al.* 1976: 39) where legitimation, power and ideology are produced within a parliamentary regime 'characterized by the combination of force [or the potential for force held by the police, the army or other means of order] and consent ... without force predominating excessively over consent' (Gramsci 1971: 80; quoted in Clarke *et al.* 1976: 39).

This Gramscian framework needs to be extended, however, in (at least) three ways. First, as Hall and Jefferson (1976) show, hegemony is seldom monolithic. Recursivity needs to be established not only between state and civil society, between domination and subservience, but also between control and resistance. Just as 'maps of meaning' are the result of differential 'cultural power' working to classify, rank and 'insert' cultures (groups and classes) in hierarchical relationship (Clarke *et al.* 1976: 11), the maps are redrawn, and sometimes the hegemonic regimes are toppled, by subcultures (or counter-cultures) that resist subordination in various ways. Although their life chances are structured and their actions often reproduce the structures that make them, cultural groups oppose, resist, make a difference. And, as will be shown below with respect to the Canadian institution of multiculturalism, the means of resistance are seldom expected.

This oppositional theme is taken up in a second theoretical extension by Joan Cocks, who rejects Gramsci's ultimately idealist notion of collective consciousness leading to a new hegemonic regime, for Foucault's (1978) 'anti-hegemonic politics'. 'Gramsci lived,' she claims, 'at the very last moment of Western history – really at the moment after the last moment ... – in which one could be so sanguine about the nature of emergent movements and formations.' This is to recognize not only the 'mechanisms of power that function outside, below and alongside the State apparatuses, on a much more minute and everyday level', but also the necessity and permanence of these mechanisms, that is, the constancy of transformation itself, which 'no mere replacement of one top-down leadership with another will do anything to transform' (Cocks 1989: 74). This double dialectic of transformation sets the agenda for contingent analysis, and justifies emphasis on cultural pluralism not only as an empirical condition at any one moment, but as a necessary condition, the 'impulse to power ... in civil society' (Cocks 1989: 75) through which a multiplicity of interests emerges. Cultural geographers, therefore, should initiate small-scale analyses of very particular sites of struggle, moments in time/place at which a change in power relationships can be identified.

To go thus far with Foucault, however, is not to impose an agenda for constitutional anarchism, or to accept a postmodern condition in which social power exists only as its negation, the mediation of the state lost in a 'justice of multiplicities' (Lyotard 1984) with the similarly idealist implication that all individuals might become equally 'empowered'. Against such a vision (and it is nothing more than that in the modern *or* postmodern context) it is necessary to establish a critical methodology for understanding the differential forms and

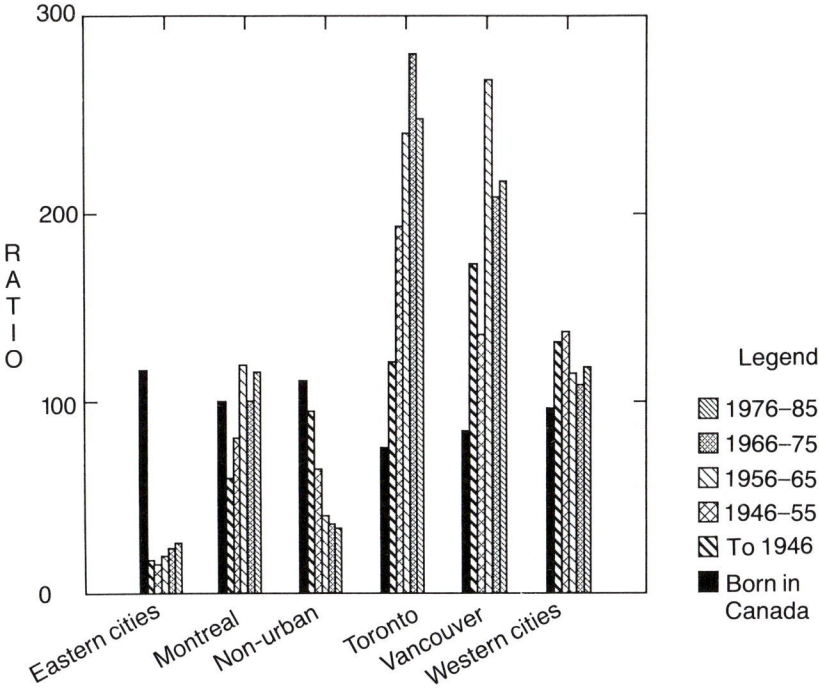

Figure 12.1 The geographical distribution of immigrants to Canada: ratio of actual to expected population

Source: Health and Welfare Canada 1989

stages of cultural relations, and for understanding the specific conditions that make some sites of struggle more important (or less) than others.

The third extension is the geographical. To ground the evocative, but none the less misleading, spatial metaphors of the cultural studies approach in real land-scapes is as much a project of demystification as it is a form of critical resistance (Soja 1989: 5). It is to *map* the meanings, to *site* the struggles, to *negotiate* the terrain so that cultural materialism becomes more than just a guiding theory. And it is not only to reassert space as a quality of social relations, it is to uncover the specific conditions in which cultural circumstances *take place*. For human beings share not only systems of production, values, ideas and political apparatuses, they share ground, as common ground, upon which their coming together and moving apart, and the conditions under which they do so, constitute the history of common life.

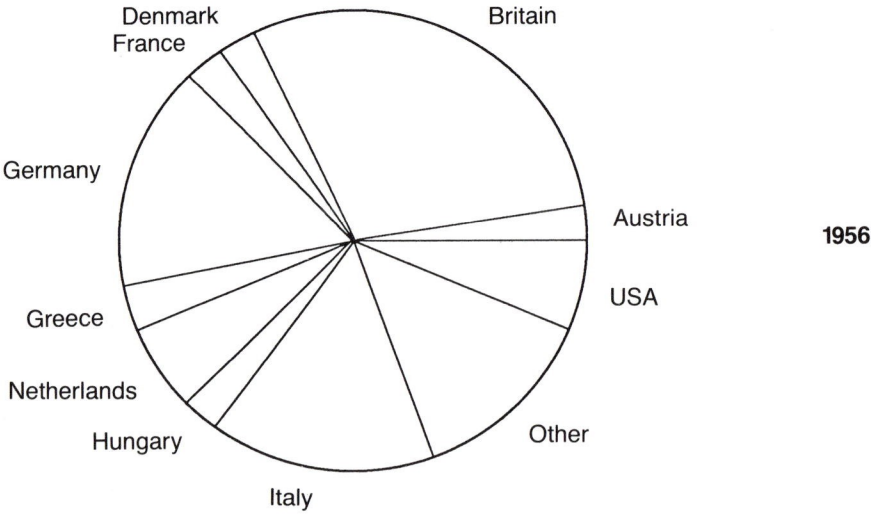

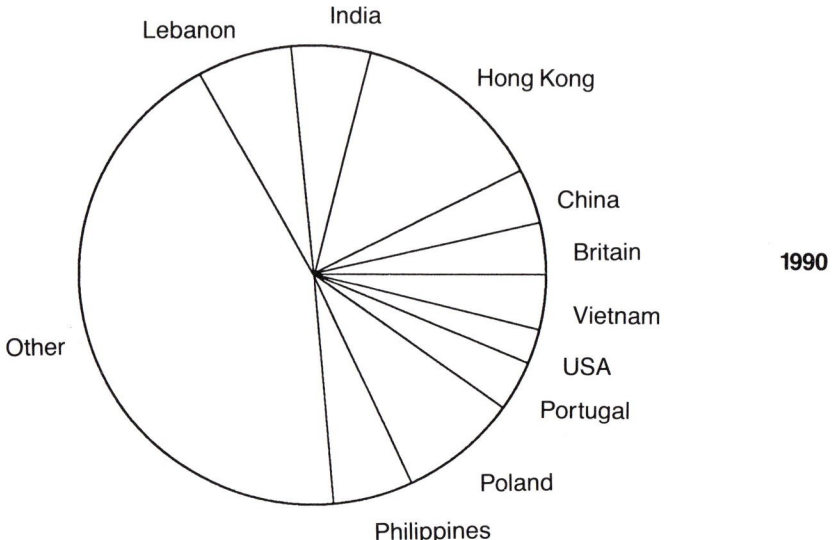

Figure 12.2 Immigrant arrivals to Canada by country of former residence, 1956 and 1990

Sources: Department of Citizenship and Immigration 1957; Employment and Immigration Canada 1990

THE HISTORICAL CONTEXT: ANGLO-CONFORMITY

'Anglo-conformity' is both a theoretical construct, based on Gordon's definition (1964) and extended in the Canadian context to mean 'that immigrants admitted to the country or their descendants would assimilate to the British group' (Burnet and Palmer 1988: 223), and an ideology that provided until fairly recently a justification for the cultural dominance of those of British ancestry (Palmer 1976).

The British North America Act (BNA Act) of 1867 established Canada as a dominion and set out the relationship between Canadians of French and British background, the 'charter' groups. At that time, the Canadian population was 61 per cent of British background, 31 per cent of French background (almost entirely concentrated in Quebec) and 8 per cent other.[9] The provisions of the BNA Act were a grudging acceptance of French-Canadian existence, however, and the first eight decades of Canada's national history were both officially and demographically Anglo-conformist, a national ideology assured by the dominance of immigrants from the United Kingdom.

A map of Canada today varies in degree of diversity as well as in ethno-cultural composition. Figure 12.1 shows stronger diversity in those provinces – Ontario, British Columbia and Quebec – with the highest proportions of urban dwellers. Figure 12.2 shows the chronological effects of changing immigration patterns over time. The make-up of the population began to shift in the 1890s, as immigration source countries expanded to include Eastern Europe and Asia. There was very little new immigration to the Atlantic provinces or Quebec, which remained overwhelmingly of British and French ethnic backgrounds respectively, with significant enclaves in Montreal of people from other backgrounds, including British (Scottish), Irish and Jewish. Ontario received a much higher level of immigration, dominated by British immigrants but including others, notably Germans in the rural areas of south-western Ontario, and Italians and Jews in Toronto. Most rapid population growth occurred in the prairie provinces, which received the largest numbers of Eastern European immigrants. British Columbia was dominated by British immigrants, with a significant stream of Asian immigrants, beginning in the 1880s with the importation of Chinese labourers to build the transnational railway and, during the first decade of the twentieth century, Japanese contract labourers to work in the primary industries.[10] In British Columbia a white supremacist movement pressured the provincial legislature to exclude completely all 'non-white' groups (Roy 1989; Ward 1978) in a context where employers sought to exploit Asian labour at lower wages than their 'white' counterparts while labour sought to remove the 'threat' of such workers against their own jobs (Phillips 1967); and where the clash of political interests between federal and provincial governments resulted in battles over legislative jurisdiction (Kobayashi 1990).

Accompanied by prejudice and racism, it remained the dominant ideology in Canada until after the Second World War. Even in 1947, Mackenzie King made a now famous speech in which he claimed that 'There will, I am sure, be general agreement with the view that the people of Canada do not wish, as a result of mass immigration, to make a fundamental alteration to the character of our population.'[12]

DEMOGRAPHIC MULTICULTURALISM

The pattern changed after the Second World War, initially with increases in immigration levels from Eastern and Central Europe, although immigration remained strongly tied to the notion of assimilability, defined mainly according to skin colour (Hawkins 1988). This change resulted from a larger and more diversified stream of immigrants. The immigrants were better educated, from more urban backgrounds, and they concentrated overwhelmingly in the large cities, primarily Toronto, followed by Vancouver and Montreal. The older pattern of greater diversity in the west than in the east was overlain by a newer pattern of metropolitan diversity. These cities have since become characterized by strong ethno-cultural concentrations, social structures organized along ethno-cultural lines and distinctive patterns in the work-force. The political significance of this post-war 'cosmopolitanization' of the large cities is important in understanding later events, because the urban ethnic groups developed a level of institutional organization, especially in Toronto, that had not occurred, or that had occurred in a much more limited way, in the rural areas (Kobayashi and Olson 1993).

Another development after the Second World War was the 'world revolution' in human rights (Humphrey 1970), prompted by the proceedings of the Nuremberg Trials, and codified in the United Nations Convention signed in 1948 (Davies 1988). Despite Canada's leading role in drafting the policy, and despite rapid changes in the composition of the population, the early post-war years saw slow change at the institutional level. The concept of human rights was none the less established at this time and gradually took hold over the next several decades. The process began with a chipping away at the hard edges of Anglo-conformity, replacing it with a dominant notion not of pluralism but of assimilation. Although the metaphor of Canadian 'mosaic' had been advanced by a few individuals since the 1920s (Palmer 1976), it was slow to achieve popular appeal. Rather, in the post-war popular rhetoric, the term 'Canadian' is substituted for the term 'British' (Buchignani 1982), but the fundamental values deemed to constitute Canadian do not shift very much.

The 1960s were a decade of transition. The demographic shift quickened as immigration policy was liberalized (Hawkins 1988). Although the privileged status of immigrants from the United Kingdom was to last for another decade,[13] in 1967 the federal government implemented a 'race-blind' immigration policy using a 'point' system that established nine criteria founded on personal attributes as a basis for eligibility. This system removed, at least in principle, barriers to immigration from particular countries or 'racial' groups, and allowed Canada to respond positively to sudden movements of people from new immigrant sources, such as Uganda in 1971.[14] Subsequent change was slow, and was not accompanied by greater public tolerance for people from 'non-traditional' (a euphemism for 'non-white') sources, but over the past three decades, as figure 12.2 suggests, there has been a significant shift, and the majority of immigrants are now visible minorities.

The 1960s also saw the social maturity of a number of non-charter ethno-cultural groups made up not only of immigrants but of successive Canadian-born generations, matched by the tentative flexing of political muscles. Although non-charter Canadians had been represented by only about 100 elected federal politicians since 1867 and were extremely under-represented within the power structure of Canadian society including the civil service (Canada, B & B Commission 1969), their political influence was growing slowly. Some, such as Jewish Canadians, concentrated in distinctive enclaves within the major cities of Toronto and Montreal and represented by the formidable organizational power of the Canadian Jewish Congress, continued and intensified the movement for human rights (Glickman 1981). Others, such as Ukrainian Canadians, concentrated in the prairie provinces, worked to develop an explicitly 'ethnic' political base (Yuzyk 1967) and to lobby for a multiculturalism policy (Canada, B & B Commission 1969; Lupul 1978). It was their increasingly visible political actions that led to speculation about the emergence of a 'third force'[15] in Canadian politics, reflecting the interests of the 'other ethnic groups'. Their interests were also for the first time part of the general sense of 'Canadian-ness' that arose during the Pearson administration, focusing on the adoption of the maple leaf flag as a symbol of national identity. There also developed a strong sense of the Canadian 'mosaic', a repudiation of the American 'melting pot'. The media at the time could not get enough of discussions of what makes a Canadian.

The academic discipline of 'ethnic studies' also came into its own during the 1960s. Scholarly debate concerning the viability of such communities centred mainly upon the theory of 'institutional completeness' (Breton 1964) which suggests that 'minority' groups assure their viability by developing a social system that includes functioning and internally controlled religious, educational and community welfare institutions. Social area analysis became fashionable in attempts to show a correspondence between institutional completeness and residential segregation in cities (e.g. Driedger and Church 1974). In opposition to the new fascination with ethnicity, John Porter published *The Vertical Mosaic* (1964), painting a disturbing picture of a Canadian society deeply split along class lines, with ethnicity one of several mechanisms for sorting social groups within a vertical hierarchy. Against the celebration of diversity by the pluralists, Porter's work supported assimilationists, who could conclude that the most effective way to dismantle the vertical mosaic would be to diminish ethnic affiliations, which were an artificial result of class divisions. Historians, on the other hand, tended to take an uncritical view of ethnicity, and focused on empirical studies of individual groups.[16]

Government bureaucracy began to recognize cultural pluralism. Within a global context that included the coming-of-age of the American civil rights movement, the growth and legitimation of the women's movement and the 'Quiet Revolution' in Quebec,[17] Anglo-conformity was being challenged, albeit in ways that would now be considered patronizing, within both the civil service and Parliament. In 1967, amidst the heady celebration of Canada's centenary, *The Canadian Family Tree* was published by the Citizenship Branch of the Secretary of State, which had by this time a handful of individuals to deal with 'ethnic' issues. As the first collection of empirical information on the backgrounds of diverse ethno-cultural groups, this book is a valuable resource. It is more interesting, however, as a statement of the dominant ethic; no heed is paid to the history of subordination of minority groups, to murmurs of political unrest or to the contentious issues of ethnicity, class and status emerging in the academic literature. Instead, it presents a 'whitewashing' (Kostash 1977) and romanticization of Canada as a 'nation of immigrants and descendants of immigrants' (p. 7), referring to individual groups in the most stereotypical terms.

The official tone began to change as a direct result of the Pearson government's Royal Commission on Bilingualism and Biculturalism, established in 1963 on the assumption that Canadian society was based on two languages and two cultures. The commission was to clarify the relationship between the two groups, and to begin the legislative process that would lead, over the next several years, to greater recognition of French-speaking Canadians. Almost as an afterthought, the commission produced Volume IV of its findings, to address the 'other ethnic groups'. The commission reiterated the 'basically bicultural character' of the country, but took 'into account the contribution made by the other ethnic groups to the cultural enrichment of Canada, and the measures that should be taken to safeguard their contribution'. It became a 'study of the way they have taken their place within the two societies that have provided Canada's social structures and institutions'. To summarize the findings, published in 1969: while the population of British ancestry was declining, its economic and political dominance continued; dominance was especially strong within the government-controlled areas of the military and civil service, where British Canadians also received higher average salaries. The effects of the commission were threefold: first, it was a public announcement of the economic and social subordination of minority ethno-cultural groups; second, through its utilization of public hearings, it established a dialogue between government and minority groups that launched the participation of these groups within the political arena; third, it encouraged an intensification of theoretical and philosophical discussion of the issues that have come to define multiculturalism: pluralism vs assimilation, individual vs group rights, language and culture, the relationship between tradition and social change, and the role of the state in promoting social change and in preserving social values. The term 'multiculturalism' arose within this context.

SYMBOLIC MULTICULTURALISM

The policy announced in October 1971 had four major objectives:

1   '[R]esources permitting, the government will seek to assist all Canadian cultural groups that have demonstrated a desire and effort to continue to develop a capacity to grow and contribute to Canada, and a clear need for assistance, the small and weak groups no less than the strong and highly organized.'

2   '[To] assist members of all cultural groups to overcome cultural barriers to full participation in Canadian society.'

3   '[To] promote creative encounters and interchange among all Canadian cultural groups in the interest of national unity.'

4   '[To] continue to assist immigrants to acquire at least one of Canada's official languages in order to become full participants in Canadian society.'

The first Minister of State for Multiculturalism, the Hon. Stanley Haidasz, was appointed in 1972. In the following year (1973–4) expenditures on multicultural programmes were $2.7 million. The budget included grants to communities, a 'Canadian Identities' programme to support the arts, folk festivals and ethnic history projects (including a commissioned series of books on major ethno-cultural groups), a multicultural studies programme that included teaching aids as well as research and multicultural conferences.[18] The Canadian Consultative Council on Multiculturalism was established, made up of political appointees from a variety of ethno-cultural backgrounds, and a programme to monitor Canadian broadcasting was implemented.[19]

Emphasis throughout the 1970s was clearly upon the Anglophone ethno-cultural groups (with funding to support non-official language projects) and decidedly upon the past, promoting what Kelner and Kallen (1974) have called 'museum culture'. The broad ranging, loosely focused objectives did not provide a clear or effective multiculturalism programme. The programme lurched forward, lacking a theoretical rudder and strongly susceptible to the winds of political opportunism. A series of Liberal Ministers during the 1970s were followed by a series of Conservative Ministers during the 1980s, all of whom made extensive and rhetorical speeches celebrating cultural diversity, and emphasizing that multiculturalism is for all Canadians.[20] Examination of their speeches reveals little departure from the sentiments expressed by Trudeau in 1971.

A collection of speeches published as a result of the Second Canadian Conference on Multiculturalism (Canadian Consultative Council on Multi-culturalism 1976) provides a sense of the official policy during the symbolic period. A 'bearpit' session invited a series of comments on multiculturalism from prominent public figures. The Hon. John Munro, then Minister, indicates the logic that informs a social construction of political interest groups, by claiming that official bilingualism was instituted to meet the demands of French

215

Canadians, while multiculturalism (despite the rhetoric of its being for all Canadians) was directed to the 'ethnic minorities':

> Two policies were adopted to give concrete recognition and government support to their realization. Bilingualism and multiculturalism were a recognition of the fact that a democracy such as ours has an obligation to assist its citizens to full participation in the national life of their country. If a Canadian identity is worth fostering and protecting against powerful outside influences; if a Canadian identity is a prerequisite of national unity, then it is equally important that the individual parts that form that identity also be encouraged and buffered against powerful forces which they are hard-pressed to resist unaided.
>
> (ibid.: 14)

Munro's statement hints at the discomfort that had existed since the beginning between bilingualism and multiculturalism, viewed as policies to represent separate constituencies of diverse ethno-cultural backgrounds in different regions.

A year later, the Hon. J.P. Guay, speaking to a meeting of the Canadian Consultative Council on Multiculturalism in Vancouver, made it clear that the issue was not even so simple as separate constituencies:

> If we accept our cultural pluralism then we assure our Canadian unity. It is as simple and as complicated as that.... Some Quebecers feel that they must stand alone in their efforts to maintain their language and culture. This should not be the case. As we recognize their just aspirations, so we must give them our whole-hearted support. In fact, the idea of multiculturalism would probably not exist in Canada today, if not for the cultural survival of French Canadians.... They have shown us that assimilation is not an option we Canadians want or choose.
>
> (speech notes)

These and other official statements of the period of symbolic multiculturalism show that, despite the constant reiteration of the inseparability of multiculturalism and bilingualism as set forth in Trudeau's original statement, no further attempt is made to clarify the exact nature of this relationship, or to justify the logical separation of language and 'culture' for policy purposes. To do so, of course, would have exposed the contradictions – political, social, geographical – upon which the 'twin pillars' rested.

Another feature is a clean distinction drawn between issues of multiculturalism and other issues, and the avoidance of statements that might link multiculturalism policy to employment, legal issues, the concerns of other 'minority' groups (except in the context of bilingualism) or international affairs. This avoidance is not surprising given the tone of the four objectives outlined above, which avoid reference to economic or political factors, to programmes that would bring about social change beyond the confines of designated groups or to issues of racism and other forms of discrimination. Language is identified as the most important means for immigrants to be 'full participants', and the entire statement uses terms that emphasize 'culture'. The political discourse, if not explicitly written to maintain a status quo in nearly every realm, certainly provides no means of changing it.

Academics were not disposed to generosity in their assessment of symbolic multiculturalism. The programmes were seen as demeaning the status of non-charter groups by relegating them to 'second-class status of "foreigners" or immigrants whose historical way of life has been transplanted intact in Canadian soil' (Kelner and Kallen 1974: 30), resulting in a 'we and they' definition of Canadian society despite its claims to the contrary (Peter 1981). To Brotz (1980) it was an analytical 'muddle' that made little sense of 'the real meaning of the term culture underneath all the rhetorical ambiguities' (p. 41) and that failed either to initiate a political dialogue concerning the role of Quebec in Canada, or to recognize the variety of national roles played out in different 'localities' (p. 46). To Perin (1983: 449) the emphasis on heritage, and in particular the collection of commissioned histories produced over the first decade, represented an 'eviscerated' kind of history, and a squeamishness at dealing with difficult social issues. In a mid-decade appraisal, Burnet (1976: 206) summed up the unfulfilled challenges of multiculturalism policy as

> how to persuade French Canadians that multiculturalism does not threaten, and indeed may be used to enhance, the position of the French language in Canada; how to ensure that governmental support does not strengthen the more conservative factions and the older generation within particular ethnic groups at the expense of the more venturesome and the younger, and large and well-organized ethnic groups at the expense of small and weak ones; how to provide opportunities for teaching the official languages and for providing other services to immigrants without appearing to alienate them from their ethnic groups; and how to make the policy serve the interests of the 'visible minorities', whose numbers are increasing, as well as the interests of white ethnic groups.

Taken together, these statements show up the weak, even contradictory, conceptual basis of symbolic multiculturalism and its limited power to affect the broader sphere of Canadian politics. They represent a cultural construction and a negotiation of ethnicity which take place most directly through the government-appointed Canadian Ethnic Studies Advisory Committee (which advises on policy matters as well as on funding of research projects) and indirectly through the funding of conferences as well as of research results published in scholarly books and journals. Furthermore, as Buchignani (1982: 16) has pointed out, research is 'a social process *par excellence*, subject to quite the same social forces as other human activity'.

Critiques of multiculturalism restricted to its own terms, however, which have dominated the academic literature as well as popular and political responses, fail to take account of the ethno-cultural groups where the effects of multiculturalism are generated. While the community groups funded by the Multiculturalism programmes are keenly aware of the importance of state funding, analysis of the relationship between Multiculturalism and its constituency is virtually non-existent (Lupul 1989). Still, Buchignani (1982), a keen observer of the effects of the policy, has noted both that 'the original charter of multiculturalism was so vague that actual programmes implemented in its name were at first largely determined by ethnic interest groups rather than by government', with the result that 'Policy and programmes soon diverged.' Another, perhaps unintended result is that 'It encouraged the creation of ethnic-based associations to the point where they are now endemic' (p. 17). By strengthening the base of community support, especially among the non-charter groups, multiculturalism had become a self-fulfilling prophecy. But it is among just such groups that we should seek the means by which resistance to Anglo-conformity and the movement for equality rights has been generated.

The Canadian Ethnocultural Council (CEC), founded in 1980 as an umbrella group representing the political interests of 38 national organizations and over 2,000 local and provincial associations, is today one of the most important products of multiculturalism policy, and one of its most effective and politically powerful critics. Its major objective is described in each issue of *Ethno Canada* (a semi-annual newsletter):

> to secure equality of opportunity, rights and dignity for ethnocultural communities in Canada. The CEC membership works by sharing informa-tion so as to develop a consensus on issues of concern to its membership and by advocating for changes on behalf of ethnic and visible minority groups.

This organization bears investigation as one of the most effective lobby groups to emerge in the last decade's negotiation of Canadian identity. While some may doubt its level of effectiveness because it receives funding from Multiculturalism (Lupul 1989), it may also be argued that its level of resistance has been raised by just such funding having been converted to its own purposes. As both the product and the nemesis of multiculturalism, the CEC played a major role in drafting the policy changes that led to the 1988 Multiculturalism Act, and has reached beyond the limited jurisdiction of multiculturalism policy to present briefs on the implications for minority groups of a wide range of legislative policies and changes, including employment equity, the Meech Lake Accord, immigration policy, broadcasting, national museums, the (US) Free Trade Agree-ment, government appointments and the criminal justice system (for a review see *Ethno Canada*, various issues).

The CEC, in coalition with a broad range of human rights organizations, has been especially strong in encouraging a shift from heritage issues to equality rights issues. Although many factors were involved in this shift, the establishment in 1983 of the Special Committee on Visible Minorities in Canada (SCVM) stands out. The committee consulted with over 1,000 groups and individuals across Canada before producing the report, *Equality Now* (Canada, SCVM 1984), containing eighty explicit recommendations for the elimination of racial discrimination. Following release of this report, the parliamentary Standing Committee on Multiculturalism was formed in 1984 (a seven-member committee made up of Members of Parliament from the three major parties), set with the task of providing the guidelines for structural multiculturalism. The extent of public involvement in these projects is indicative of the growing importance in Canada of parliamentary committees in establishing direct links between government and the grass roots.

As a representative of civil society, the CEC also mediates between the grass roots and state institutions. Whereas its initiatives are generated within its council, it represents the communities, regionally and sectorally, through a hierarchy of spokespersons, to allow both the coalescence and coalition of interests. If a 'third force' in Canada exists, it is manifest through this organization. The actual process of representation, as well as the dispersal, concentration or transformation of political power that occurs through its collective actions, requires major study. It is at this level of analysis that the 'opposition imagination' of the Canadian populace can be discovered.

STRUCTURAL MULTICULTURALISM

In a recent analysis of the policy, Burnet and Palmer (1988: 223) state that

> Canadian society no longer consists of a British mainstream and variegated tributaries: it is now essentially ethnically diverse. Its institutions, although based on British models, are uniquely Canadian, and within them people of many origins take their places without a sense of strangeness or inferiority. Discrimination still exists, especially against recently arrived groups, but there is an array of formal and informal means of combating it, and few Canadians hesitate to employ those means. Canada has become multicultural.

Although many would challenge its optimism, this progressive model of multiculturalism, assuming a direct and beneficial correspondence between policy and social change, is adopted in a recent government assessment of the policy's past achievements and future needs (Canada, Standing Committee on Multiculturalism 1987: 27). According to the report, the initiative of the 1971 policy adoption prodded the country from *demographic multiculturalism* to *symbolic multiculturalism*. What was now needed was a Canadian Multiculturalism Act which would establish *structural multiculturalism*, wherein systematic legal and bureaucratic mechanisms are in place to ensure the efficacy of the multiculturalism policy and to incorporate advances in human rights made over the past decade. The objective was to *multiculturalize* those activities within the jurisdiction of the federal government, as a basis for addressing a larger social context in which

The Multiculturalism Policy of 1971 is clearly insufficient and out of date. It does not have the ability to respond to the needs of today's multicultural society. There is a sense that this 15-year-old policy is floundering. It needs clear direction.

The cultural industries and government programs are not doing enough to preserve and enhance our multicultural reality. Ethnocultural and visible minorities continue to face varying degrees of discrimination in employment. Various government policies and departments do not pay adequate attention to multiculturalism and Canada's ethnocultural and racial diversity, be it in matters of health and welfare, justice, youth issues, women's issues, or trade. With a budget of only $23.6 million which is minimal in comparison with other programs, Multiculturalism can only be seen as something of a marginal policy. Acceptance and support of Multiculturalism is carried out more in a fringe or peripheral sense. The mainstream of Canadian society and institutions have yet to be 'multiculturalized'.

> (Canada, Standing Committee on Multiculturalism 1987: 'Executive summary')

In sharp contrast to the rhetorical tone of 1971, the eighteen recommendations document the need for strengthened policy, a separate and restructured Department of Multiculturalism and new programmes to create equality according to eight principles: 'multiculturalism for all Canadians, advancement of multiculturalism within a bilingual framework, equality of opportunity, preservation and enhancement of cultural diversity, elimination of discrimination, establishment of affirmative measures, enhancement of heritage languages and support for immigrant integration' (ibid.: 3). Since then, substantial actions have been taken to implement the majority of the recommendations. It is beyond the scope of this essay to outline the differences between the recommendations and the actual legislation. They are, however, summarized as follows:[21]

- A new policy of multiculturalism was enshrined through the Canadian Multiculturalism Act 1988.
- A separate Department of Multiculturalism is being created.
- Legislation was promised to create a Heritage Languages Institute.
- A budget increase of $62 million over the next five years was announced in May 1988, but later reduced to $52, including a cut to community programmes but maintaining operational levels.
- A new Broadcasting Bill contains several provisions for multiculturalism policy.
- A new Multiculturalism Advisory Council has been established.
- The Canadian Race Relations Foundation has been created to fulfil a commitment made as part of the Japanese–Canadian redress settlement of September 1988.

A basic problem of the present policy is to go beyond contradictions already institutionalized in the past. The dual objective of heritage and equality multiculturalism may not be possible, in the same way that individual and collective rights are often in conflict, but that duality is given, and must form the basis for future policy initiatives. Similarly, a legacy of the 'muddled' concepts of the symbolic period is an edifice of representation in which socially constructed categories such as 'ethnic', 'race' and even 'multiculturalism' have defined social relations. Their effects are difficult to overcome despite changing analytical notions.

The issue of 'race relations' provides an important example. Almost no one, within government or the interest groups concerned, disputes the fact that racism is a serious and growing problem (see Barnett 1987; Henry 1986; Kobayashi 1990; Naidoo 1989; *The Network* 1989). The Standing Committee makes 'race relations' one of its most important priorities, and recommends a number of ways in which anti-racist policies might be adopted. In response to the 1984 report, racism has recently been moved to the front burner of priorities within the Multiculturalism Directorate, with the rejuvenation in 1988 of the Race Relations Directorate (established in 1982). Its mandate is primarily educational, and includes funding for programmes in the communities (with police, business, schools, the media, various levels of government and social services). Recent actions to heighten public awareness of racism come under the slogan 'Together we're better!' and include a nationwide student contest to mark International Day for the Elimination of Racial Discrimination (21 March 1990), distribution of media packages, advertisements in all community newspapers across the country and publication of a booklet on racism, which claims:

Racism and racial discrimination are facts of life in Canada.

They exist openly and blatantly in the attitudes and actions of individuals. They exist privately in the fears, in the prejudices and stereotypes held by many people, and in plain ignorance. And they exist in our institutions.

(Canada, Secretary of State of Canada and Minister of State Multiculturalism and Citizenship 1989: 3)

Effective implementation of this anti-racist policy is inhibited, however, by lack of a clear definition of the premise 'race', which is only an indicator of how much needs to be done if the process of racism as a whole is to be understood. Nearly all government documents continue to treat 'race' as an unproblematic category, naturally given rather than socially constructed. The objective, therefore, becomes that of overcoming discrimination, fostering equality *regardless* of 'race'. Such an objective only perpetuates the separation of human beings according to arbitrary phenotypical characteristics, instead of addressing the social processes by which races are created through an ideology of physical (or cultural) difference as the product, not of nature, but of racism itself (Anderson 1987; Jackson 1987; Kobayashi 1990; Miles 1989). The ground is prepared for future assumptions about human difference. The conceptual confusion is only enhanced by the notion of 'race relations', a term adopted following the British model (Commission for Racial Equality 1987). In both Britain and Canada the term 'race relations' euphemistically projects an image of racial harmony, rather than the more distressing image of racism occurring as a social relationship of dominance and subordination, created by and engendering structural inequality. Furthermore, while academic discussions have the luxury of deep analysis, and can accept or reject terms on the basis of reasoned (if not reasonable) argument, governments must face the much more difficult task of wresting a notion from common discourse once it has become public domain. The notion of 'race' is a product of colonialism and past prejudices as much as present practices, and will not go away because it is reconceptualised.

A second problem, likely to emerge in the coming months as a crisis in Canadian confederation, is the contradiction between multiculturalism and other declared interests. The claim since 1971 that no ethnic group takes precedence over another has been challenged in a number of ways. The conceptual separation of the charter groups ('ordinary Canadians') and others ('ethnics') has been maintained despite policy to the contrary and, in the absence of effective affirmative action, is likely to remain as long as there are prejudicial attitudes and social inequalities. The relationship between French Canadians and the 'other' ethnic groups has never been an easy one, and the opinion that multiculturalism undermines the claims of Francophone Quebecers is widely held. The notion of a 'third force' only supports this view, because it challenges the notion that current political controversy can be understood according to a perspective of 'two solitudes'. Therefore, as long as popular notions of ethnicity create minority status, and until the issue of Quebec's status is resolved, it is unlikely that multiculturalism will become a policy for all Canadians.

The recent controversy over the Meech Lake Accord brings these contradictions to the top of the public agenda and brings to the surface old hatreds and prejudices. In August 1987, the CEC made a presentation to the Parliamentary Committee on the Constitutional Amendment (CEC 1987), claiming that multiculturalism is as important as bilingualism as a 'fundamental characteristic' of Canadian society.[22] They questioned the meaning of 'distinct society' applied to Quebec in relation to other social groupings that might also be labelled 'distinct'. More importantly, they expressed concerns that provisions within the accord would supersede fundamental equality rights now guaranteed in the charter. They pointed out, furthermore, that the claims made on behalf of ethno-cultural minorities at the time of drafting the 1982 Constitution had not yet been addressed, and that the present amendment process provided an opportunity to redress this situation. Evelyn Kallen (1988: S107) writes:

> Throughout both 1980s constitutional debates, amendments designed to further entrench the special and dominant status of Canada's two founding peoples – English/Protestant and French/Catholic 'charter groups' – have

assumed top priority. Accordingly, the collective (linguistic, religious and broader cultural) rights of this country's majority ethnic groups have consistently taken precedence over the corresponding rights of Canada's ethnic minorities.

> From a human rights perspective ... constitutional amendments specified in provisions of the Accord do not serve to promote national unity and group equality across the country. Rather ... [they] serve to ossify long-institutionalized status inequalities among Canada's ethnic and non-ethnic minorities.

There is no simple or clear solution to this dilemma, however, given Kallen's later classification of collective claims: claims between French and English Canadians; claims of multiculturalism (and recent immigrants); and claims of native (Indian, Inuit and Métis) peoples. The claims of the first order are geographical (the status of the province of Quebec), of the second dispersed (the rights of minorities throughout the country) and of the third national (nationhood land claims). Not only are the principles upon which these claims are put forward at odds, but they overlap upon the very landscape in which Canadians must coexist. The eventual failure to ratify the Meech Lake Accord (see Postscript) was as much a result of the intractability of these contradictory claims as of the negotiation process through which the contradictions seem to be entrenched rather than overcome.

Finally, there exist contradictions between the interests of the state and the interests of civil society. This issue, too, arose with respect to the Meech Lake Accord which Cairns (1988) claimed would fail in part because the constitution makes 'spectators' of rights-bearing citizens whose needs are not considered in an amending formula that places priority on relations between federal and provincial levels of government. Similarly, Multiculturalism is *at least* as concerned with its power *vis-à-vis* other arms of government as it is with addressing the problems of a multicultural society. This point was clear in the presentation of the Bill (C-18), which was initiated in part as a response to recommendations of the so-called Nielsen Task Force, formed in 1986 to conduct an unprecedented and comprehensive review and analysis of the entire Canadian bureaucracy. The Task Force produced a 21-volume report finding that the federal bureaucracy as a whole was insufficiently 'multiculturalized', but making specific recommendations that would have seen Multiculturalism cut back in size, function and budget, with many of its responsibilities dispersed through other departments. While the recommendations were subsequently abandoned, they took their course in prompting changes to multiculturalism policy, as outlined above, justified in terms of the need to promote social change, but at least partially motivated by the need to protect Multiculturalism interests.

These three examples expose part of the process by which institutional relations are worked out and interests of state and civil society mediated. To understand this process it is necessary to go beyond the formal institution, to situate Multiculturalism within a national cultural context and to recognize multiculturalism as a web of complex social relationships, a multiplicity of powers and constraints, a shifting composition of cultural representation.

CONCLUSION

The term 'multiculturalism' carries a sense of dynamism and diversity. Any representation of multiculturalism, however, also carries the contradictions inherent in cultural processes in general, and in Canadian culture in particular. There is a contradiction between the 'multicultural' composition of the population and multiculturalism policy. Within the population, difference is socially constructed, geographically diverse and unequal. The streets of Canada's cities are indeed sites of struggle where difference is acted out, creating landscapes that include racism, residential segregation, ethno-cultural variations in access to work and its rewards, as well as street festivals, 'ethnic' restaurants and school-yard games. The interests of Canadian citizens are as diverse as the landscapes of francophone Trois Rivières, the Kahnawake Indian Reservation, Vancouver's Chinatown and Toronto's old elite district of Rosedale. Such places must be seen, therefore, not simply as cultural landscapes, but as manifestations of political objectives that cannot be accommodated without struggle and contradiction. The Meech Lake Accord was the latest, and perhaps the most divisive, example of geographical contradiction. Multiculturalism as a policy, on the other hand, is formulated against conflict. It upholds not only the value of diversity but the concept of equality rights. Its challenge is not only to create the conditions – through legislation, policy and programmes of affirmative action – in which equality rights can be realized, but also to recognize that equality is not possible except as an abstract and idealized concept. Present actions respond differentially to a differential past and create a differential future. The issue is whether, aware of such fundamental contradictions, we might structure a future in which contradiction is ameliorated at specific times and places.

Further contradiction exists between the interests of civil society and those of the state. Multiculturalism policy, and the web of cultural relationships that it engenders, is an important and effective means of mediation, and ties Canadian government to popular interests in ways seldom achieved in other countries. None the less, the Department of Multiculturalism is organized as a balancing act between effecting social policies and maintaining a place within a complex bureaucratic structure. The halls of government and bureaucracy (the House of Commons and the building at 15 Eddy Street, Hull, where Multiculturalism is located) are sites of struggle tied, in concrete as well as figurative terms, to the streets of Vancouver, Toronto and Montreal.

Finally, the analysis of relations between Multiculturalism and its constituency shows that rights awareness has become a significant (though far from dominant) dimension of Canadian cultural politics, marked by the institutional shift from multiculturalism as heritage to multiculturalism as equality rights. If this tendency to participatory democracy represents, as Smith (1989a) suggests, an indication of the 'new times', and if it suggests a framework for geographical analysis centred on institutional relationships, it also demands a politicized cultural geography in which rights are not (like 'race') taken as morally given, but analysed as the outcome of cultural process, socially constructed within a web of cultural relationships that define places and structure practice. It sets the need to explore the sites of struggle within which social constructions occur.

POSTSCRIPT

The Meech Lake Accord was not ratified on 23 June 1990. There are many reasons, only a few of which have been mentioned above. But the unexpected resistance came from the aboriginal peoples of Canada through Elijah Harper, a New Democratic Party member of the Manitoba Legislature, who used procedural tactics to stall a legislative vote on the Accord until past the deadline for ratification. The unresolved issues were taken up again (Canada 1992), when the First Ministers negotiated a more extensive set of constitutional amendments that included, *inter alia*, Quebec's becoming a signatory to the Constitution, reform of the Senate and of Parliamentary voting rights, and recognition of the principle of self government for aboriginal peoples, but no specific additional commitments to promote multiculturalism or strengthen equality rights. The new Accord was abandoned after a national referendum on 26 October 1992 indicated a minority of Canadian people in favour of its adoption.

NOTES

1 The term 'Multiculturalism' in upper case is used to refer to the Multiculturalism Directorate of the Secretary of State; 'multiculturalism' in lower case refers to the policy. The term 'multicultural' is a descriptive term which refers to demographic diversity.

2 The term 'institution' is used here as a 'noun of action or process' (Williams 1976: 168–9) to denote the *emplacement* of cultural norms, a historical process whereby social relationships are established and organized, institutionalized at all levels, but also a geographical process.

3 These three stages are identified but not defined by the Standing Committee on Multiculturalism (Canada, Standing Committee on Multiculturalism 1987: 27).

4 The Constitution Act 1982 established Canada's right to frame its own Constitution, thus superseding the British North America Act 1867, which could hitherto be amended only by an Act of the British Parliament. Part I of the new Constitution consists of the Canadian Charter of Rights and Freedoms, which supersedes (but does not cancel) the Canadian Bill of Rights 1960. The Charter affirms religious and linguistic rights established in the BNA Act and entrenches certain rights of aboriginal Canadians and equality rights of men and women. Protection of minority ethno-cultural groups is not entrenched, although Section 15(1) provides equality rights for individuals under the law, and Section 27 provides that the Charter 'shall be interpreted in a manner consistent with the preservation and enhancement of the multicultural heritage of Canadians'. Quebec is not a signatory, having refused in 1981 to join the other nine provinces in ratifying a Constitution in which it did not

have a special veto power over future changes.

5 The accord came about as the result of a meeting held at Meech Lake, Quebec, of the First Ministers (the Premiers of the ten provinces plus the Prime Minister) in June 1987, to work out the terms under which Quebec would become a signatory to the Constitution. The accord was signed by the eleven participants, subject to ratification in three years.

6 Under the former structure the Multiculturalism portfolio was held by a Minister of State subordinate to the Secretary of State. A separate Ministry within the Department of the Secretary of State was formed in 1988, and legislation (Bill C-18) has been introduced to create a separate Department of Multiculturalism and Citizenship.

7 The term 'cultural studies' is used to mean the work, especially, of Stuart Hall and his colleagues through the Centre for Contemporary Cultural Studies at Birmingham. Although the present discussion focuses upon a theoretical overview by Clarke *et al.* (1976) contained in Hall and Jefferson (1976), this Centre has produced a large quantity of empirical work that uncovers the working of cultural processes which affect subordinate groups in British society. More generally, 'cultural studies' refers to the theoretical contributions of Raymond Williams (especially 1961, 1977, 1980, 1981) and Antonio Gramsci (especially 1971). A recent book by Joan Cocks (1989) takes cultural studies considerably further by providing a feminist analysis of culture and political power. The most complete incorporation of a cultural studies perspective within geography is by Jackson (1989).

8 Mannette's discussion of hegemony derives from Clarke *et al.* (1976: 38–41).

9 The designation of ethnic groups is socially constructed but not thereby arbitrary. It is limited by the notions of ethnicity encoded within the Census of Canada, the major source of demographic data. In fact, French Canadians could be identified as the largest 'ethnic' group, if the 'British' category is divided into 'Irish', 'Scottish' and 'English'. Whereas there is considerable historical justification for doing so, the relative closeness of these three groups *vis-à-vis* other Canadians, and their position (especially the latter two) at the top of the social hierarchy, leads me to place them together.

10 There is a vast literature on the history of immigration and ethno-cultural groups. The most comprehensive single work is Burnet with Palmer (1988), while Palmer (1976) provides a context for understanding immigration and multiculturalism. Journals such as *Polyphony* and *Canadian Ethnic Studies* contain case studies.

11 The zenith of Canadian racism was reached with the uprooting, internment, dispossession and partial deportation of Japanese Canadians from 1941 to 1949 (Adachi 1976; Kobayashi 1987a, 1987b; Sunahara 1981). This was, however, only the culmination of a history in which white supremacism had been a major force (Ward 1978); and in which racism was institutionalized in the legal system by controlling immigration, and by the differential rights of citizens to work, vote and hold public office or property (Anderson 1987; Kobayashi 1990; Roy 1989), as well as rights of access to public places (Kerr 1969).

12 *House of Commons Debates*, 1 May 1947: 2644–6.

13 The special status of British immigrants was implicitly reduced in the Canadian Human Rights Act 1960 and explicitly removed in the Citizenship Act 1977. See McChesney (1987).

14 In that year, Canada received over 5,000 persons of South Asian origin, who fled the new Ugandan administration of Idi Amin.

15 The term 'third force' is attributed to Senator Paul Yuzyk, who introduced it in his first speech before the Senate on 3 May 1964 (Kelner and Kallen 1974; 33).

16 No attempt is made here to review either theoretical or empirical studies of ethnicity

in Canada. The most comprehensive review of historical studies is Burnet with Palmer (1988; but cf. Perin 1983). Anderson and Frideres (1981) provide a detailed account of theories of ethnicity.

17 The 'Quiet Revolution' refers to the period following the election of a Liberal government in Quebec in 1962. Congruent with a rise in living standards, modernization of industry, and the movement of Québecois into the professions and the civil service (encouraged by the federal government's programme to promote bilingualism within the federal civil service), Quebec nationalism took hold during the 1960s. By 1970, the actions of separatists led to sporadic violence, culminating in the 'October Crisis' (Gwyn 1980). The power of Quebec nationalism was reflected in 1976 by the election of the Parti Québecois, under René Lévesque as Premier. (See, for example, Lee 1979; Lévesque 1968; McRoberts and Posgate 1976; Morris and Lanphier 1977; *Canadian Review of Sociology and Anthropology* 15(2), 1978).

18 Various publications of the Multiculturalism Directorate provide details of programmes. Each year a list of funded projects is published. The most succinct description of the early programmes, however, is Kelner and Kallen (1974). A general history of the multiculturalism policy is provided in Canada, Standing Committee on Multiculturalism (1987).

19 A Broadcasting Bill, introduced in the House of Commons in fall 1989, incorporates suggestions from successive rounds of consultations and studies over the past fifteen years.

20 Texts of speeches presented by the successive Ministers are held in the Library of the Department of the Secretary of State, 15 Eddy, Hull, Quebec (Sect. SOS Cit. Br. Multi.).

21 A review of legislative and policy changes, as well as response to the changes by the CEC, is contained in the Fall 1989 issue of *Ethno Canada*. The most controversial difference lies in the failure to create a Commissioner of Multiculturalism to police the enactment of provisions in the Multiculturalism Act.

22 There had earlier been an extensive debate as to whether bilingualism should be identified as '*a* fundamental characteristic' or '*the* fundamental characteristic' (emphasis added). The final wording settled on 'a' but no further characteristics are identified in Section 1.

ACKNOWLEDGEMENTS

I wish to thank Peter Jackson and the editors for comments on drafts of this essay, and Jan Penrose and Susan Smith for discussions which have helped clarify some of the ideas. Funding was provided by the Secretary of State, Multiculturalism, for a larger project on racism in Canada for which this essay forms part of the context. The errors and opinions are entirely my own.

REFERENCES

Adachi, K. (1976) *The Enemy That Never Was: A History of Japanese Canadians*, Toronto: McClelland & Stewart.

Agnew, J.A. (1989) 'The devaluation of place in social science', in J.A. Agnew and J.S. Duncan (eds) *The Power of Place: Bringing Together Geographical and Sociological Imaginations*, London: Unwin Hyman, 9–29.

Althusser, L. (1971) 'Ideology and ideological state apparatuses', in L. Althusser *Lenin and Philosophy and Other Essays*, London: New Left Books, 121–76.

Anderson, A.B. and Frideres, J.S. (1981) *Ethnicity in Canada: Theoretical Perspectives*, Toronto; Butterworths.

Anderson, K. (1987) 'The idea of Chinatown: the power of place and institutional practice in the making of a racial category', *Annals of the Association of American Geographers* 77(4): 580–98.

—— (1988) 'Cultural hegemony and the race-definition process in Chinatown, Vancouver: 1880–1980', *Environment and Planning D: Society and Space* 6: 127–49.

Barker, E. (1961) *Principles of Cultural Pluralism*, New York: Oxford University Press.

Barnett, S. (1987) *Is God a Racist? The Right Wing in Canada*, Toronto: University of Toronto Press.

Berger, R. (1982) *Fragile Freedoms: Human Rights and Dissent in Canada*, Toronto: Irwin.

Breton, R. (1964) 'Institutional completeness of ethnic communities and personal relations to immigrants', *American Journal of Sociology* 70: 193–205.

Brotz, H. (1980) 'Multiculturalism in Canada: a muddle', *Canadian Public Policy*, 6 (Winter): 41–6.

Buchignani, N. (1982) 'Canadian ethnic research and multiculturalism', *Journal of Canadian Studies* 17(1): 16–34.

Burnet, J. (1975) 'Multiculturalism, immigration, and racism: a comment on the Canadian immigration and population study', *Canadian Ethnic Studies* VII(1): 35–9.

—— (1976) 'Ethnicity: Canadian experience and policy', *Sociological Focus* 9(2): 199–207.

—— (1981) 'Multiculturalism: 10 years later', in J. Elliot (ed.) *Two Nations, Many Cultures: Ethnic Groups in Canada*, Scarborough: Prentice-Hall, 235–42.

—— with Palmer, H. (1988) *Coming Canadians: An Introduction to a History of Canada's Peoples*, Toronto: McClelland & Stewart.

Butters, S. (1976) 'The logic of participant observation: a critical view', in S. Hall and T. Jefferson (eds) *Resistance through Rituals: Youth Subcultures in Post-War Britain*, London: Unwin Hyman, 253–73.

Cairns, A.C. (1988) 'Citizens (outsiders) and governments (insiders) in constitution-making: the case of Meech Lake', *Canadian Public Policy* XIV: S121–45.

Canada (1971) *House of Commons Debates*, 8 October.

—— (1987a) *Strengthening the Canadian Federation: The Constitution Amendment, 1987*, Ottawa: Dept of Supply and Services.

—— (1987b) *Multiculturalism ... being Canadian*, Ottawa: Dept of Supply and Services.

Canada, Department of Health and Welfare (1989) *Charting Canada's Future: A Report of the Demographic Review*, Ottawa: Dept of Supply and Services.

Canada, Department of the Secretary of State, Citizenship Branch (1967) *The Canadian Family Tree*, Ottawa: Queen's Printer.

Canada, House of Commons (1985) *Minutes of the Proceedings of Evidence of the Standing Committee on Multiculturalism*, Ottawa: Dept of Supply and Services.

Canada, Parliamentary Committee on Equality Rights, J. Patrick Boyer MP, Chairman (1985) *Equality for All*, Ottawa: Queen's Printer.

Canada, Royal Commission on Bilingualism and Biculturalism (1969) *The Cultural Contribution of the Other Ethnic Groups*, Vol. IV, Ottawa: Queen's Printer.

Canada, Secretary of State and Minister of State Multiculturalism and Citizenship (1989) *Eliminating Racial Discrimination in Canada*, Ottawa: Dept of Supply and Services.

Canada, Special Committee on Visible Minorities in Canadian Society, Bob Daudlin MP, Chairman (1984) *Equality Now!*, Ottawa: Dept of Supply and Services.

Canada, Standing Committee on Multiculturalism, Gus Mitges MP, Chairman (1987) *Multiculturalism: Building the Canadian Mosaic*, Ottawa: Queen's Printer.

Canada, Special Joint Committee on a Renewed Canada, Hon. Gérald A. Beaudoin and Dorothy Dobbie MP, Joint Chairpersons (1992) *A Renewed Canada*, Ottawa: Supply and Services Canada.

Canadian Consultative Council on Multiculturalism (1976) *Conference Report: Second Canadian Conference on Multiculturalism*, Ottawa: Minister of Supply and Services.

Canadian Ethnocultural Council (1987), 'To the back of the bus', brief presented to the Parliamentary Committee on the Constitutional Amendment Act 1987.

Clarke, J., Hall, S., Jefferson, T. and Roberts, B. (1976) 'Subcultures, cultures and class: a theoretical overview', in S. Hall and T. Jefferson (eds) *Resistance through Rituals: Youth Subcultures in Post-War Britain*, London: Hutchinson/Centre for Contemporary Cultural Studies, 9–74.

Cocks, J. (1989) *The Oppositional Imagination: Feminism, Critique and Political Theory*, London and New York: Routledge.

Commission for Racial Equality (ed) (1987) *Race and Ethnic Relations in Britain: Past, Present and Future*, London: CRE; reprinted from *New Community* XIV: 1–2.

Cosgrove, D.E. and Jackson, P. (1987) 'New directions in cultural geography', *Area* 19: 95–101.

Davies, P. (ed.) (1988) *Human Rights*, London and New York: Routledge.

Driedger, L. and Church, G. (1974) 'Residential segregation and institutional completeness: a comparison of ethnic minorities', *Canadian Review of Sociology and Anthropology* 11(1): 30–52.

Ethno Canada, newsletter of the Canadian Ethnocultural Council.

Foucault, M. (1978) *The History of Sexuality*, Vol. 1: *An Introduction*, trans. R. Hurley, New York: Vintage.

Furnivall, J.S. (1956) *Colonial Policy and Practice*, New York: New York University Press.

Glickman, Y. (1981) 'Political socialization and the social protest of Canadian Jewry: some historical and contemporary perspectives', in J. Dahlie and T. Fernando (eds) *Ethnicity, Power and Politics in Canada*, Toronto: Methuen, 123–50.

Gordon, M. (1964) *Assimilation in American Life*, New York: Oxford University Press.

Gramsci, A. (1971) *Selections from the Prison Notebooks*, London: Lawrence & Wishhart.

Gregory, D. and Ley, D. (1988) 'Culture's geographies', *Environment and Planning D: Society and Space* 6(2): 115–16.

Gwyn, R. (1980) *The Northern Magus*, Toronto: McClelland & Stewart.

Hall, S. and Jefferson, T. (eds) (1976) *Resistance through Rituals: Youth Subcultures in Post-War Britain*, London: Hutchinson/Centre for Contemporary Cultural Studies.

Hawkins, F. (1988) *Canada and Immigration: Public Policy and Public Concern*, Montreal: McGill-Queen's University Press.

Henry, F. (1986) *Race Relations Research in Canada Today: A 'State of the Art' Review*, Ottawa: Canadian Human Rights Commission.

Humphrey, J.P. (1970) 'The world revolution and human rights', in A. Gotlieb (ed.) *Human Rights, Federalism and Minorities*, Toronto: Canadian Institute of International Affairs, 157–79.

Inglis, F. (1977) 'Nation and community: a landscape and its morality', *The Sociological Review* 25(3): 489–514.

Jackson, P. (1987) 'The idea of "race" and the geography of racism', in P. Jackson (ed.) *Race and Racism: Essays in Social Geography*, London: Allen and Unwin, 3–22.

—— (1989) *Maps of Meaning: An Introduction to Cultural Geography*, London: Unwin Hyman.

Kallen, E. (1982) *Ethnicity and Human Rights in Canada*, Toronto: Gage.

—— (1988) 'The Meech Lake Accord: entrenching a pecking order of minority rights', *Canadian Public Policy* XIV: S107–20.

Kelner, M. and Kallen, E. (1974) 'The multicultural policy: Canada's response to ethnic diversity', *Journal of Comparative Sociology* 2: 21–34.

Kerr, R.W. (1969) *Legislation against Discrimination in Canada*, New Brunswick: Human Rights Commission.

Kobayashi, A. (1987a) 'From tyranny to justice: the uprooting of Japanese Canadians after 1941', *Tribune Juive* 5: 28–35.

—— (1987b) 'Real or apprehended? The Japanese–Canadian redress issue and human rights', *The Human Rights Advocate* 3: Rights and Freedoms Section, 1–3.

—— (1989) 'A critique of dialectical landscape', in A. Kobayashi and S. Mackenzie (eds) *Remaking Human Geography*, London: Unwin Hyman, 164–84.

—— (1990) 'Racism and law in Canada: a geographical perspective', *Urban Geography* 11: 447–73.

Kobayashi, A. and Olson, S. (forthcoming) 'Ethnicity in Canadian cities', in D. Ley and L. Bourne (eds) *The Changing Social Geography of Canadian Cities*, Montreal: McGill-Queen's University Press.

Kostash, M. (1977) *All of Baba's Children*, Edmonton: Hurtig.

Lee, D.J. (1979) 'The evolution of nationalism in Quebec', in J. Elliot (ed.) *Two Nations, Many Cultures*, Scarborough: Prentice-Hall, 60–74.

Lévesque, R. (1968) *An Option for Quebec*, Toronto: McClelland & Stewart.

Ley, D. (1984) 'Pluralism and the Canadian state', in C. Clarke, D. Ley and C. Peach (eds) *Geography and Ethnic Pluralism*, London: Allen & Unwin, 87–110.

Lupul, M.R. (ed.) (1978) *Ukrainian Canadians, Multiculturalism, and Separatism: An Assessment*, Edmonton: University of Alberta Press.

—— (1989) 'Networking, discrimination and multiculturalism as a social philosophy', *Canadian Ethnic Studies* XXI(2): 1–12.

Lyotard, J.-F. (1984) *The Postmodern Condition: A Report on Knowledge*, trans. G. Bennington and B. Massumi, Minneapolis: Minnesota University Press.

McChesney, R.A. (1987) 'Canada', in J. Donnelly and R.E. Howard (eds) *International Handbook of Human Rights*, New York: Greenwood Press, 29–47.

McRoberts, K.H. and Posgate, D. (1976) *Quebec: Social Change and Political Crisis*, Toronto: McClelland & Stewart.

Mannette, J.A. (1988) '"A trial in which no one goes to jail": the Donald Marshall Inquiry as hegemonic renegotiation', *Canadian Ethnic Studies* XX(3): 166–80.

Miles, R. (1989) *Racism*, London: Routledge.

Morris, R.N. and Lanphier, C.M. (1977) *Three Scales of Inequality: Perspectives on French–English Relations*, Toronto: Longman.

Naidoo, J. (1989) 'Canada's response to racism: visible minorities in Ontario', background paper, Third International Symposium at Oxford, The World Refugee Crisis: British and Canadian Responses.

The Network (1989) Newsletter of the Race Relations Committee, Canadian Ethno-cultural Council 2(1), Summer.

Palmer, H. (1976) 'Reluctant hosts: Anglo-Canadian views of multiculturalism in the twentieth century', in Canadian Consultative Council on Multiculturalism (CCCM) *Conference Report: Second Canadian Conference on Multiculturalism*, Ottawa: Dept of Supply and Services, 81–118.

Perin, R. (1983) 'Clio as an ethnic: the third force in Canadian historiography', *Canadian Historical Review* LXIV(4): 441–67.

Peter, K. (1981) 'The myth of multiculturalism and other political fables', in J. Dahlie and T. Fernando (eds) *Ethnicity, Power and Politics in Canada*, Toronto: Methuen, 56–67.

Phillips, P. (1967) *No Power Greater: A Century of Labour in B.C.*, Vancouver: BC Federation of Labour and the BOAG Foundation.

Porter, J. (1964) *The Vertical Mosaic: An Analysis of Social Class and Power in Canada*, Toronto: University of Toronto Press.

Roy, P.E. (1989) *A White Man's Province: British Columbia Politicians and Chinese and Japanese Immigrants 1858–1914*, Vancouver: University of British Columbia Press.

Sartre, J.P. (1963) *Search for a Method*, trans. H. Barnes, New York: Vintage.

—— (1976) *Critique of Dialectical Reason*, trans. A. Sheridan-Smith, ed. J. Ree, London: New Left Books.

Sigler, J.A. (1983) *Minority Rights: A Comparative Analysis*, Contributions in Political Science No. 104, Westport: Greenwood Press.

Smith, M.G. (1965) *The Plural Society in the British West Indies*, Berkeley: University of California Press.

Smith, S.J. (1989a) 'Society, space and citizenship: a human geography for the "new times"?', *Transactions of the Institute of British Geographers* (new series) 14: 144–56.

—— (1989b) *The Politics of 'Race' and Residence*, Oxford: Polity.

Soja, E.W. (1989) *Postmodern Geographies: The Reassertion of Space in Critical Social Theory*, London: Verso.

Sunahara, A.G. (1981) *The Politics of Racism: The Uprooting of Japanese Canadians during the Second World War*, Toronto: Lorimer.

Ward, P. (1978) *White Canada Forever: Popular Attitudes and Public Policy toward Orientals in British Columbia*, Montreal: McGill-Queen's University Press.

Weinfeld, M. (1981) 'Myth and reality in the Canadian mosaic: "affective ethnicity", in R.M. Bienvenue and J.E. Goldstein (eds) *Ethnicity and Ethnic Relations in Canada*, 2nd edn, Toronto: Butterworths, 65–86.

Williams, R. (1961) *Culture and Society 1880–1950*, Harmondsworth: Penguin.

—— (1976) *Keywords: A Vocabulary of Culture and Society*, London: Fontana.

—— (1977) *Marxism and Literature*, Oxford: Oxford University Press.

—— (1980) *Problems in Materialism and Culture*. London: Verso, New Left Books.

—— (1981) *Culture*, London: Fontana.

Yuzyk, P. (1967) *Ukrainian Canadians: Their Place and Role in Canadian Life*, Toronto: Ukrainian Canadian Business and Professional Federation.

13

REPRESENTING POWER
The politics and poetics of urban form in the Kandyan Kingdom
James Duncan

INTRODUCTION

In this essay I explore the relationship between representation, power and landscape in a particular place and time: the Kandyan Kingdom in early nineteenth-century Sri Lanka. In focusing upon the landscape, I situate my interpretation within a long tradition of artefactual analysis within American cultural geography. And yet in other respects the type of analysis undertaken here constitutes a radical departure from past practice in that the cultural is conceived of as inextricably intertwined with the political, and the landscape is seen not as reflective of socio-cultural processes but as constitutive of them.

In studying the relationship between socio-cultural processes and landscape I draw heavily upon the concepts of discursive field, discourse and narrative, for a culture's signifying system can be thought of as composed of what Foucault (1970) has termed discursive fields containing discourses which are in turn composed of narratives (Geertz 1983). Discussions of discourse have usually been applied to Western societies and texts constructed around rational enlightenment precepts. However, the same strategy is equally, and perhaps more relevant to a highly ritualized society, even if it is largely pre-literate. Here, as Geertz (1980) has shown within the context of Bali, landscape can play its part among a diversity of texts to consolidate a culture and a politics.

The story I tell is of a group of Kandyans living in the central mountains of Sri Lanka in the early nineteenth century – a weak and vain king, his scheming nobles and a long-suffering peasantry who were exploited both by the king and by the nobles. At the heart of the political drama is the city of Kandy, the royal capital. In order to set the context for this political drama, I will first show how the landscape of the city is inter-textually connected to the larger political discourse of kingship in the Kandyan Kingdom. As such, the layout of the city is an important part of the way in which this discourse is represented to the citizenry at large. I will then focus upon a massive city-building programme which the king undertook in the early nineteenth century, both to celebrate his army's victory over an invading British force, and to heighten his charismatic appeal. Finally, I will examine the way various interest groups within the kingdom, the king, the nobles and the peasants, represented this building programme – how a politics of interpreting the landscape was at once rooted within the larger political discourse of power and yet ultimately undermined that very discourse. The landscape of the city, as we shall see, is not merely the site where the political struggle takes place, it becomes the *means* by which each party attempts to defeat the other. As such the landscape is an important part of the practice of power.

DISCOURSES OF POWER

The representations of kingship and landscape are part of a larger unified set of practices. Representations always operate within discursive fields. The latter are usually focused around what might be broadly termed institutions. For example, in any society there are discursive fields around law, science or politics; these in turn contain a range of competing discourses constituted by a set of narratives, concepts and ideologies relevant to a particular realm of social practice. Some of these discourses may be hegemonic, in that they reinforce an existing power structure, while others are at least potentially contestatory. At times there may be a relatively stable discursive order in which competing discourses coexist, while at other times they may come into open conflict. Discourses have a dual nature in that they simultaneously define the social framework of intelligibility within which practices are communicated and negotiated, while they serve as resources to be used in the pursuit of political power. The representation of power is culturally variable. It is articulated through different symbols within different social formations. Within the Kandyan Kingdom in the early nineteenth century the discursive field of political power was centred around the institution of kingship, within which there were two distinct yet intertwined discourses on kingship, the Asokan and the Sakran. Both of these models are common throughout Indian Asia. Each specified the reciprocal obligations and duties of ruler and ruled.

The Asokan discourse was based upon accounts of the great third-century BC Indian Buddhist monarch, Asoka. This model is of a righteous ruler devoted to fostering the Buddhist religion and the welfare of his people.

The Sakran discourse drew upon texts of Sakra, the king of the gods. Accordingly, the king was expected to model himself upon Sakra and become a *cakravarti*, a universal monarch who rules over his people just as Sakra rules over the kingdom of the gods. Although the king was also expected to be just and benevolent, greatest emphasis was placed upon the glorious and divine qualities of kingship. He was to be thought of as a god-king.

These discourses had political and religious texts associated with them which provided precedent for and thereby legitimated the practice of kingship. More-over, each discourse had a landscape model associated with it. The Asokan emphasized the production of a landscape dominated by public works for the benefit of the citizens and religious structures such as monasteries and *dagobas* containing religious relics. The Sakran, on the other hand, produced landscapes dominated by great palaces and monumental public spaces modelled upon the texts of the city of the king of the gods in heaven. The concretization of these discourses became an important political issue in a poor kingdom such as the Kandyan one, for building programmes of either the Asokan or Sakran variety were costly both in terms of resources and human labour. The desire of Kandyan kings to be great builders was always fraught with the political risk of bankrupting the country or alienating the people due to great demands for *rajakariva* (corvée labour).

Although the Sakran and Asokan discourses differed in their models of kingship, they coexisted in syncretic form within the discursive field of politics throughout Kandyan history as kings adopted elements of each model. The *Mahavamsa* (1950: 55), the great chronicle of kingship written by Buddhist monks over the centuries, attempted to reconcile these discourses. The king in Kandy as the earthly representative of Sakra was expected to protect the religion and the people. The kings of Kandy possessed what is alleged to be a relic of the Buddha, snatched from his funeral pyre in the sixth century BC. This relic legitimized the king's rule, for its possession showed that the king had obtained the Buddha's favour and the kingdom thus received an aura of moral excellence and supernatural power (Malalgoda 1976: 14; Seneviratne 1978: 182–4; Smith 1978a: 87; 1978b: 52). One could see this syncretism represented in the landscape of Kandy as a balance was achieved between religious structures and palaces. Although sanctioned by texts, this syncretism was, nevertheless, a fragile entity, based upon the willingness of all parties concerned to overlook any contradictions between the discourses. It was precisely this structural instability which allowed for a vibrant politics to take place. For when political disputes erupted, all parties attempted to justify their claims in terms of one or the other of the discourses. Inevitably, then, on such occasions the differences between them were brought to the fore for strategic reasons.

The city of Kandy became the capital of an independent state in the mountainous core of the island in the late fifteenth century. The palladium of kings, the Tooth Relic, was brought to the city in the late sixteenth century, lending prestige to the capital and the king who occupied the throne (De Silva 1981: 201). In the early eighteenth century the last of the ethnic Sinhalese kings of Kandy died without legitimate offspring and the kingdom passed through his wife's line to a South Indian Tamil dynasty known as the Nayakkar. From that date on the Kandyan Kingdom had Tamil kings ruling a Sinhalese nobility and populace. The Nayakkars' Tamil ethnicity was always a potential source of political instability in the kingdom, for Tamils from South India were the traditional enemies of Sinhalese rulers. Throughout the eighteenth and early nineteenth centuries there was a constant struggle between the king and the Sinhalese nobles for control over the administrative bureaucracy of the kingdom.

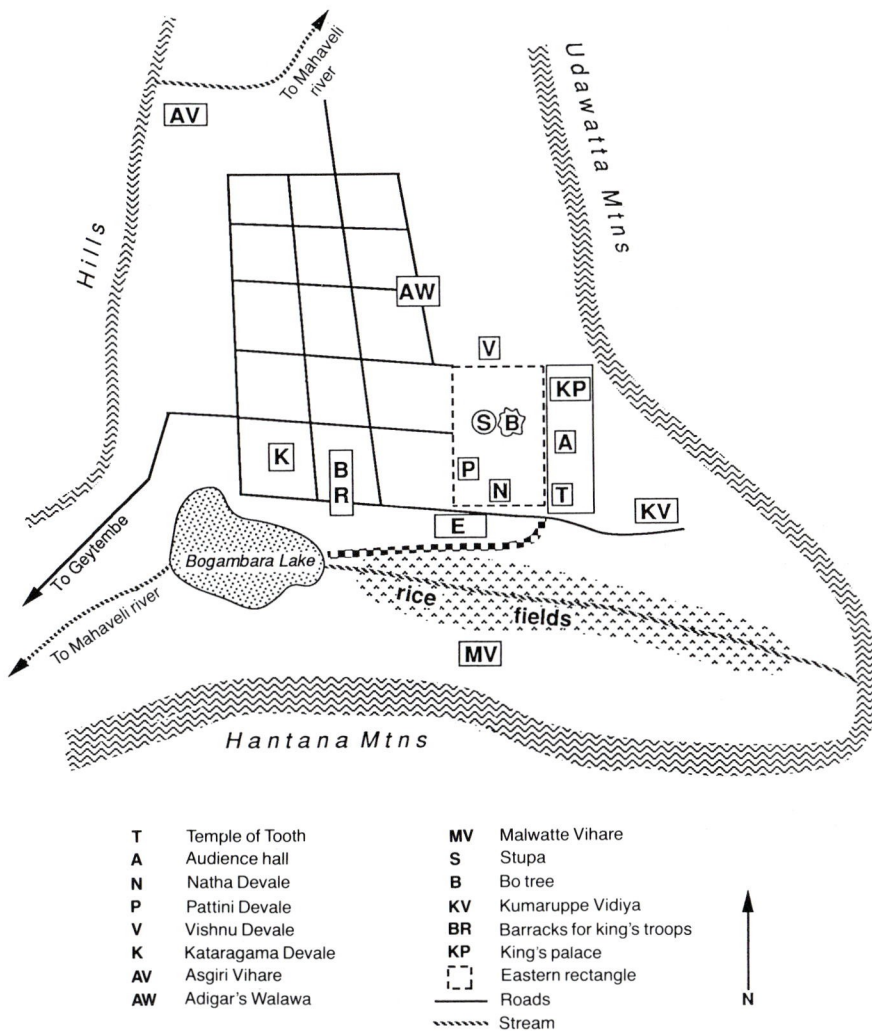

Figure 13.1 Map of Kandy in 1800

Source: Duncan 1990: 75. Reproduced with permission from Cambridge University Press.

Increasingly throughout this period as the kings' control over the bureaucracy weakened they sought to counterbalance this loss by increasing their political legitimacy. They did this by simultaneously conforming to both the Asokan and Sakran discourses of kingship. They made an effort to be, or appear to be, fervent Buddhists. They demonstrated this fervour by building religious structures for the Buddhist clergy and, in the words of Seneviratne (1978: 182), conspicuously venerating the Tooth Relic of the Buddha. They also conformed to Sakran ideals by greatly elaborating the symbolism of the god-king in court ritual and ambitious programmes of palace building and city beautification.

Before exploring the role that the city played in the political process I outline the morphology and symbolism of the city as it existed at the beginning of the nineteenth century. I then examine the ambitious building programme of the last king of Kandy, which I argue contributed greatly to his downfall. The very form of the city itself suggests that it was built as a cosmic capital, a representation of the city of the king of the gods descended to earth (Figure 13.1).

The city was composed of two rectangles, the sacred shape of the cities of the gods (Maimataya n.d.; Mus 1937: 69). The western rectangle comprised the city proper. It was here that the nobles had their residences and where the common people had their shops and houses. The western rectangle was divided into four sections by two major avenues running north–south and east–west (Keppitipola 1918). The number four and multiples of four were highly symbolic as representing totality, the four quarters of the world which the king as a universal monarch symbolically sought to control.

It was the eastern rectangle containing the palace of the king and the temples of the gods that was the true locus of ritual power in the kingdom and which played a key role in the legitimation of political power. It was here that one could see most clearly the attempt to mirror the city of the king of the gods in the Kandyan capital. According to the religious texts which inform the Sakran discourse on kingship, the gods live on the top of Mount Meru which lies at the centre of the universe and connects the world to the heavens. Mount Meru, the central cosmic mountain, is represented as having a central peak flanked by four buttress peaks, the eastern being the location of Sakra's palace. Here we have a parallel between the palace of Sakra containing a shrine for the Tooth Relic on Mount Mandara, the eastern peak of Meru, and the palace-temple complex with its audience hall and Temple of the Tooth Relic on the eastern side of Kandy. There are shrines to the gods on the northern, southern and western sides of the rectangle as well. These shrines were thought of as the actual abodes of the gods in the capital (Obeyesekere 1984: 37). At the northern boundary of the sacred rectangle lies the Temple of Visnu. Post-Vedic mythology places Visnu's paradise, which is called Vaikuntha, on the *northern* peak of Mount Meru (Stutley and Stutley 1984: 316).

At the southern edge of the sacred rectangle is the temple of the god Natha who is also thought of as the *Maitreya* (future Buddha) whose heavenly location according to the *Satara Devala Devi Puwata* (1954: 158) is in the Tusita heaven directly above Mount Meru.

At the western edge of the sacred rectangle lies the Temple of Pattini who is a Bodhisattva (future Buddha) and therefore dwells in the Tusita heaven over Mount Meru (Obeyesekere 1984: 59). The temples to the four gods, therefore, mark out the four cardinal directions of the sacred rectangle and render it a representation of the heavens on the top of the cosmic mountain at the centre of the universe.

At the centre of the sacred rectangle there is a sacred Bo tree which is said to come from a shoot of the great Bo tree under which the Buddha was enlightened in the sixth century BC in northern India. The Bo tree like Mount Meru is a cosmic axis that unites the earth and the heavens (Coomaraswamy 1956: 32, 97). Next to the Bo tree in the centre of the sacred rectangle sits a dagoba which contains the begging bowl of the Buddha. Dagobas are symbolic of Mount Meru and are thought to have cosmic axes running through them (Mabbett 1983: 75–7). Here Asokan iconography is linked to the Sakran in that Sakra is reputed to have built a dagoba on the top of Mount Meru for a relic of the Buddha (Ariyapala 1956: 372).

Drawing upon the landscape models associated with the Sakran and Asokan discourses on kingship, Kandy served as a stage upon which a god-king who was also Buddhist monarch could display both his benevolence and ritual power to his nobles and commoners. This was the city which the last king of Kandy, Sri Vikrama, inherited when he became king in 1798.

SRI VIKRAMA'S BUILDING PROGRAMME

In 1803 an English army of three thousand men arrived in Kandy in order to capture the king and insert a puppet ruler who would be subservient to the British. The king torched a portion of his palace as well as the city's major temples so that they would not be desecrated by the invaders. He then melted away into the surrounding mountains with his army. The main British force ransacked the city, retreated and left a garrison to hold it. The Kandyans, who were adept at guerrilla warfare, gradually choked off the supply lines connecting the garrison to the coast and waited in the mountains surrounding the city for the European troops to be weakened by fever and hunger. After holding the city for several months, the British garrison was captured and summarily executed by the Kandyans. This represented a great victory for the Kandyan king as the British were reluctant to undertake another invasion.

The king used this period of peace to attempt to reassert control over the nobility, who for a century had gradually been growing in power. However, during the next five years he had very limited success in bringing his nobles into line, and they continued to plot among themselves and with the British. Increasingly, the king withdrew, surrounding himself with a circle of Tamil advisers, thereby further alienating his Sinhalese nobles and setting off a new round of plotting against him. He responded to these threats from the nobles by seizing the property of families who were implicated in the unrest and executing suspected leaders. Frustrated by the continual intrigues of his nobles, Sri Vikrama turned to a time-honoured tradition of South Asia kings, a magnificent building programme in the capital to demonstrate his greatness both to his supporters and his enemies. Such a building programme was prescribed by both the Sakran and Asokan models. However, the former required palaces or allusions to the world of the god-king, while the latter expected stupas and monasteries. The king

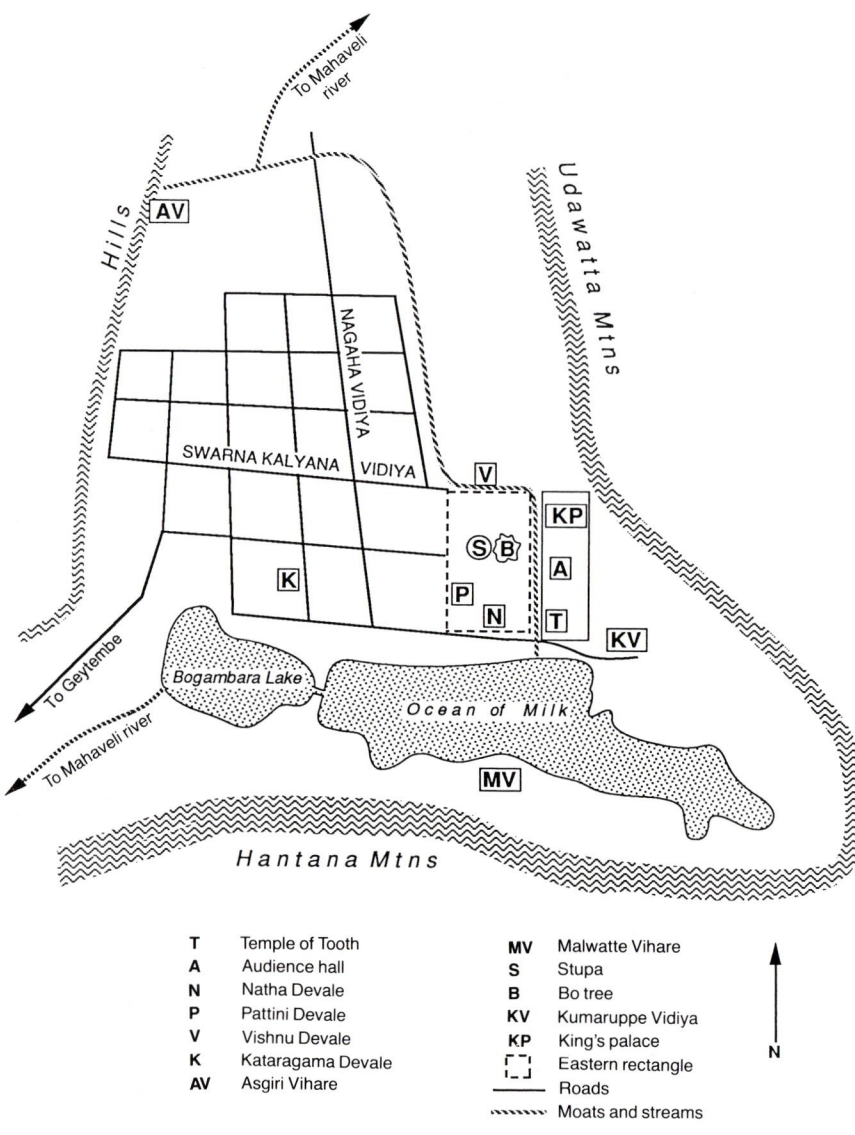

T	Temple of Tooth	**MV**	Malwatte Vihare
A	Audience hall	**S**	Stupa
N	Natha Devale	**B**	Bo tree
P	Pattini Devale	**KV**	Kumaruppe Vidiya
V	Vishnu Devale	**KP**	King's palace
K	Kataragama Devale	⌐⌐	Eastern rectangle
AV	Asgiri Vihare	——	Roads
		⌁⌁⌁	Moats and streams

Nagaha Vidiya – N–S divide of city
Swarna Kalyana Vidiya – E–W divide of city

Figure 13.2 Map of Kandy in 1815

Source: Duncan 1990: 83. Reproduced with permission from Cambridge University Press.

emulated Sakra, who after the defeat of the *asuras* (demons) rebuilt his palace and capital with the help of Visvakarma the divine architect. At the same time the king also emulated the earlier hero-kings of Lanka, who after military victories engaged in major public works projects. Between 1809 and 1812 there was nearly continuous rebuilding and enlarging of the city, the palace and the royal gardens (Figure 13.2). By the end of 1810 the renovations on the palace were completed and a *pattirippuwa* (octagonal tower) from which the king could address his subjects had been added to the Temple of the Tooth Relic (Plate 13.1). This octagonal structure was of great symbolic significance, for when the king stood in this tower he stood at the centre of the world with the eight points of the compass radiating out around him, symbolizing and magically reinforcing his power. The temple-palace complex was surrounded by a moat and double wall. The outer wall around the moat was topped by a series of triangular waves and each wave was incised with four triangular niches (Plate 13.1). The wave pattern symbolizes the heavenly Ganges river which falls from the sky on to the top of Mount Mandara and runs down its side into the cosmic ocean at the mountain's foot. The triangular niches represent both the triangular leaves of the Bo tree (a cosmic axis) and, since the triangle is the symbol for fire in Hindu iconography, the thunderbolt of Sakra the rain god. The higher, inner wall was called the Celestial Rampart and was topped by an undulating pattern representing clouds. Each cloud in this wall was incised by four niches in the shape of a trefoil which is the symbol for Mount Meru. The palace-temple complex rose above this Meru wall just as Sakra's palace rose from the top of Mount Meru.

During the next year the king expanded the western rectangle by adding five new city blocks. This extension of the city was undertaken in order to solve some of the problems besetting the state, through the magic of parallelism. The *Ingrisi Hatana* (1906: verse 249) a poem commissioned by the king in 1812, claims that 'After measuring the length and width of the city, he [the king] made it into a perfect square in order that the expenditure and income of the kingdom would be equal.' Furthermore by adding five squares to the western rectangle he brought their number to twenty-one, which was the number of provinces in the kingdom. By reproducing within the western rectangle, through the parallelism of like numbers, the provinces in the kingdom, he could magically control the provinces from his palace at the centre of the world.

Sri Vikrama's greatest building project was the creation of a huge artificial lake to the south of the city, by constructing a great dam at the west end of the rice fields belonging to the major temples and monasteries in the city (Figure 13.1). Here he created a lake over two miles in circumference. Between 1810 and 1812, 2,000 to 3,000 labourers were forced to work on the construction of this dam and the walls surrounding the lake (Figure 13.2). This project caused serious disruption to the city and entailed the confiscation of temple lands as well as the property of private citizens. The king called this lake the Kiri Muhuda (Ocean of Milk), which was the name of the cosmic ocean which lay at the foot of Mount Meru. The king also ordered a canal cut around the perimeter of the city (Figure

Plate 13.1 Kandy: the palace-temple complex in 1815

Source: Duncan 1990: 81. Reproduced with permission from Cambridge University Press.

13.2). Such a moat was too small to serve defensive purposes and served only as an allusion to the annular ocean surrounding Mount Meru.

We can understand this building programme, therefore, as an attempt by the last king to create a more perfect reproduction of the world of the gods within his capital and thereby to approximate more closely the glory and the power of Sakra, the king of the gods. He wished to emulate not only the king of the gods, but also the Lankan hero-kings of old who ruled from monumental capitals in the rich irrigated lands in the north of the island.

Unfortunately, however, the Kandyan Kingdom was poor, due in part to the limited amount of productive rice land and to the policy of the Europeans who controlled the coasts. First the Portuguese, then the Dutch and finally the British pursued a policy of economically weakening the Kandyan state in order to control it more effectively. As a result of this penury, the king was unable to pay for his construction projects and thus relied entirely upon corvée labour. However, the normally accepted amount of labour due the king was greatly exceeded. The question was also raised as to whether so much labour should be devoted to Sakran constructions, for it was noted that this king had failed in his duties as an Asokan monarch as he built no monasteries or stupas. This can in part be explained by the king's antipathy to the nobility who controlled the top positions in the Buddhist clergy and who were implicated in some of the plots against him. Instead of supporting the monkhood, the king in fact seized temple lands and dragooned peasants whose labour was supposed to be under the control of the temples. Sri Vikrama's unwillingness to engage in even token Asokan projects during these years was a great political miscalculation on his part, for it allowed his enemies to argue that he had violated his duties as king.

THE POLITICS OF INTERPRETATION

Let us now turn to a consideration of the politics of interpretation that surrounded the king's building project. As we shall see, not only was the project conceived within the discursive field of Kandyan kingship, but the political debate over the programme was structured by this same discourse as well. Because the landscape of the city was seen by all parties as a representation of political power, criticism of the landscape and the process by which it was built inevitably was seen as a criticism of the practice of power itself. Kandyan society was highly textualized in that all official action depended upon a series of authoritative historical texts. As we shall see, various groups within the kingdom – the king and his supporters, rebellious nobles and disaffected peasants – each clustered around a particular reading of kingly behaviour and urban form. These readings were then employed in attempts to gain political advantage over their opponents.

The king's interpretation

There are a number of documents written during and immediately after the building programme was completed that further explain the king's actions and attempt to answer the criticisms that had been launched against him. A major document is the *Ingrisi Hatana* (1906), a poem written in 1812 at the king's behest to commemorate his victory over the English in 1803 and celebrate the glory of his rebuilt capital. Other accounts such as the *Dalada Vittiya* (1974) written in 1812 and several *sannasas* (royal grants usually inscribed on copper) are interesting because they describe the city building within the context of religious models of city buildings and the city building of the hero-kings of Lanka.

Since one of the charges against the king was that his construction violated the traditional laws, he responded in the *Ingrisi Hatana* (1906: v. 82) that he had 'summoned the ministers, capable of protecting the world, knowing all of the rules and regulations, and what should and should not be done'. In this way, the king no doubt wished to place the burden of responsibility upon the shoulders of his Sinhalese advisers. Sri Vikrama stated as his goal to create a capital 'filled with all the wealth and requisites of cities, and which flourishes with the prosperity of the fortunate city of Indra [Sakra] (as if) reflected on the mirror of the earth' (*Moligoda Sannasa* 1814).

In order to have his capital truly mirror Sakra's capital, changes to the fabric of the city had been made. The king, therefore, had 'brought master builders of all types from the three areas of the island who were like the god Visvakarma in their ability' (*Ingrisi Hatana* 1906: v. 249). Here we can see the king specifically alluding to the myth of the divine architect who rebuilt Sakra's city after his great victory over the demons. He justified his great wall and moat as 'resembling Sakra's palace'. He claimed that he planted trees so that 'the city was made to resemble the heavenly city of Sakra' (ibid.: vv. 250–3, 255).

It is curious that only one verse was devoted to his greatest building project, the artificial lake which he named the Ocean of Milk after the primordial ocean lying at the foot of Mount Meru. Perhaps little mention is made of the lake because of the political tension that was running high among the aristocracy, the clergy and the peasants over what these groups saw as appropriations of lands and demands for corvée labour.

The *Ingrisi Hatana* (1906: v. 203) also is at pains to compare Sri Vikrama's building programme to that of Parakramabahu, one of the hero-kings of medieval Lanka. One discrepancy that the poem fails to mention, however, is that whereas Parakramabahu built many religious structures in addition to his palace and artificial lakes, Sri Vikrama completely ignored the former. In fact the *Ingrisi Hatana* stands as a remarkable document not merely because it provides insight into the king's justifications, but because of its near total absence of Asokan imagery. The document points to the grave imbalance that was developing between the Sakran and Asokan discourses on kingship. It was an imbalance that his enemies would be quick to exploit.

The nobles' interpretation

Throughout the eighteenth century the Nayakkar kings had maintained power by pursuing three principal strategies. They tried to limit the power of the nobles by turning different factions against one another. They had embraced the Asokan model of kingship, becoming lavish patrons of Buddhism. They increasingly yielded control over the provincial bureaucracy and retreated into a Sakran display of pomp and ceremony. Through this threefold strategy they were able both to placate the powerful nobles and to appear to the populace as powerful, charismatic rulers. One source of tension between the Nayakkar kings and the Sinhalese nobles was the growing community of Tamil relatives and retinue who came to form an inner circle within the palace bureaucracy and who were granted special economic privileges by the kings. In the early nineteenth century, powerful Sinhalese nobles found themselves not only excluded from the king's inner circle, but frequently in debt to the king's relatives as well. Sri Vikrama, installed by the nobles in 1798 as a young king whom they thought would serve their interests, became intent after the defeat of the British in 1803 on reclaiming some of the power that the Nayakkars had lost to the Sinhalese nobles. As a result, the nobles plotted with the British, and the king responded by punishing members of the leading families and beheading the Prime Minister who had installed him.

In 1814 the new Prime Minister by the name of Ahalepola also revolted against the king but his revolt was suppressed and he fled to British territory on the coast. The ferocity of the king's response to Ahalepola's faction among the nobility and the increasingly indiscriminate revenge taken upon noble families finally united the nobles in opposition against him. In failing to maintain competing factions among the nobility, Sri Vikrama neglected to employ one of the central strategies of Nayakkar rule.

The king's political problems were greatly exacerbated by his increasingly poor relations with the monkhood, whose interests as we have seen had been neglected. He even went so far as to execute a monk whom he suspected of being involved in Ahalepola's abortive 1814 revolt. Sri Vikrama, therefore, by abandoning the Asokan model of kingship, abandoned the second political stratagem of the Nayakkar kings. Although during the first decade of the nineteenth century the king had retained the support of the common people, he soon lost this as well by greatly exceeding the customary demands for corvée labour. Thus ironically he lost the support of the people which the building project was expected to win.

Ahalepola's revolt against the king in 1814 was simply the first move in his campaign to become king himself. From exile in Colombo he co-ordinated the British invasion of Kandy and convinced the other factions among the Sinhalese nobility to offer no resistance to the British army. Ahalepola returned to Kandy with the British in 1815 and attempted to convince both the British and the Sinhalese nobles and peasantry that he should be the next king of the Kandyans.

He commissioned two heroic poems to be written about himself in a style reserved for kings. The *Kirala Sandesaya* (1958) and the *Ahalepola Varnanava* (n.d.) contain interesting information on the nobles' interpretation of the king's building programme. As one would expect, the argument in these texts was anti-Sakran and was phrased within the Asokan discourse on kingship. The textual basis of Ahalepola's claims were derived from the *Mahavamsa* (1950), a largely Asokan document presenting the history of early Lankan kingship from the perspective of the clergy. The two texts first document the injustices committed by the king against the clergy and the people, and then present Ahalepola as a hero-king who has driven the wicked Tamil oppressors away and returned to claim his rightful place as the head of a Sinhalese dynasty. The two texts perform different political tasks. The *Kirala Sandesaya*, although it alludes in general terms to the king's construction activity, focuses attention primarily upon the king as a Tamil oppressor. The *Ahalepola Varnanava*, on the other hand, discusses the king's city building in great detail. Both texts rhetorically situate the king's building programme at the centre of the discourse of kingship by citing violations of the codes of kingship which occurred. The poems specifically accuse the king of sacrilege, in cutting two Bo trees and demolishing a preaching hall (*Ahalepola Varnanava* n.d.: v. 22). The king was accused of destroying the Buddhist religion by confiscating temple lands and killing monks (*Kirala Sandesaya* 1958: v. 37–9).

Whereas the king's account skirts the politically sensitive issue of the construction of the great artificial lake, the nobles' account focuses in upon it. He was accused of having supervisors who were so cruel that they made peasants work without food all day and beat them like animals (*Ahalepola Varnanava* n.d.: v. 26–8).

The *Ahalepola Varnanava* (n.d.: vv. 29–35, 43–4, 47) also attacked the building programme for bringing in large numbers of people to the city from the provinces and removing houses and trees in order to build new streets. The poem claims that the king stood like a devil on the *pattirippuwa* (octagonal tower) and shouted at the people to work harder. The poem claims that great mountains of earth were built chaotically throughout the city. Street patterns were disrupted and a great lake was constructed to no apparent purpose. Furthermore, the king is also accused of torturing those who criticized his destruction of a Buddhist temple.

The *Ahalepola Varnanava*, then, is a political tract which drew its power from the Asokan discourse on power. It is interesting, however, that there is no direct attack on the Sakran model of the god-king. This becomes especially puzzling when the whole point of the king's city building was to enhance his claim to be a god-king. Why is it that the *Ahalepola Varnanava* portrays the building project as irrational? The reason, I believe, is that Ahalepola wished to become king himself and expected to adopt elements of the Sakran model as all kings of Kandy had done. He therefore embraced the Asokan discourse while remaining silent about the Sakran.

In fact, the changes which the king made to the city did not violate tradition. Rather they were part of an ongoing process of urban development in Kandy which had been continuing for centuries. What violated tradition was, however, the king's excessive demands for peasant labour and his failure to negotiate a settlement with the *sangha* (monkhood) for religious property. It was the king's ever worsening relations with the nobles, however, which ultimately propelled the building programme to the centre of a political crisis.

The peasants' interpretation

We know from the diary of the English master-spy John D'Oyly (1975: 74–5) that a great deal of unrest was developing in the southern provinces which were administered by the Prime Minister Ahalepola. Peasants in this region were reputed to have fled into the hills to escape the king's agents while the latter destroyed their houses. There is, however, little direct evidence of the peasants' views of the king's city-building projects. What we have seen so far is filtered through the lens of Ahalepola's attack on the king or of the reports of D'Oyly's spies. There is one source available, however, a series of ritual texts performed during the Kandyan period. These texts took the form of a series of allegories about two mythical kings, the good king of Soli and the wicked king of Pandi. They continued to be performed in particular villages in the southern Kandyan province which had rebelled against the king in 1814 until the 1950s when they were recorded by the anthropologist Gannanath Obeyesekere (1984: 326–33). One text which is called the *pataha* (pond ritual) spoke of the building programme of the evil Tamil king of Pandi. This ritual appears to be an extremely detailed allegorical attack upon Sri Vikrama's projects. This allegory corresponds closely to Ahalepola's critique and yet it diverges from the noble's critique in one crucial respect. Obeyesekere's account of the ritual is interspersed with my own commentary in light of what we know of Sri Vikrama's building programme.

The narrator of the ritual tells the audience that the king of Pandi had a wondrous city that resembled the city of the king of the gods built by Visvakarma the divine architect. But this evil king was not content and asked himself, 'In what respect is our city different from heaven?' The narrator adds, 'It is a thought that will bring disaster on him.' So far the story of the king of Pandi corresponds to that of Sri Vikrama in Kandy whose city was like that of the king of the gods, but who vainly wished to achieve an even closer correspondence. The chief minister reads the king of Pandi's mind and says:

> O mighty one, lord of the seven world systems
> (*sakvalas*)
> O warrior powerful as Sakra himself
> Like Sakra himself possessing three eyes
> Is it your pleasure that we build a pond?

The king then commands his people to begin digging a lake 'huge in both length and breadth' and to complete it within seven days. The ritual describes the suffering of the labourers:

> He [the king] summons Brahmans, watchers, and
>   supervisors
> To break the laziness in the camp
> Young lads were tied up together
> And beaten to make them work.

The bad king is accused of grabbing 'hold of idlers' and beating 'out their brains'. They say 'he cuts their bodies and slaughters the lads'. The poem continues:

> A foolish king in spite of his broad forehead
> To please him we carry large baskets on our heads
> We suffer a thousand sorrows and misfortunes
> Our heads are bald from carrying these baskets.
>
> We dig up stones and trees, and heap the earth
> But these efforts are worthless, like husking a
>   coconut without the kernel.
>
> People gather to work like a sheaf of reeds
> Their mouths were so parched that they even forgot
>   their suffering
> Like bulls they bit their lips and bore it all
> Who can escape the sins of past births?

In spite of all this labour, the king says, 'These are idle workmen, they should be impaled on an *ula* [sharpened stake]. No inquiry is necessary.'

The parallel between the peasants' allegorical account of the building of the lake and Ahalepola's is striking. In both cases it is argued that the king oppresses his subjects by greatly exceeding the customary demands for corvée labour and engaging in massive works of no religious merit or public utility. The two accounts differ in one key respect, however – their explanation of the king's motivation for the building project. Ahalepola's account, while arguing from an Asokan point of view, portrays the king as evil and irrational. However, because of his own desire to become king, Ahalepola does not directly attack the Sakran model of kingship. The peasants, on the other hand, with no such strategic goals in mind adopt a much more radically Asokan position. Rather than obscuring the king's reasons as Ahalepola does they argue that the object of the building programme was to produce a closer parallel between the landscape of the city of the gods and the capital of the god-king. I would argue that the building project and the excessive labour demands by the king pushed the peasants into an interpretative stance which had historical precedent but had rarely been adopted, a radical Asokanism that denied the historical syncretism of the Sakran with the Asokan model of kingship.

CONCLUSION

The landscape of the city of Kandy since its founding has been an inter-textual creation; it has acted as both a concretization and a reaffirmation of the Kandyan discursive field of kingship. While composed of both Asokan and Sakran discourses and inherently unstable, this field was held together by the willingness of all parties to overlook major differences in these two discourses. The struggle for power between the nobles and the king came to a head over the building programme because the nobles saw that the king had alienated the peasants. What ensued was a struggle over the interpretation of the landscape, which ultimately became a struggle over the nature of political authority. Each of the three interpretations which I have examined is politically inspired. Each reflected the perceived interests of the group that espoused it. While all were confined within the larger discursive field of which they were a part, they were also shaped in opposition to the discourses of opposing groups. The king's interpretation was relentlessly Sakran, the nobles' was Asokan with a tacit acceptance of Sakran claims, while the peasants' was uncompromisingly Asokan. Ultimately the king's interpretation could no longer prevail, his political legitimacy being so undermined that he was betrayed to the British by his own people.

No reading is innocent; each of the readings I have examined had a political agenda. In Kandyan society, both the urban landscape and the way people respond to it were shaped by texts. The fact that the form of protest was highly inter-textual, employing allegorical rituals that had been used in earlier centuries, is evidence that the whole culture, even for illiterate peasants, was highly textualized. There are, however, as we have shown, multiple and contestatory texts that comprise a discursive field, so that this field may become a site of struggle for power. To struggle for and against political legitimation in the Kandyan Kingdom was in large part a matter of bringing certain texts and particular readings to the forefront of political discourse. In this sense one would have to agree with Jacob Neusner (1975: 195) that 'the point and purpose of tradition is not to pass on historical facts but both to create and to interpret contemporary reality, to intervene in history'.

ACKNOWLEDGEMENTS

I am indebted to Nancy Duncan for comments on an earlier draft. An extended version of the essay appears in Duncan (1990).

REFERENCES

Ahalepola Varnanava/Vaduga Hatana (n.d. [original early nineteenth century]) ed. B. Gunasekara, Colombo.
Ariyapala, M.B. (1956) *Society in Medieval Ceylon*, Colombo: K.V.G. De Silva.
Coomaraswamy, A.K. (1956) *Medieval Sinhalese Art*, 2nd edn, New York: Pantheon.
Dalada Vittiya (1974) (Original 1812) ed. Puchibandara Sanasgala.

De Silva, K.M. (1981) *A History of Sri Lanka*, Berkeley: University of California Press.

D'Oyly, J. (1975) 'A sketch of the constitution of the Kandyan Kingdom', *Ceylon Historical Journal* 24, Colombo: Tisara Prakasakayo.

Duncan, J.S. (1990) *The City as Text: The Politics of Landscape Interpretation in the Kandyan Kingdom*, Cambridge: Cambridge University Press.

Foucault, M. (1970) *The Order of Things*, New York: Random House.

Geertz, C. (1980) *Negara: The Theatre State in Nineteenth Century Bali*, Princeton: Princeton University Press.

—— (1983) 'Centers, kings and charisma: reflections on the symbolics of power', in C. Geertz *Local Knowledge: Further Essays in Interpretive Anthropology*, New York: Basic Books, 121–46.

Ingrisi Hatana (1906 [original 1812]) Colombo: privately published.

Keppitipola, T.B. (1918) 'Map of Kandy town about the year 1815 AD', *The Ceylon Antiquary* 4: 75.

Kirala Sandesaya (1958 [original 1816]) ed. C. Godakumbure, Colombo: privately published.

Mabbet, I.W. (1983) 'The symbolism of Mount Meru', *History of Religions* 23: 64–83.

Mahavamsa: Or the Great Chronicle of Ceylon (1950) trans. and ed. W. Geiger, Colombo: Ceylon Government Information Service.

Maimataya (n.d.) ed. H.P. Perera Appuhamay, Colombo: privately published.

Malalgoda, K. (1976) *Buddhism in Sinhalese Society 1750–1900: A Study of Religious Revival and Change*, Berkeley: University of California Press.

Moligoda Sannasa (1904) [original 1814]) reproduced in H.C.P. Bell *Report on the Kegalla District*, Colombo: Government Printer; reprint of *Archaeological Survey of Ceylon* 19 (1892).

Mus, P. (1937) 'Angkor in the time of Jayavarman VII', *Indian Arts and Letters* 11: 65–75.

Neusner, J. (1975) 'The study of religion as the study of tradition', *History of Religions* 14: 191–206.

Obeyesekere, G. (1984) *The Cult of the Goddess Pattini*, Chicago: University of Chicago Press.

Satara Devala Devi Puwata (1954 [original eighteenth century]) in H. Nevill (coll.) and P.E.P. Deranyagala (ed.) Colombo: Ceylon National Museums Manuscript Series Vol. 5, pp. 158–9.

Seneviratne, H.L. (1978) 'Religion and legitimation of power in the Kandyan Kingdom', in B.L. Smith (ed.) *Religion and Legitimation of Power in Sri Lanka*, Pennsylvania: Anima, 177–87.

Smith, B.L. (1978a) 'Kingship, the Sangha, and the process of legitimation in Anuradhapura, Ceylon: an interpretive essay', in B.L. Smith (ed.) *Religion and Legitimation of Power in Sri Lanka*, Pennsylvania: Anima, 73–95.

—— (1978b) 'The ideal social order as portrayed in the chronicles of Ceylon', in B.L. Smith (ed.) *Religion and Legitimation of Power in Sri Lanka*, Pennsylvania: Anima, 48–72.

Stutley, M. and Stutley, J. (1984) *Harper's Dictionary of Hinduism*, New York: Harper & Row.

Part IV
ON REPRESENTING CULTURAL GEOGRAPHY

14

REPRESENTING SPACE
Space, scale and culture in social science
John Agnew

Representations of geographical space have not elicited much attention in the social sciences as non-geographers have by and large adopted specific representations through tacit assumption rather than explicit adoption. The purpose of this essay is to open up debate over how geographical space is regarded in contemporary social science by identifying some dominant conceptions of space and relating these to the dominant conceptions of scale and culture which condition them. The essay is organized as follows: first, the broad connections between conceptions of space, scale and culture are outlined; second, some specific representations of space are described; third, a 'counter-representation' of space in the form of a concept of place based on recent work in cultural geography is briefly described as an alternative to dominant representations. Some possible reasons are suggested for its absence from contemporary social science.

SPACE, SCALE AND CULTURE

Representations of space are not isolated, idiosyncratic or marginal aspects of the social sciences as alleged by those commentators who confuse the absence of *their* representation of space from social sciences with the absence of all representations. Rather, representations of space are embedded in all social sciences but intertwined in complex ways with representations of scale and culture. In this essay space is taken to refer to the presumed effect of location, or *where* social processes are taking place, upon those processes; scale refers to the spatial *level*, local, national, global, at which the presumed effect of location is operative; and culture refers to the matrix of socially constructed practices and ideas that 'mediate' between location and social processes.

In contemporary social science two 'interpretative communities' relate space, scale and culture in distinctive ways, although, as the second section shows, different emphases are apparent within the two groups. The first group sees space as, or becoming, *national*. In other words, space is viewed in terms of a national-state unit of historical study in which a 'modern' national culture increasingly displaces 'traditional' or residual local ones. This representation of space has long been dominant in such fields as political sociology, political history, macro-economics, international relations and public administration. It perhaps reaches an extreme in the 'nationalization thesis' of political sociology in which homogeneous levels of support for political parties and uniformity of response to political forces are seen as producing an increasingly 'national' politics in which national social cleavages and national cultures slowly replace those at more local scales (Lipset and Rokkan 1967).

A second interpretative community views space as *structural*. From this point of view, the spatial effects of particular units such as nodes, districts or regions are fixed and constant due to their relationships with one another. Thus an industrial core area (or city) is contrasted with a resource periphery (or hinterland) in terms of a structural social/cultural relationship of superiority/inferiority. This representation of space is characteristic of much academic geography, including both spatial analysis and regional geography, as well as regional anthropology, economic history and neo-Marxist sociology (especially world-system theory). There are important differences in the scale at which 'core–periphery' structures are regarded as operating. For example, city–hinterland relationships, central city–outer city relationships and city hierarchies are the focus of much urban geography, whereas global core–periphery–semi-periphery distinctions are key to world-system theory in sociology. In each case, however, culture is viewed as an inert product of the dominant spatial-structural relationships that provide the framework for study.

Neither of the two 'interpretative communities' has demonstrated much self-consciousness about its representations of space, scale and culture. They are 'hidden' geographies. To some commentators this is tantamount to a declaration of 'spacelessness', i.e. if it is not written about explicitly it does not exist. This would seem to be an element in Soja's (1989) allegation of a 'lack' of space in contemporary social science. However, this essay will show that social science is filled with assumptions and presumptions about space. What is in doubt is their adequacy.

In the first place neither interpretative community sees space and society as other than 'externally related'. By this I mean that there is a tendency to view space either as a board or backdrop across which social processes 'move' and are 'imprinted' or as a set of fixed 'containers' at particular scales for cultures and social processes. In this regard, Soja's (1989) argument on behalf of an alternative 'internally related' conception of space and society is very much to the point. Second, scale is almost always treated in terms of either the fixed or the emerging dominance of one level. So rather than relating scales to one another, spatial effects are regarded as the product of one scale with other scales at best viewed as residual or emergent. The national and the global have thus achieved privileged status in different genres of social science. Third, the representations of culture which tend to correlate with these conceptions of space and scale are largely inert and superorganic. Culture (a black box of practices, values and beliefs) is a *function* of the structural position of a particular spatial unit (core or periphery) or a *product* of the emergence of a dominant scale (national), rather than a changing matrix of practices and ideas that *actively* mediates scale relationships in the constitution of social processes, as argued in much recent theoretical work in cultural geography. The third section of this essay returns to these criticisms as a basis for constructing an alternative representation of space.

REPRESENTATIONS OF SPACE

National representations of space

Nationalization

In the field of international relations the basic unit of analysis is the state construed as a territorial or geographical container. Ideas of organic nationhood and sovereignty are realized in a political geography of sharply delimited and inviolable spaces. This is a concept of space totally separated from concepts of time and change. Changes in military technology, worldwide communication systems and an increasingly interdependent world economy have made the idea of rigid territorial boundaries problematic, but the dominant 'realist' tradition in international relations is locked into an essentially eighteenth-century conception of national territorial spaces as homogeneous and exclusive (Walker 1984). One major consequence of this viewpoint is its restriction of authentic politics within territorial boundaries. 'External' to national boundaries are 'relations', 'admitting at best only of order and rules of accommodation' (ibid.: 532). A particular 'spatial metaphor', therefore, lies at the heart of contemporary international relations as an academic field. Indeed, state-centric international relations have undergone something of a renaissance in recent years (e.g. Hall and Ikenberry 1989). Alternative ways of understanding the territorial state as a *particular* spatial form have not received much attention (except see Cox 1981).

Much of 'mainstream' sociology equates society with the boundaries established by national states. Irrespective of their other differences, sociologies inspired by Marx, Durkheim and Weber have all come to accept the social divisions (and census categories) of the national state as coincident with their operational definitions of society. Abstract discussions of society give way to empirical examination of national societies, a tendency especially apparent in the field of political sociology where there is a history of seeing state and society in a one-to-one relationship, thereby legitimizing socially the political claims of the state. In the late 1960s this perspective was institutionalized, if it is today challenged by some of the structural representations of space identified later. In a major survey of the field Lipset and Rokkan (1967) declared that Western European political alignments had stabilized around national social cleavages with the coming of mass suffrage. Rose and Urwin (1969) presented a similar picture of enduring and stable national alignments. The 1950s and 1960s suggested that with the industrializing and urbanizing impulses of modernization, citizens transfer their allegiances from the local community to national social groups and 'conflict is no longer between constituent territorial units of the nation, but between different conceptions of the constitution and organization of the national party' (Lipset and Rokkan 1967: 23).

In the United States, geographical mobility is the mechanism that has 'weakened the attachments of Americans to home and place; for increasing numbers ... a state or a city is only a location where one happens to lie at the moment' (McWilliams 1972: 32). National types of economic organization, consumption and leisure activities have come to predominate, and brought even the US South, the bastion of territorial distinctiveness, into the national main-stream (McKinney and Bourque 1971).

Following on from the nationalization of social life has been the nationaliz-ation of political life (Campbell *et al.* 1966). In what proved to be an influential book Schattschneider (1960: 87) wrote: 'elections since 1932 have substituted a national political alignment for an extreme sectional alignment everywhere in the country except in the South'. The central theme of much writing on politics in the countries of Western Europe likewise has been increasing national homogeneity. To Blondel (1963: 26), for example, 'Britain is essentially a homogeneous nation in which the major distinctions are not based on geography, but on social and economic conditions ... in Britain, national class differences are the main divisions of society.' Even local conditions are viewed as much the same everywhere: 'the conditions and preoccupations of life are much the same in Bristol and Bradford' (Pulzer 1967: 43). That this viewpoint is not just a curiosity of the 1960s is illustrated by such recent work as Bogdanor (1983) and McAllister (1987).

In a widely hailed book on the 'modernization' of rural France, Weber (1976) argues that between 1870 and 1914 peasants became French. From diversity and particularity, not to say barbarism, emerged homogeneity and civilization. This occurred because as France grew more prosperous, roads, railways, markets, schools, national newspapers and military conscription penetrated the country-side. These rationalizing and nationalizing trends undermined rural particular-ism, opened the countryside to new ideas, goods and practices and tied previously isolated communities into the national culture and social life. Finally, some writers on Italian politics have argued for a 'territorial homogenization' in levels of support for the major Italian political parties (see Agnew 1987). This view became especially popular in the 1970s when, as will be argued in a later section of the essay using this case, a temporary trend towards nationalization was widely seen as a permanent secular trend.

Over the past ten years the nationalization representation of space has been increasingly criticized (Agnew 1987). Local and regional patterns of political competition and dominance have not only persisted but strengthened. Increasing homogeneity of response to political parties is either lacking or a product of peculiar electoral and political circumstances. However it is defined, therefore, 'nationalization' does not seem to make much sense as the spatial foundation for social science. But rather than abandoning it a number of *ad hoc* responses to its failures have been proposed. Three stand out: the effects of population composition, the neighbourhood effect and local culture representations are held to account for residual local and regional spatial effects in the face of nationalization.

Population composition

This approach proposes that geographical variation in political and social phenomena is a function of population mix or 'composition effects', often misleadingly called contextual effects. In particular, social groups are not evenly distributed. There are two major problems with this argument. First, it presumes that the national-level designations used to label groups have the same significance everywhere, i.e. that terms such as 'working class' or 'Catholic' have spatially invariant meanings. Second, empirical research in political sociology, for example, suggests that differences in support for political parties between regions are due more to variations in class support than to differences in class composition. To account for this, processes relating to the influence of local political environments through information flow and issue orientation are sometimes invoked: the so-called neighbourhood effect.

Neighbourhood effect

The evidence for neighbourhood effects is often persuasive. Numerous authors have shown that the more dominant numerically a given social class is in an electoral district the greater the support not only from the majority class but also from minority ones. Favoured explanations for this include the existence of locally oriented voters who 'switch' from their national class position, and local 'information effects' which favour numerical majorities. Much of modern electoral geography has been concerned with attempting to demonstrate the neighbourhood effect as the fundamental contribution of 'geography' to social science (see Agnew 1989a). However, the concept of the neighbourhood effect rests upon a number of dubious premises and is also now subject to considerable empirical dispute. In particular, local outcomes in terms of bias towards the numerically preponderant group's presumed 'natural' political orientation do not offer much information as to how they occur. Indeed for the 'local voter' thesis to work a number of implausible assumptions must be invoked: e.g. that voters are equally susceptible to conversion irrespective of strength of party identification, the equal probability of sending and receiving partisan messages, the random mixing of majority and minority voters in electoral districts, and clearly distinguishable national and local voters (see Bodman 1983).

Local cultures

Local-level effects are sometimes interpreted as the product of local subcultures or political cultures. These are widely regarded as residual effects, 'left over' after national-level 'variables' have provided the bulk of explanation. In recent British electoral studies Johnston (e.g. 1986) has stressed the role of local political party organizations in setting local political agendas, in particular through control over local government, and thus becoming 'part of the local culture, participating in the socialization of individuals ... and creating a base that can be mobilized and drawn upon at election times' (Johnston 1986: 52). Similar arguments have been made by some students of Italian politics. For example, Baccetti (1987) and Trigilia (1986) base their conceptions of Italian political life upon the long-run effects of distinctive and fixed political traditions associated with different Italian regions. Putnam *et al.* (1985) see the relative success of different Italian regional governments since 1970 in terms of 'regional' political cultures. The problem here is twofold. Conventional statistical models show little in the way of effects that can be labelled as endogenous 'local cultures' (e.g. Warde *et al.* 1988), although these models are themselves based upon assumptions that are problematic (Agnew 1989a). More importantly, the term 'culture' as used in this literature is little more than a conceptual 'black box' for practices and ideas whose nature and political and social effects cannot change and are set for all time in permanent counterpoint to one another and to those of 'creeping' nationalization (e.g. Cartocci 1987).

Structural representations of space

In the past two decades the *core* (or *centre*)–*periphery* couplet has become central to much work in the social sciences. The distinction between geographical core regions and geographical peripheries has been especially important for some students of world political economy, such as Wallerstein and his world-system perspective, and for some students of political development and 'integration' within states (e.g. Bensel 1984; Rokkan and Urwin 1983; Tarrow 1977). The concepts of centre and periphery can also be defined in an abstract or semiotic sense. They then signify the Sacred and the Profane or the socially Fundamental and the Marginal. This is the usage one finds in comparative studies in religion (Eliade 1959) and structural-functional sociology (Shils 1975) respectively. But in much recent social science research, particularly that operating under the label of the world-system perspective, centre and periphery have a primarily geographical definition. However, centrality is not *merely* locational. Clearly, the designation of a place as a centre or part of a core implies a relatively dominant position for that place *vis-à-vis* all other places.

Wallerstein's basic research, as reflected in *The Modern World-System*, Volumes I (1974) and II (1980), involves detailing the process whereby some places in Western Europe became the core of a world capitalist economy by 'peripheralizing' the rest of the known and accessible world. For some places to be dominant others had to be dominated. The origin of this process lay in the struggle for dominance between states *within* the core which initially extended out beyond Western Europe into Eastern Europe and the Americas and later into the rest of the world. The period 1450–1650 was a time of expansion, whereas 1600–1750 was one of 'consolidation' or involution. Outside the core, several types of state are identified: those in the periphery, characterized by an absence of industry, lack of local political control, but a flow of resources to the core states; those in the semi-periphery, marked by 'proto-industrialization' and an ability to repel domination from the core states; and those in the external arena, the part of the world external to the integrated world economy. It is not clear, however, that it is appropriate to refer to 'states' in the periphery and external arena in the sense that it is appropriate to use the term in connection with the core and semi-periphery. The European state, as Wallerstein stresses, was a major element in Europe's exceptionalism.

Wallerstein emphasizes that 'the concept of spatial boundaries [is] a central axis of the analysis of this book' (1980: 245). Yet it is also apparent that by and large Wallerstein views these spatial boundaries and their associated spatial concepts as the outcome of a historical process of European inter-state competition and expansion. Other work under the world-system perspective, however, uses the spatial concepts in a different way. For example, Bergesen (1980: 7) asserts that 'core, semi-periphery, and periphery are properties of the world-system as a whole and this creates a set of structural relations at the distinctly world level of analysis'. Wallerstein and Hopkins (1977: 137) are drawn on for support. They appear to contradict Wallerstein's previous emphasis on inter-state competition as the basis for the growth of a world system when they write:

> If there is one thing which distinguishes a world-system perspective from any other, it is its insistence that the unit of analysis is a world-system defined in terms of economic processes and links, and not any units defined in terms of juridical, political, cultural, geological, etc., criteria.

Thus, according to Bergesen (1980: 8), 'The constituent parts [of the world system] – core and periphery – which exist *prior* to the world division of labor come first.' Interaction between these different areas comes only after they have been defined. But Bergesen wants to go even further than this. For him a 'world mode of production' rather than a global division of labour among societies is the *cause* of spatial differentiation (ibid.: 10–11). Spatial differentiation, in its turn, *causes* interaction.

By way of example for the second type of literature on political development, Rokkan and Urwin (1983) focus on the territorial structuring of Europe. They stress the multi-dimensionality of centre–periphery relationships. There are three sets of relationships: military-administrative, economic and cultural. Peripheries can be conquered and incorporated by centres in all ways or only one way. But changes in one type of relationship affect the others. For example, increasing economic dependence affects cultural distinctiveness. Rokkan and Urwin identify three regions at the scale of Europe as a whole: a central European city belt, the core areas of state building, and peripheries which have resisted incorporation into modern states. The objective is to construct a typology of 'peripheral predicaments': political reactions of peripheries to processes of subordination and incorporation. Tarrow (1977) likewise distinguishes between cultural, economic and administrative dimensions of centre–periphery links in his study of the roles of local politicians in Italy and France. He argues that in political systems in which political representatives at the centre disagree fundamentally 'integration at the periphery and integration between centre and periphery have ... become arenas of reconciliation compensating in part for the policy paralysis at the top' (ibid.: 253). Finally, Bensel (1984) sees the political development of the United States in terms of a process of 'sectional alignments' around fundamental economic interests which are defined geographically in terms of a national core (the north-east) and a national periphery (the rest of the country). He traces the voting behaviour of congressional delegations over the period 1880–1980 in terms of a basic core–periphery dichotomy; although he notes a waxing and waning of its intensity throughout the period under review.

The use of the spatial couplet 'core–periphery' carries a number of potentially misleading and dangerous implications. Three come to mind: the hypostatization and ascription of causality to spatial categories, the pattern-process inference problem and the single scale of analysis problem. Much of the writing in the world-system tradition is ambiguous concerning the causative role of spatial concepts such as core and periphery. For example, Chase-Dunn and Rubinson (1977: 475) assert that 'the exploitation of the periphery by the core enables core capital to coopt core labor into a national alliance'. The writing of Frank, if it can be subsumed under the world-system label given its influence on Wallerstein, contains many examples of statements to the effect that spatial location has a determining effect on economic development. For example, in his widely referenced *Capitalism and Underdevelopment in Latin America*, Frank writes:

> The metropolis [core] expropriates economic surplus from its satellites [periphery] and appropriates it for its own economic development. The satellites remain underdeveloped for lack of access to their own surplus and as a consequence of the same polarization and exploitative contradictions which the metropolis *introduces* and *maintains* in the satellite's domestic structure.

> (Frank 1969: 9)

Of course, the language used in these quotations can be viewed as 'shorthand' for a more complex argument. But it is precisely here that the danger arises. One can start out using spatial concepts as shorthand for complex sociological processes but slip easily into *substituting* the spatial concepts for the more complex argument. This is often what appears to happen to advocates of a world-system perspective although, as previously noted, it is certainly not intrinsic to the perspective. Further, writing such as that of Bergesen, with its emphasis on the determining nature of the world system as a whole, is much more likely to involve potential hypostatization and ascription of causality to spatial categories than are the careful historical analyses conducted by Wallerstein. Such reification also occurs in the political development literature, although here the usage is often heuristic or metaphorical. In Tarrow (1977), for example, the term periphery is not defined. At different points Tarrow seems to equate the periphery with small towns and rural areas, or alternatively with everywhere outside central government institutions. The danger here is of confusing an evocative metaphor with an analytic concept (Pezzino 1980).

A major inference problem in the social sciences concerns the extent to which one can infer a specific social process from a geographical pattern (Olsson 1969). Thus, one might agree that the modern world system is geographically differentiated into a core, semi-periphery and periphery, but argue that the process of differentiation involves principles of initial and cumulative advantage for 'core' states and the role of transportation cost minimization within a world market for spatial differentiation elsewhere. A danger inherent to the world-system perspective as elaborated by Wallerstein and others is that, knowing the contemporary geographical structure of the world system, they have inferred from this pattern, and its historical accretion, a specific and historically transcendental process of genesis. To quote Chase-Dunn and Rubinson (1977: 457), 'The basic structure of the modern world-system is reproduced as the system moves through a series of geographical expansions to encompass the whole globe.' It is not surprising, therefore, that people and places can be seen as exhibiting a 'posterior destiny', to use a phrase of Runciman's (1980); their present positions within the world system determined by the historic geographical roles they have performed for the system. It appears that sociological functionalism is not dead after all, but alive and well in geographical disguise.

Although social scientists are often constrained by the necessity to move from a contemporary geographical pattern to a generative process, it is important that they keep an open mind concerning any inferences that are made. Some advocates of the world-system perspective, however, do no such thing. They see the geographical structure of the modern world system and their inferred process as inextricably intertwined. But alternative processes have been proposed and they should be accorded serious attention even if they trace the origins of the 'modern world system' (global market economy) to the eighteenth and nineteenth rather than the fifteenth and sixteenth centuries (Polanyi 1944). Likewise on the national scale, core–periphery differences in wealth and well-being cannot be reduced a priori to a single dimension such as state redistributive policies (as in Rokkan and Urwin 1983). Other processes relating to private industrial investment and international economic linkages may well be of greater importance.

A unique feature of the world-system perspective is its emphasis on the preeminence of a single scale of analysis. Usually this is the global scale, but even when, as in Hechter's (1975) writing, the national scale is the context of analysis, the single scale suffices for explanatory purposes. It is this feature of the perspective that has been the subject of most criticism even though both Wallerstein (1974) and Frank (1979) have emphasized the need to combine scales of analysis. For some critics it is the lack of attention to the *internal* social structures of core and peripheral states that constitutes the major lacuna (e.g. Evans 1979; Skocpol 1977). For others it is the emphasis on homogeneity within the spatial categories, particularly the periphery (e.g. Smith 1979). Finally, some writers take exception to the *passive* or reactive role accorded to 'the periphery' or, more appropriately, states outside the core (e.g. Gran 1979).

As it stands, however, and perhaps inevitably given its avowed 'holism', the world-system perspective tends to overstress the supra-state determinants of development and thus the central role of geographical differentiation on a world scale. In particular, factors of growth and change internal to states in either core or periphery are downplayed. Indeed the autonomy of states or other geographical parts as against the whole is often denied (Bergesen 1980). Perhaps this results from a perception of overemphasis on states in much previous writing. But it is nevertheless unbalanced. Social, economic and political activities certainly can have causes on scales other than that of the world system and its spatially differentiated geographical 'parts'.

Authors from a variety of disciplines have proposed that the way in which regional settlement systems organize and structure social relationships also shapes cultural and economic institutions. Locational analysis of *city–hinterland* links has been especially marked among a group of anthropologists associated with G.W. Skinner (see e.g. Smith 1976) who have drawn upon economic geography, in particular central-place theory, and economic historians interested in the spread of industrialization and its association with state development (e.g. de Vries 1984; Pollard 1981). Skinner (1977) provides a way of integrating the geography of the economy with that of the state. He does so by describing two hierarchical orders: one constituted from the base up and resulting from economic transactions, which through the definition of market areas created a hierarchy of central places, and one imposed from above resulting from imperial control and the imposition of administrative functions. The match/mismatch between these hierarchical orders is used as a base or template for answering research questions about economic growth, political rebellion and such anthro-pological issues as kinship networks, ethnicity and religiosity. The approach is a combination of Von Thünen's regional core–periphery model of rent as a direct function of transport costs, and Christaller's model of a central-place hierarchy within a region. Agrarian societies such as late Imperial China, late Tokugawa/ early Meiji Japan and early nineteenth-century France are viewed as sets of macro-regions functionally integrated and differentiated by core–periphery divisions largely of a physiographic nature and internal urban systems providing a 'skeletal structure' for each region's economy and administrative system. Each macro-region is seen as a largely autonomous and integrated socio-economic system.

Of course, it is not clear that Skinner's model, developed for the case of China, can be exported elsewhere without considerable elaboration. For example, the history of the Chinese state and its impress upon East Asia is totally different from the process of state development in Western Europe. More fundamentally, as Hobsbawm (1989: 235) points out, the role of the city–hinterlands and regions changes as the nature of the economic systems in which they are embedded changes. Today, economies are less organized in sectoral-regional terms than was the case before the Second World War. Political movements have become national rather than remained regional or become international in organization and objectives. International contexts and national units, therefore, can be more important for analysis than city–hinterlands. Most of all, however, city–hinterlands and the systems of which they are part change over time. For example, today it makes little sense to view city hierarchies in exclusively national terms. Not only do different city–hinterlands or regions have distinctive links to the world economy, they are also now parts of an increasingly integrated global urban system.

PLACE AS A COUNTER-REPRESENTATION

What all the representations of space reviewed share is an 'abstract attitude'. Spatial metaphors are used for categorizing and containing observations without much attention to their impact on the selection and ordering of the 'concrete particulars' themselves. Modern social science suffers from a sort of 'agnosia' (or disorder of perception) in which representations of space set boundaries for non-spatial processes rather than provide an understanding of space and society as inextricably intertwined.

One approach to providing a way out of the impasse reached by conventional representations of space is to focus on a central concept of cultural geography, the concept of place. This provides the possibility of addressing the three inadequacies of dominant representations: space as backdrop, single-scale exclusivity and a superorganic conception of culture. Place is one of those 'contestable' concepts (Gallie 1955–6) whose application is a matter of dispute. In geography, it is sometimes used synonymously with location, point, area or space. This confusion has led Giddens (1983: 79) to suggest that the term 'locale' substitute for place to indicate the physical settings in which social relations are constituted. The concept of place as context for social relations has suffered particularly from its assimilation in sociological discourse to the concept of community. But an emphasis on places as the physical and social contexts for action has long been characteristic of microsociology and much humanistic geography (Ley and Samuels 1978). This perspective represents the position that in order to *explain* human behaviour one must deal with the 'micro-episodes' of everyday life and their embeddedness in concrete milieux or contexts. To

Giddens (1983: 79), therefore, 'Locales are not just points in space in which action occurs, any more than time is a series of intervals into which action is somehow inserted.'

An emphasis on the contextuality of action and practices, a notion central to the definition of place adopted here, has been a feature of such culture-sensitive approaches as symbolic interactionism, cognitive sociology, ethnomethodology, social phenomenology, ethogenics and ethnoscience. This literature stresses that people do not experience life in the abstract context of 'mass society'. Their knowledge is acquired, and they live their lives, in the context of 'social worlds' dominated by the perspectives of different 'reference groups', in which meaning is attributed to acts and events through communication and interaction with limited numbers of people (Shibutani 1955; Smith 1984). In everyday life such social worlds provide the boundaries for social learning and interpretation. This is as true for 'cosmopolitans', people with an orientation towards a wider world, as it is for the mass of people, 'locals', people whose interests and definition of life are locally oriented. Even for 'jet-setters', reference groups and locale-specific 'significant others' define the rhythm of their movements from Acapulco to Aspen to Gstaad to Cannes. When in a locale, so to speak, they follow the routines and rituals of that locale. Of course, most cosmopolitans have rather more limited geographical itineraries and are tied to a few dominant reference groups and one set of locales at a time. Locals are even more socially, and spatially, constrained (Dahmann 1982).

But as Giddens and others have pointed out, microsociology often misses the impact on the constitution of action of the *longue durée* of structured social practices. There is a sense in which 'locales' could be anywhere. However, they are not. They are *located* according to the demands of a spatially extensive division of labour, the global system of material production and distribution, and variable patterns of political authority and control. The 'face-to-face society' of the locale in which action is embedded is in its turn embedded within a wider 'territorial society'. This is important because, as Pred (1984: 283) argues:

> In industrialized countries the spatial and social division of labor occurs at a macro-level within a system of places while retaining a local component. Thus, especially in capitalist countries, but also in command-economy countries, the production and distribution projects occurring within a local area are directly or indirectly connected to the dialectics of more macro-level structuration processes.

Therefore, place is not just locale, as setting for activity and social interaction, but also location. The reproduction and transformation of social relations must take *place* somewhere. Pred puts it as follows (ibid.: 279):

> Place ... always involves an appropriation and transformation of space and nature that is inseparable from the reproduction and transformation of society in time and space. As such, place is not only what is fleetingly

262

observed on the landscape, a locale, or setting for activity and social inter-action (Giddens 1979: 206–8; Giddens 1981: 39, 45). It also is what takes place ceaselessly, what contributes to history in a specific context through the creation and utilization of a physical setting.

But place is also more than an 'object'. Concrete, everyday practices give rise to a cultural mediation or 'structure of feeling', to use Williams's (1977) phrase, or 'felt sense of the quality of life at a particular place and time' (Pred 1983: 58). This sense of place reinforces the social-spatial definition of place from *inside*, so to speak. The identification with place that *can* follow contributes yet another aspect to the meaning of place: one place or 'territory' in its differentiation from other places can become an 'object' of identity for a 'subject'. This is *not* the same as community in the sense of a way of life based on a high degree of personal intimacy and sociability. But of course this could be present also.

Interwoven in the concept of place suggested here, therefore, are three major elements: *locale*, the settings in which social relations are constituted (these can be informal or institutional); *location*, the effects upon locales of social and economic processes operating at wider scales; and *sense of place*, the local 'structure of feeling'. Or, by way of example, home, work, school, church, and so on, form nodes around which human activities circulate and which *in toto* can create a sense of place, both geographically and socially. Place, therefore, refers to discrete if 'elastic' areas in which settings for the constitution of social relations are located and with which people can identify. The 'paths' and 'projects' of everyday life, to use the language of time-geography, provide the practical 'glue' for place in these three senses (Pred 1984). To the extent that places are similar in these respects, interconnected and contiguous, one can refer to a 'region' of places (Cox 1969). In that situation the sense of place can be *projected* on to the region or a 'nation' and give rise to regionalism or nationalism. The sense of place need not be restricted to the scale of the locality.

The question of how to define place has exercised some geographers and others for many years. In their approaches to it, one or other of the three elements has tended to predominate. For example, economic geographers have tended to emphasize location; cultural geographers have been centrally concerned with sense of place; and a few humanistic geographers have concerned themselves with locale. Rarely have the three aspects been brought together. A key tenet in the present definition is that the local social worlds of place (locale) *cannot* be understood apart from the *objective* macro-order of location and the *subjective* territorial identity of sense of place. They are all related; if ultimately locale is the most central element sociologically it must be grounded geographi-cally. In other words, locale is the core geo-sociological element in place, but it is structured by the pressures of location and gives rise to its own sense of place that may in certain circumstances extend beyond the locality.

Rather than a 'metric' space, divided into compact areas, place involves a conception of 'topological' space in which diverse scales are brought together through networks of 'internal' and 'external' ties in defining geographical variation in social phenomena. This geographical variation responds to changes in the interaction of the networks that interweave the internal and the external, i.e. locale and location. In other words, geographical variation cannot be 'read off' from one geographical scale. It is the necessary concomitant of the inter-relation of social processes on different scales that 'come together' or are mediated through the cultural practices of particular places. Geography, there-fore, is implicated *in* social processes rather than being a 'backdrop' or a 'board' *upon* which social processes are inscribed.

This conception of place may be illustrated by way of example rather than further abstract discussion. The geography of Italian electoral politics since 1945 includes three political-geographical 'regimes' that have dominated at different periods. The first regime, dominant from 1947 to 1963, is a clear regional (meso-scale) pattern of support for the major political parties based upon social-economic relationships and local cultures that clustered regionally. The second regime (1963–76) witnessed the expansion of the Communist Party (PCI) out of its regional stronghold into a competitive position with the Christian Democratic Party (DC). This expansion had different sets of causes in different places, but the net effect was to suggest a nationalization of the two major parties. The third regime, since 1976, has seen increased support for minor parties, the geographi-cal 'retreat' of the PCI and a more localized pattern of political expression in general, reflecting the increased 'patchiness' of Italian economic growth, social change and cultural practices.

Rather than transcendental areal differentiation on a regional scale, the geography of Italian electoral politics represents areal variation on a variety of scales that switch in importance over time. Different scales have dominated expression as place-based causal processes have brought about a changing balance of geographical sameness and difference. The period 1947–63 is that of the 'classical' electoral geography of Italy established most definitively by Galli and his colleagues (Galli and Prandi 1970). They divided the country into six zones on the basis of levels of support for the three major parties, the PCI, DC and the Socialists (PSI), and the strength of the major political subcultures, the socialist and the Catholic. There was a strongly rooted 'cultural hegemony' in only two of these zones: 'la zona bianca' and 'la zona rossa' (Muscara 1987; Stern 1975). But in electoral terms, support for specific political parties was remarkably clustered regionally in 1953 (Rizzi 1986): the PCI in the centre, the PNM (monarchists) and MSI (neo-Fascists) in the south and Sicily, DC in the north-east and the south. In the 1950s, Italian politics followed a regional 'regime' reflecting a similarity at the regional scale of place-based social, economic and political relationships.

The second period, 1963–76, marks a break with the regional pattern characteristic of the 1950s. Two electoral shifts were especially clear: the expansion of support for the PCI outside 'la zona rossa' (along with its consolidation inside), particularly in the industrial north-west and parts of the south, and the breakdown of 'la zona bianca' as a number of small parties made inroads into the previously hegemonic support for DC in parts of the north-east. The net effect of these changes was a *seeming* nationalization of the major parties, even though they still maintained traditional areas of strength. In addition to the geographical expansion of the PCI and its overall increasing share of the national vote, the other major feature of the period 1963–76 was the so-called 'breakdown' of the Catholic subculture or hegemonic position in 'la zona bianca' or north-east and the loss of voters to DC subsequent to this. The argument is that DC, being largely an electoral rather than a mass party with a large membership, had relied heavily on affiliated organizations, many of a religious nature, to mobilize its support. But in the 1960s, as a result of heavy outmigration from rural areas in the Veneto, Trento and Friuli, the constituent regions of 'la zona bianca', and the growing industrialization of some areas, such as Venice, Treviso, Trento and Pordenone, the traditional social networks and communal institutions upon which DC hegemony was based began to collapse (Parisi 1971; Sani 1977).

The nationalizing political-geographical regime peaked in 1976 when DC and PCI together accounted for 73 per cent of the national vote. Although this trend had distinctive causes relating to the geographically differentiated social and economic impacts of the 'economic miracle' and their interplay with political and organization traditions, it was widely interpreted as a permanent 'nationalization' of political life. DC and the PCI were now *national* political parties. The election of 1976 seemed to seal it once and for all as the PCI expanded in constituencies where it had previously been weak or where its previous support had stagnated (but not much in the north-east): +10.6 per cent in Naples–Caserta; +10.9 per cent in Cagliari–Nuoro (Sardinia); +9.6 per cent in Rome–Viterbo–Frosinone; and +9.2 per cent in Turin–Novara–Vercelli.

The 1979 election indicated a much more complex geography of political strength and variation than had been characteristic previously. Since then all parties have been less regionalized than in the past (Rizzi 1986). The 1983 and 1987 elections suggest a trend towards a localization or increased differentiation of political expression. In 1983, DC lost 5.4 per cent nationally. But the PCI was not the beneficiary. Rather, it was smaller parties such as the PSI and the Republicans (PRI) in the north and the MSI in the south that gained most. In 1987, DC recovered somewhat from 1983 but without a major geographical expansion. The major loser this time was the PCI, which lost ground not only in the north-east and the north-west but also in some provinces of 'la zona rossa' to the PSI and a variety of smaller parties including the Radicals (PR), the 'Greens' and Democrazia Proletaria (DP) (Leonardi 1987).

How can this localization be explained? One element is the increasingly differentiated pattern of economic change after a previous era of concentration. While the economic boom of the early 1960s concentrated economic growth increasingly in the north-west, by the late 1960s there was considerable decentralization of industrial activity out of the north-west and into the north-east and the centre. This 'new' economic geography must be placed in a macro-economic context (Santagata 1981). The big firms that were important in the economic boom of the early 1960s operated in sectors that were in continual international crisis after the early 1970s. This encouraged strategies of adaptation such as vertical disintegration and the stimulation of investment in the non-unionized small firm sector. Causes other than economic restructuring have also contributed to the contemporary localizing political geography. One of these has been the failure of parties to adapt successfully to recent social and economic change. In Trento and Udine, for example, DC has had problems adapting to the new economy. In large parts of the south and the north-west, the PCI has been unable to capitalize on earlier successes mainly because, in the south, it has neither had control over the state resources that lubricate the politics of many parts of that region, nor been able to build a cultural hegemony. In the north-west, its major 'vanguard' of unionized workers has been much reduced in economic importance, while at the same time other parties have become better organized and the particular problems of southern immigrants have largely receded from the political agenda (Pasquino 1985; Sassoon 1981).

The emergence of effective regional-level governments since 1970 has also reinforced the localization of interests and 'sense of place'. Where parties have achieved some strength and legitimacy through control over regional govern-ments, they have been able to build local coalitions for national politics based upon the pursuit of local interests. The PCI, for example, has benefited from its control of or participation in the regional governments of Emilia-Romagna, Toscana and Umbria. But it has suffered elsewhere, and other parties such as DC or the PSI have benefited, because of lack of control over patronage jobs and inability to write regional political agendas (Putnam *et al.* 1985).

Finally, to the extent that the former successes of DC and the PCI in, respectively, 'la zona bianca' and 'la zona rossa' rested on the 'total' social institutions with which they were affiliated (unions, co-operatives, clubs, etc.), as well as social isolation, the shifting orientations of these institutions and the rise of the consumer society have opened up possibilities for the smaller parties. There is some evidence that, since the late 1960s, the ties between DC and the PCI and their supportive organizations, especially the unions, have weakened (Hellman 1987; Mershon 1987). The parties themselves are responsible for some of this. In order to expand nationally, they have often had to abandon or at least limit the ideological appeal that served so well in areas of traditional strength. They have also had to respond in some areas to 'new' movements (e.g. the 'Greens'), which has opened them up for both factionalism and essentially localized forms of organization and ideology (Amyot 1981). More generally, parties do not always 'travel well'. Thus, in comparing north-east with central Italy, the question of compatibility between party 'style' and local 'style' arises. Stern (1975: 223) notes the evolution of two very different forms of political hegemony, each with distinct characteristics that necessitate sharply contrasting forms of maintenance. The Christian Democratic variety that flourishes in northeastern Italy is fueled efficiently by a stable social organization that deemphasizes the place of politics in community life. In comparison the Communist variant thriving in central Italy … accents the urgent attention that political matters should command among the local citizenry and thereby constantly reaffirms the relatively recent sense of legitimacy that underlies PCI control.

Of course, these hegemonies have always had local roots and, in some localities, their power is still quite visible, as recent studies of Bologna and Vicenza suggest (Allum and Andrighetto 1982; Kertzer 1980). There is persistence in place as well as change. At present, not only is support for the parties more obviously localized, so are the parties themselves. In post-war Italian electoral politics, therefore, thinking in terms of place and changing scales of social-cultural differentiation appears more fruitful than alternative approaches to spatial representation such as nationalization or a fixed core–periphery structure.

CONCLUSION

If the place perspective has merit why has it not received the attention of other representations of space? I can only speculate here but I believe there are four elements to an answer. First, as I have argued elsewhere (Agnew 1989b) the concept of place became fatefully identified with that of community in eighteenth- and nineteenth-century social science. As community was viewed to be in eclipse under the onslaught of industrialization and urbanization, place was eclipsed too. A 'methodological nationalism' came to prevail in many quarters in which society, rather than remaining an abstraction or ideal type, became coterminous with the boundaries of national states. Culture followed. Second, the abstract, homogeneous, boundary-defining views of space characteristic of dominant representations of space in social science have been part and parcel of the identity that grew up in the late nineteenth century between abstraction and scientific validity (Kern 1983). Local context and culture have been regarded as antithetical to the enterprise of social science with its 'imposition of uniformity as a means to universality' (Ley 1989). Third, the distinction between scales or 'levels of analysis', initially a taxonomic device to distinguish areas of study (international relations vs domestic politics; micro- vs macroeconomics, etc.) and levels of generalization and causality (ecological vs individual inference), has come to 'masquerade as a theoretical principle. For classification often leads to reification' (Walker 1984: 539). In this way the possibility of integrating scales of analysis has become either institutionally difficult (because different disciplines 'specialize' on different scales) or intellectually heretical. That the process of analysis (or reduction) necessarily assumes a correlate of synthesis has not been of much concern to modern social science. Fourth, representations of space are not 'merely' epistemic; functions of how we 'just' happen to think. They are related to the dominant political and material conditions of different eras (Williams 1977). Much of contemporary social science and associated representations of space are historical products of the late eighteenth and nineteenth centuries. The 'founding fathers' (Marx, Weber, etc.) still cast long shadows. Abstraction, uniformity, scale exclusivism and superorganic culture are all parts of their legacy. Yet, rather than starting over we continue to search for insight on space (and social process) where it cannot be found in works scarcely relevant to the conditions of life in the late twentieth century.

Throughout contemporary social science, representations of space are important elements of major theoretical perspectives. There is no spaceless social science. Dominant perspectives, however, see space and society as externally related, view scale in terms of one fixed or emerging level of 'greatest' significance and tend to have superorganic and inert conceptions of culture. Different representations exhibit different forms of these drawbacks. In the end, however, each *mis*represents the role of space in social life. One purpose of this essay has been to open up discussion of implicit (and explicit) representations of space very much in the spirit of the examination of previously unexamined assumptions that masquerades today under the label of 'poststructuralism'. But I am not only interested in exposing or deconstructing. I am also interested in the possibility of representing space in a way that overcomes the limitations of dominant representations. This is why I have discussed a cultural representation of place as a counter-representation of space.

It is popular these days to proclaim the increasing 'spatiality' of social science (e.g. Dear 1988; Soja 1989) as if social science had hitherto been 'spaceless'. However, if there is a conclusion to be drawn from a survey of contemporary social science it is that change and time have been obscured in favour of stasis and fixed spatial metaphors. It is what we mean by spatiality that counts, not merely proclaiming any old version of it.

REFERENCES

Agnew, J.A. (1987) *Place and Politics: The Geographical Mediation of State and Society*, London: Allen & Unwin.

—— (1989a) 'From political methodology to geographical social theory: a critical review of electoral geography 1960–1987', in R.J. Johnston, P. Taylor and F. Shelley (eds) *Developments in Electoral Geography*, London: Routledge, 15–21.

—— (1989b) 'The devaluation of place in social science' in J.A. Agnew and J.S. Duncan (eds) *The Power of Place: Bringing Together Geographical and Sociological Imaginations*, London: Unwin Hyman, 9–29.

Allum, P.A. and Andrighetto, T. (1982) 'Elezioni e elettorato a Vicenza nel dopoguerra', *Quaderni di Sociologia* 30: 335–97.

Amyot, G. (1981) *The Italian Communist Party: The Crisis of the Popular Front Strategy*, New York: St Martin's Press.

Baccetti, C. (1987) 'Memoria storica e continuità elettorale. Una zona rossa nella Toscana rossa', *Italia Contemporanea* 167: 7–30.

Bensel, R.F. (1984) *Sectionalism and American Political Development, 1880–1980*, Madison: University of Wisconsin Press.

Bergesen, A. (1980) 'From utilitarianism to globology: the shift from the individual to the world as a whole as the primordial unit of analysis', in A. Bergesen (ed.) *Studies in the Modern World-System*, New York: Academic Press, 1–12.

Blondel, J. (1963) *Voters, Parties, and Leaders*, Harmondsworth: Penguin.

Bodman, A.R. (1983) 'The neighbourhood effect: a test of the Butler-Stokes model', *British Journal of Political Science* 12: 124–31.

Bogdanor, V. (1983) *Multi-Party Politics and the Constitution*, Cambridge: Cambridge University Press.

Campbell, A. *et al.* (1966) *Elections and the Political Order*, New York: Wiley.

Cartocci, R. (1987) 'Otto risposte a un problema: la divisione dell'Italia in zone politicamente omogenee', *Polis* 1: 481–514.

Chase-Dunn, C. and Rubinson, R. (1977) 'Toward a structural perspective on the world-system', *Politics and Society* 7: 453–76.

Cox, K.R. (1969) 'On the utility and definition of regions in comparative political sociology', *Comparative Political Studies* 2: 68–98.

Cox, R.W. (1981) 'Social forces, states and world orders: beyond international relations theory', *Millenium* 10: 126–55.

Dahmann, D.C. (1982) *Locals and Cosmopolitans: Patterns of Spatial Mobility during the Transition from Youth to Adulthood*, Geographical Research Paper No. 204, Chicago: University of Chicago.

Dear, M. (1988) 'The postmodern challenge: reconstructing human geography', *Transactions of the Institute of British Geographers* (new series) 13: 262–74.

de Vries, J. (1984) *European Urbanization, 1500–1800*, Cambridge: Cambridge University Press.

Eliade, M. (1959) *The Sacred and the Profane*, New York: Harper & Row.

Evans, P. (1979) 'Beyond center and periphery: a comment on the contribution of the world-system approach to the study of development', *Sociological Inquiry* 49: 15–20.

Frank, A.G. (1969) *Capitalism and Underdevelopment in Latin America*, Harmondsworth: Penguin.

—— (1979) *Dependent Accumulation and Development*, New York: Monthly Review Press.

Galli, G. and Prandi, A. (1970) *Patterns of Political Participation in Italy*, New Haven: Yale University Press.

Gallie, W.B. (1955–6) 'Essentially contested concepts', *Proceedings of the Aristotelian Society* 56: 167–98.

Giddens, A. (1979) *Central Problems in Social Theory: Action, Structure, and Contradiction in Social Analysis*, London: Macmillan.

—— (1981) *A Contemporary Critique of Historical Materialism*, Berkeley: University of California Press.

—— (1983) 'Comments on the theory of structuration', *Journal for the Theory of Social Behavior* 13: 75–80.

Gran, P. (1979) *Islamic Roots of Capitalism: Egypt 1760–1840*, Austin: University of Texas Press.

Hall, J.A. and Ikenberry, G.J. (1989) *The State*, Minneapolis: University of Minnesota Press.

Hechter, M. (1975) *Internal Colonialism: The Celtic Fringe in British National Development, 1536–1966*, Berkeley: University of California Press.

Hellman, J.A. (1987) *Journeys among Women: Feminism in Five Italian Cities*, New York: Oxford University Press.

Hobsbawm, E. (1989) 'La dimensione statale come fondamento delle articolazioni regionali', in F. Andreucci and A. Pescarolo (eds) *Gli spazi del potere*, Florence: Usher, 185–97.

Johnston, R.J. (1986) 'The neighbourhood effect revisited: spatial science or political regionalism?', *Environment and Planning D: Society and Space* 4: 41–55.

Kern, S. (1983) *The Culture of Time and Space, 1880–1918*, Cambridge, Mass.: Harvard University Press.

Kertzer, D.K. (1980) *Comrades and Christians: Religion and Political Struggle in Communist Italy*, Cambridge: Cambridge University Press.

Leonardi, R. (1987) 'The changing balance: the rise of small parties in the 1983 elections', in H.R. Penniman (ed.) *Italy at the Polls, 1983*, Durham, NC: Duke University Press, 100–19.

Ley, D. (1989) 'Modernism, postmodernism and the struggle for place', in J.A. Agnew and J.S. Duncan (eds) *The Power of Place: Bringing Together Geographical and Sociological Imaginations*, London: Unwin Hyman, 44–65.

Ley, D. and Samuels, M. (eds) (1978) *Humanistic Geography*, Chicago: Maaroufa.

Lipset, S.M. and Rokkan, S. (1967) 'Cleavage structures, party systems, and voter alignments', in S.M. Lipset and S. Rokkan (eds) *Party Systems and Voter Alignments: Cross-National Perspectives*, New York: Free Press, 3–64.

McAllister, I. (1987) 'Social context, turnout, and the vote: Australian and British comparisons', *Political Geography Quarterly* 6: 17–30.

McKinney, J.C. and Bourque, L.B. (1971) 'The changing south: national incorporation of a region', *American Sociological Review* 36: 399–412.

McWilliams, W.C. (1972) 'The American constitutions', in G. Pomper *et al.*, (eds) *The Performance of American Government*, New York: Free Press.

Mershon, C.A. (1987) 'Unions and politics in Italy', in H.R. Penniman (ed.) *Italy at the Polls, 1983*, Durham, NC: Duke University Press, 120–45.

Muscarà, C. (1987) 'Dalla geografia elettorale alla geografia politica: il caso italiano delle aree bianca e rossa', *Bollettino della Società Geografica Italiana* 4: 269–302.

Olsson, G. (1969) 'Inference problems in locational analysis', in K.R. Cox and R.G. Golledge (eds) *Behavioural Problems in Geography*, Evanston: Northwestern University Studies in Geography, 14–34.

Parisi, A. (1971) 'La matrice socio-religiosa del dissenso cattolico in Italia', *Il Mulino* 21: 637–57.

Pasquino, G. (1985) 'Il partito communista nel sistema politico italiano', in G. Pasquino (ed.) *Il sistema politico italiano*, Bari: Laterza, 126–68.

Pezzino, P. (1980) 'Sistemas politico e clientelismo in Italia', *Italia Contemporanea* 140: 123–7.

Polanyi, K. (1944) *The Great Transformation*, Boston: Beacon.

Pollard, S. (1981) *Peaceful Conquest: The Industrialization of Europe, 1760–1970*, Oxford: Oxford University Press.

Pred, A. (1983) 'Structuration and place: on the becoming of sense of place and structure of feeling', *Journal for the Theory of Social Behavior* 13: 45–68.

—— (1984) 'Place as historically contingent process: structuration and the time-geography of becoming places', *Annals of the Association of American Geographers* 74: 279–97.

Pulzer, P.J. (1967) *Political Representation and Elections in Britain*, London: Allen & Unwin.

Putnam, R. *et al.* (1985) 'Il rendimento dei governi regionali', in G. Pasquino (ed.) *Il sistema politico italiano*, Bari: Laterza, 345–83.

Rizzi, E. (1986) *Atlante geo-storico, 1946–1983: le elezioni politiche e il parlamento nell' Italia repubblicana*, Milan: GSI.

Rokkan, S. and Urwin, D.W. (1983) *Economy, Territory, Identity: Politics of West European Peripheries*, London: Sage.

Rose, R. and Urwin, D.W. (1969) 'Social cohesion, political parties and strains in regimes', *Comparative Political Studies* 2: 7–67.

Runciman, W.G. (1980) 'Comparative sociology or narrative history? A note on the methodology of Perry Anderson', *European Journal of Sociology* 21: 162–78.

Sani, G. (1977) 'Le elezioni degli anni settanta: terremoto o evoluzione?' in A. Parisi and G. Pasquino (eds) *Continuità e mutamento elettorale in Italia*, Bologna: Il Mulino, 67–102.

Santagata, W. (1981) 'Ciclo politico-economico: il caso italiano, 1953–79', *Stato e Mercato* 2: 257–99.

Sassoon, D. (1981) *The Strategy of the Italian Communist Party*, New York: St Martin's Press.

Schattschneider, E.E. (1960) *The Semisovereign People*, New York: Holt, Rinehart & Winston.

Shibutani, T. (1955) 'Reference groups as perspectives', *American Journal of Sociology* 60: 562–9.

Shils, E.(1975) *Center and Periphery: Essays in Macrosociology*, Chicago: University of Chicago Press.

Skinner, G.W. (ed.) (1977) *The City in Late Imperial China*, Stanford: Stanford University Press.

Skocpol, T. (1977) 'Wallerstein's world capitalist system: a theoretical and historical critique', *American Journal of Sociology* 84: 1075–90.

Smith, C. (ed.) (1976) *Regional Analysis*, Vol. II: *Social Systems*, New York: Academic Press.

Smith, S. (1984) 'Practicing humanistic geography', *Annals of the Association of American Geographers* 74: 353–74.

Smith, T. (1979) 'The underdevelopment of development theory: the case of dependency theory', *World Politics* 32: 247–88.

Soja, E.W. (1989) *Postmodern Geographies: The Reassertion of Space in Critical Social Theory*, London: Verso.

Stern, A. (1975) 'Political legitimacy in local politics: the Communist Party in northeastern Italy', in D. Blackmer and S. Tarrow (eds) *Communism in Italy and France*, Princeton: Princeton University Press, 221–58.

Tarrow, S. (1977) *Between Center and Periphery: Grassroots Politicians in Italy and France*, New Haven: Yale University Press.

Trigilia, C. (1986) *Grandi partiti e piccole imprese*, Bologna: Il Mulino.

Walker, R.B.J. (1984) 'The territorial state and the theme of Gulliver', *International Journal* 39: 529–52.

Wallerstein, I. (1974) *The Modern World-System*, Vol. I: *Capitalist Agriculture and the Origins of the European World-Economy in the Sixteenth Century*, New York: Academic Press.

—— (1980) *The Modern World-System*, Vol. II: *Mercantilism and the Consolidation of the European World-Economy, 1600–1750*, New York: Academic Press.

Wallerstein, I. and Hopkins, T.K. (1977) 'Patterns of development of the modern world-system', *Review* 1: 111–45.

Warde, A. *et al.* (1988) 'Class, consumption and voting: an ecological analysis of wards and towns in the 1980 local elections in England', *Political Geography Quarterly* 7: 339–51.

Weber, E. (1976) *Peasants into Frenchmen: The Modernization of Rural France, 1870–1914*, Stanford: Stanford University Press.

Williams, R. (1977) *Marxism and Literature*, Oxford: Oxford University Press.

15

INTERVENTIONS IN THE HISTORICAL GEOGRAPHY OF MODERNITY
Social theory, spatiality and the politics of representation

Derek Gregory

Lucidity came to me when I at last succumbed to the vertigo of the modern. This last word, no sooner formulated, melts in the mouth. The same thing happens with the whole vocabulary of life.... However, the path I was following was such that I could no longer avoid consulting the map of its territory.

(Louis Aragon, *Paris Peasant*, 1980)

Aragon's dilemma is not an oddity of early twentieth-century surrealism; it is, rather, a symptomatic condition of modernity. When Baudelaire (1984) reflected on 'The painter of modern life'[1] and called for an art capable of registering the passing moment without destroying its transient passage, one could already sense the pull of what Connor (1989: 4) calls that 'irrevocable tension between the way human beings felt they lived and the forms used to render that sensation'. The tension was aggravated by the pulsating drive of technical and scientific change through the nineteenth and into the twentieth centuries, especially in Europe and North America, but it was also heightened – and, I think, generalized in significantly new ways – by the turmoil of the First World War and the explosive force of the Russian Revolution.[2] With the First World War, wrote Benjamin (1968b: 84) on the eve of the Second,

a process began to become apparent which has not halted since then. Was it not noticeable at the end of the war that men returned from the battlefield grown silent – not richer, but poorer in communicable experience? ... For never has experience been contradicted more thoroughly than strategic experience by tactical warfare, economic experience by inflation, bodily experience by mechanical warfare, moral experience by those in power. A generation that had gone to school on a horse-drawn streetcar now stood under the open sky in a countryside in which nothing remained unchanged but the clouds, and beneath these clouds, in a field of force of destructive torrents and explosions, was the tiny, fragile human body.[3]

The dilemma of resolving contradictions like these – or, at any rate, of representing them, to oneself and to others – not only called into question ordinary conceptions of time: it also shattered conventional conceptions of space.[4] By the opening decades of the twentieth century, the classical disassociation between what Jameson calls *Wesen* and *Erscheinung*, essence and appearance, structure and lived experience, had been transcoded into a radically new relation between space and place: by then, 'the truth of [daily] experience no longer coincide[d] with the place in which it takes place' (Jameson 1991: 411).[5] Early modernism was so deeply implicated in this multiple and compound crisis of representation that I do not think it possible to claim (as some commentators have claimed) that *fin-de-siècle* art or social thought – the distinctions between the two were often highly ambiguous – somehow privileged time over space.[6] One might more readily argue that the characteristically modernist gesture was to disrupt narrative sequence, to explode temporal structure and to accentuate simultaneity as a way of declaring that 'things [did] not so much fall apart as fall together' (Lunn 1985: 35). What lay behind many of these experiments, so it seems to me, was not so much an insistence on the non-representability of art or a defiant commitment to *l'art pour l'art* as the belief that the 'vertigo of the modern' could be represented through an exploration of its spatiality: through a reading of 'the map of its territory'.

I want to move across the same terrain in the present essay. I do so in four stages. I begin with some observations about the purchase of these claims on our own precarious present. I then seek to make my discussion more concrete by considering some of the recent writings of David Harvey and Allan Pred, whose reconstructions of two cities torn apart by the convulsions of modernity (Paris and Stockholm) have much to tell us about these matters. But for them to speak to one another, I think it necessary to introduce another voice to mediate between them: that of Walter Benjamin. Through his extraordinary interventions in the historical geography of modernity, one can begin to glimpse some of the deeper resonances between politico-intellectual concerns at the beginning and the end of the twentieth century. One can also start to tease out some of the more intricate relations between social theory, spatiality and the politics of representation.

NOAH'S ARK? HUMAN GEOGRAPHY AND SOCIAL THEORY

Take the construction of the Ark. What does he do? He builds it in gopher-wood. *Gopher*-wood? Even Shem objected, but no, that was what he wanted and that was what he had to have. The fact that not much gopher-wood grew nearby was brushed aside. No doubt he was merely following

273

instructions from his role-model; but even so. Anyone who knows anything about wood – and *I* speak with some authority in the matter – could have told him that a couple of dozen other tree-types would have done as well, if not better; and what's more, the idea of building all parts of a boat from a single wood is ridiculous. You should choose your material according to the purpose for which it is intended; everyone knows that. Still, this was old Noah for you – no flexibility of mind at all. Only saw one side of the question.

(Julian Barnes, *A History of the World in 10½ Chapters*, 1989)

After the Second World War a wedge was driven between the arts and the social sciences. The geometric turn in post-war human geography installed a supposedly objective 'spatial science' at the centre of the discipline, but even the early critiques of this project – which were, I suppose, the first post-war engagements between human geography and social theory – as often as not remained complicit in the austere canons of this high modernism. Typically, they proposed to substitute one set of foundational claims for another. But for the past decade or so there have been signs of a gathering revolt against high modernism across the whole field of the humanities and the social sciences. That revolt is not confined to the celebrants of a putative postmodernism, of course, since the critique of (high) modernism can also be made from quite other discursive positions. None the less, as human geography has been progressively 're-socialized', so its interest in those questions of culture that spiral around the debate over postmodernism has intensified and its engagements with social theory have become more creative in form and more critical in temper. I do not want to be drawn into any boundary disputes over the first of these terms – there have been too many deadening proclamations about the 'nature' or 'spirit and purpose' of geography – but I suspect that I need to say something about the second because I do not use it as a synonym for 'sociology' or even 'social science'. Instead, I propose to treat social theory as a series of overlapping, contending and colliding discourses that seek, in various ways and for various purposes, to make social life intelligible. This is a bare-bones characterization, but it does not limit social theory to any one discipline and it contains two words which I hope will breathe life into what follows.

In the first place, I treat social theory as a series of *discourses* because the term underlines the embeddedness of social theory in social life – those traces of its historical geography that conventional social theory seeks to suppress but which are none the less indelibly present in the very questions it asks and the answers it gives: those contexts and casements which shape our local knowledges, however imperiously global their claims to know[7] – and, just as binding, the practical consequences of understanding (and, indeed, being in) the world like this rather than like that. As I use the term, social theory is not a commentary on social life but an intervention in social life. To speak of social theory as discourse is thus to emphasize the politics of social theory which are put in place, sometimes openly and sometimes covertly, through the multiple ligatures between 'knowledge' and 'power'.[8]

And I insist that social theory is about making social life *intelligible* because this does not immediately privilege one, nominally 'scientific' way of knowing over and against another. In particular, it appears to offer some defence against the slashes of Occam's razor; I continue to believe that Lévi-Strauss was at his most insightful when he remarked that explanation was not about substituting simplicity for complexity but replacing a complexity we don't understand by one we do. Ironically, his own structural anthropology was predicated on a singular reductionism and it is not difficult to see why the development of post-structuralism should have seemed so promising precisely because it offered a more complex model of representation. For much the same reason, I am suspicious of any discourse that gathers to itself privileges and closures which sustain a supremely self-confident claim to a singular Truth somehow independent of subject-position.

I want to argue that human geographers have to work with social theory, conceived thus, for two reasons.

First, we have little choice. Empiricism is not an option, if it ever was, because the 'facts' do not (and never will) 'speak for themselves', no matter how closely (as E.P. Thompson once urged) we 'listen'. For all Thompson's worries about theory (or, rather, Theory), nobody working in the humanities and the social sciences can escape working in those media that seek to *make* social life intelligible: and that emphasis on 'making' – on the constitutive function of theory – is of the upmost importance (Thompson 1978).[9] It imposes a responsibility on all of us to make our theoretical commitments as openly as possible and to invite a critical inspection of them. More parochially, I would add that geometry is not an option either, and that any attempt to install genuinely human geographies in place of the abstract lattices of spatial science must engage seriously and vigilantly with social theory. In doing so, as Harris recognizes, the horizon of meaning is visibly enlarged and the intellectual range of enquiry considerably extended (Harris 1991).[10] Indeed, I do not think it would be difficult to show that some of the most productive, creative (and, yes, empirical) work in human geography in recent years has been brought about through such an engagement with social theory.

And yet: social theory is not a Noah's Ark that can magically save human geography (or any other discipline) from the floodwaters. Or at any rate, if it is, then its hull is so riddled with woodworm that major reconstruction is necessary to keep it afloat. Second, then, I believe that we have to work with social theory because it *needs* so much work. Here I want to pay particular attention to the development of Marxist or Marxisant theorizations – not because they are the weakest points: far from it – but because they bear most directly on the writers whose work I will discuss in due course. Three problems are particularly intrusive.

(a) **Ethnocentrism**: Even some of the most seemingly radical of writers assume a vantage-point which tacitly adopts the positions and assumptions of a

Western intellectual culture or openly privilege 'the West' as a model and mirror for other societies. Habermas is only the most obvious among them.[11] Indeed, 'Western Marxism', which once seemed so liberating in its break from the political and economic rigidities of the orthodox canon which had been imposed on Eastern Europe and much of Asia, is now seen by many (otherwise sympathetic) commentators as a distinctly Western *European* discourse which inevitably suffers from the implications of its adjective (cf. Young 1990). Ever since Said's magisterial assault on the discourse of Orientalism, it has become increasingly difficult to ignore the ethnocentrism of Western social theory (Said 1978). And yet there remains something deeply (and perhaps inescapably) troubling about attempts to undo those closures from within. Mitchell's *Colonizing Egypt*, for example, is surely unequalled in its critical force: in its disclosure of the materiality of discourse, its inscription in space and its constitution of subjectivities. It comes as something of a shock to realize that its force depends upon a remarkable appropriation of high European theory – and on the work of Benjamin, Foucault and Heidegger in particular – to unmask: what? The aggrandizing of European social thought (Mitchell 1988, 1989).[12]

(b) **Sexism**: The sexism of mainstream social theory is incontrovertible (though hardly unassailable), and so it is perhaps not surprising that human geography should have smuggled in so many of its patriarchal assumptions. A commitment to an explicitly critical version of social theory provides few safeguards: not only the classical formulations of Marx, but also the contemporary constructions of Foucault and (particularly) Habermas are vulnerable to a feminist critique.[13] I sometimes think that the price human geography has paid for being one of the last of the social sciences to take Marxism seriously has been to become one of the last to take feminism seriously. Notations of class have been foregrounded in such a way (and with such power) that notations of gender and sexuality have been more or less marginalized. On some occasions this has taken the form of outright exclusion; on others, these relations have been incorporated into discussions whose boundaries and terms continue to be policed by the primacy of class.[14] It is small wonder that so many should feel 'outside The Project' (Christopherson 1989). I know that some feminist critics would object that 'theory' is itself a peculiarly masculine way of knowing – that it is implicated in a distanced, remote and even instrumental mode of seeing and being in the world – but I hope it might nevertheless be possible to argue for the valorization of theories that are constructed out of personal experience – out of what Probyn urges as an attempt to open up 'the local' – without having them summarily dismissed as somehow 'parochial' (Probyn 1990).

(c) **Abstraction**: Modern social theory still bears the marks of its Enlightenment origins and its claims to know continue to respond (in different ways) to Kant's attempt to install reason as the undisputed arbiter in all spheres of social life: science, morality, art. Indeed, it is perhaps a characteristic of modern intellectuals – 'universal intellectuals', as Foucault once called them – to see themselves as legislators: as dealers in generalities rather than brokers in particulars, uniquely qualified to chart the course of society-in-general (or society-as-totality).[15] The discourses of modern social theory are thus driven by an assertive generality in which, as Habermas (1987: 322–3) puts it, 'the transcendent moment of universality bursts every provinciality asunder'. He acknowledges that they are inevitably 'carriers of context-bound everyday practice', embedded in a particular here and now, but insists that they also typically claim to erase all particularities and to *transcend* space and time. Yet this rebounds on Habermas's own project: in Spivak's perceptive remark (1990: 111), 'Habermas makes a lot of sense in the history of the West German political context. He makes a mistake by universalizing it.' Even so, the objection ought not to be that social theories are inescapably context-bound, but rather that the origins of 'travelling theory' need to be scrupulously acknowledged because it will always be freighted with a host of assumptions, often derived from different and even radically incommensurable sites, which may not – and sometimes should not – survive the journey intact.[16]

My purpose in the present essay is not to elaborate on these objections; I merely want to establish that they bear directly on the textual composition of our enquiries. I think it possible to show, through a consideration of these concerns, that the form in which we render our accounts is not a secondary matter – a mere embellishment or ornamentation – but that different textual strategies imply different ways of working with social theory.

These three concerns are not easily disentangled from one another. Taken together, they suggest at once the power and the vulnerability of the meta-narratives through which modern social theory has prosecuted its legislating (and legitimating) functions. But they also dovetail with a renewed crisis of representation. For if, as these critiques imply, there is no privileged vantage-point, no singular place of reflection, no unambiguous closure, no unitary logic, then how can we make the lives of other people intelligible to us – how can we bring them within the horizon of our own (limited) sensibilities and competences – without in some way being invasive, colonizing, even violent? Yet surely we are not to be condemned, in imagination or in practice, to our own eccentric worlds? I can offer no answer to what Clifford (1988: 3) calls this predicament of 'ethnographic modernity': perhaps all we can do, at present anyway, is to disclose our vulnerabilities and, as Spivak (1990: 9) puts it, 'un-learn our privilege as our loss' (see also Spivak 1988). If so, then part of this deconstructive process will entail an examination of our textual strategies – and in particular a consideration of the duplicities of narrative and image – because it is through these modes of representation that many of our most commonplace privileges are unthinkingly put in place. To put it as starkly as possible, the crisis of representation has once again brought the politics of social theory and the poetics of social enquiry *into the same discursive place.*

This is the pivot around which the rest of the essay moves. My readings of Harvey, Benjamin and Pred are intended to bring into focus the politics and the poetics of three historical geographies, each of which seeks, in somewhat different ways, to provide a 'map' of modernity. Since the metaphor of the map has become increasingly common – and increasingly contentious (Connor 1989: 227) – in discussions of this sort, I should say that I use the term with two caveats. First, I realize that the supposed objectivity of 'maps' is an effective fiction: that their texts and images are as vulnerable to deconstruction as any others. Second, the 'maps' which I discuss here are something other than metaphorical devices: their accounts of the inscription of social life in space mean that they also have a substantial materiality.

THE LITERARY DIVER: DAVID HARVEY AND SECOND EMPIRE PARIS

> I find myself most deeply impressed by those works ... that function as both literature and social science, as history and contemporary commentary.
>
> (David Harvey, *Consciousness and the Urban Experience*, 1985: xv)

Harvey's investigations of *Consciousness and the Urban Experience* (1985) are focused on a dazzling reconstruction of Second Empire Paris.[17] His interest in that city is twofold: he wants to use its turbulent historical geography to illuminate the relations between capitalism, modernity and urbanization – or, to put this the other way round, to bring the general propositions of historical materialism to bear upon the particular case of Benjamin's 'capital of the nineteenth century' – and also to clarify the politics of space or what he calls, more specifically, 'the urbanization of revolution'. Largely for this reason, Harvey's major essay in that collection moves determinedly towards (though it does not quite reach) the Paris Commune of 1871. 'We have much to learn from the study of such struggles', Harvey concludes (ibid.: 220), and it will be an important part of my argument that the shape and direction of this 'learning' – and of his politico-intellectual project as a whole – implicates him in an essentially *progressive* conception of history in which the past is able to illuminate the present precisely because each is conceived as a moment in the unfolding of a single master-narrative.

Harvey's portrayal of the events leading up to the Commune is, in part, framed by the work of Henri Lefebvre.[18] This Marxist outlaw, if I may so call him, has provided the inspiration for at least two other reconstructions which can bring Harvey's image into a sharper and more distinctive focus. For Castells, the Commune was not the harbinger of any working-class revolution; it was, rather, a protest against the instruments of speculation, 'the manipulator of the rules of exchange, not the one who appropriated the means of production', and against the institutions of the *ancien régime*, 'the accountants of the old morality'. Harvey differs from Castells not so much in the targets he identifies – on the contrary, he makes much of the ravages of speculation and the involvement of the state in underwriting the transformation of urban space – but in the theoretical armature he deploys. Castells plainly departs much further from historical materialism than Harvey would like. 'For the Commune of Paris', Castells (1983) declares, 'surplus value was an historical abstraction.' Harvey, in contrast, keeps close to the labour theory of value and so penetrates much further into the world of work and the process of production. In doing so, his trajectory is paralleled (though I think on a different level) by Ross's (1988) contextual reading of Rimbaud's poetry. She sees the Commune as a series of anti-hierarchical gestures and improvisations, as 'a revolt against deep forms of social regimentation' which (as Harvey insists) included the imposition of capitalist conceptions of 'work'; but she gives much greater prominence to the ways in which these tactics were elaborated within different *cultural* spaces. This is emphatically not to say that Harvey is indifferent to culture and consciousness, but it is to say (as I now want to show) that his presentation of these themes is made through an almost subterranean textual strategy which contrasts markedly with the way in which he presents the basal logic of the capitalist mode of production.

Marxism and meta-narrative

The master-narrative that makes his primary presentation possible is determined by what is in many ways an immensely productive re-reading of Marx. Harvey repeatedly draws attention to his previous account of *The Limits to Captial* (1982), and the containing contours of late nineteenth-century Paris are accordingly mapped by the instruments of historico-geographical materialism which are calibrated by the concept of the commodity inscribed at its centre. Harvey's leading claim, I take it, is that the rationalization of urban space depended on the mobilization of finance capital – on a new prominence for money, credit and speculation – which installed both *spaces as commodities* and *commodities in spaces*.

On one side of this coin, there was a much tighter integration of finance capital and landed property, to such a degree that 'Parisian property was more and more appreciated as a pure financial asset, as a form of fictitious capital whose exchange value, integrated into the general circulation of capital, entirely dominates use value' (ibid.: 82). Not surprisingly, as its pulverized spaces were commodified, so Paris became a divided city:

> That Paris was more spatially segregated in 1870 than in 1850 was only to be expected, given the manner in which flows of capital were unleashed to the tasks of restructuring the built environment and its spatial con-figuration. The new condition of land use competition organized through land and property speculation forced all manner of adaptations upon users. Much of the worker population was dispersed to the periphery ... or doubled up in overcrowded, high-rent locations closer to the center. Industry likewise faced the choice of changing its labour process or suburbanizing.
>
> (ibid.: 95)

On the other side, and literally so, the centre of the city was increasingly given over to the conspicuous commodification of social life:

> Hausman[n] tried ... to sell a new and more modern conception of community in which the power of money was celebrated as spectacle and display on the *grands boulevards*, in the *grands magasins*, in the cafés and at the races, and above all in those spectacular 'celebrations of the commodity fetish', the *expositions universelles....* [Those] World Exhibitions, as Benjamin put it, were 'places of pilgrimage to the fetish Commodity', occasions on which 'the phantasmagoria of capitalist culture attained its most radiant unfurling'.
>
> (ibid.: 103, 200)

Much of Harvey's essay, as he teases out and traces through one thread after another, is clearly structured by the circuits of capital that are wired together in this way to illuminate the totality of capitalist modernization in Second Empire Paris. And make no mistake: Harvey is determined to present his account *as* a totalization. 'To dissect the totality into isolated fragments', he warns, is 'to lose contact with the complex interrelations that intertwine to produce the simple narrative of historico-geographical change that must surely be our goal' (ibid.: 68).

Must it? Harvey's primary motifs give his project a certain robustness but, I think it fair to say, a certain rigidity too. Yet many of the events and themes that he describes quite clearly refuse to be 'mastered' by his leading narrative. Most recalcitrant are the images of women that occur again and again as the text unfolds: the city of Paris, which appeared to Balzac as a 'mysterious, capricious and often venal' woman and to Zola as a 'fallen and brutalized' one (ibid.: 180); the 'powerfully formed, bare-breasted' Liberty of Delacroix and the 'terrifying' prostitute posing as Liberty by the Tuileries in Flaubert (ibid.: 191–4); the 'voluptuous sensuality' of the Orient, supposedly 'the locus of irrational and erotic femininity' (ibid.: 201). These are all highly specific images of female sexuality which seem to have a double function. For contemporary men, as Harvey recognizes, they provided ways of debasing the struggles of women and so domesticating resistance to the established patriarchal order. For Harvey, himself, however, they become a licence for bringing gender and sexuality within the frame of meaning provided by the master-narrative of capitalist production and reproduction. With money as the 'universal whore', Harvey can declare that 'the city itself has become prostituted to the circulation of money and capital' (ibid.: 177). The coding is unexceptional – one can find it in Marx, in Baudelaire and Benjamin[19] – and it is not without value. Yet the iconographic tradition which these images embody cannot be assimilated as directly to the logic of capital and class as Harvey seems to think.[20] One example must suffice. When Harvey watches the way in which Frédéric Moreau, the bourgeois hero of Flaubert's *L'éducation sentimentale*, 'glides as easily from space to space and relationship to relationship as money and commodities change hands' (ibid.: 204), what he fails to notice is that Frédéric was not only capitalizing on the freedom conferred by his class position: for, as Massey (1991: 47–9) remarks, 'he did have another little advantage too'.[21]

Modernity and the midway: interiors and exteriors

Towards the closing sections of Harvey's essay, however, and overlapping with these images, glimpses of another historical geography come to the surface: one which is looser, more open in texture and perhaps even more politically charged. Many of these moments are occasioned by engagements with Baudelaire – 'that apostle of modernity' (1982: 174) – and Benjamin. In my view, these passages reveal Harvey at his very best: and a best that is, thematically and stylistically, far removed from the wooden prose that supports his earlier narrative. With Benjamin looking over his shoulder, often guiding his pen, Harvey offers a series of arresting and often moving readings of Baudelaire's poems. Slowly, he begins to inaugurate a different rhetorical and representational strategy: one which drives many of his abstract claims into the concrete of everyday experience. Modernity, so he seems to say, is like a swing-boat on the midway, poised in an agonizing stasis between past and future.

> There is ... a contradiction in Baudelaire's sense of modernity after the bittersweet experience of 'creative destruction' on the barricades of 1848. Tradition has to be overthrown, with violence if necessary, in order to grapple with the present and create the future. But the loss of tradition wrenches away the sheet anchors of our understanding and leaves us drifting, powerless. The aim of the artist, he wrote in 1860, is 'to extract from fashion the poetry that resides in its historical envelope', to understand modernity as 'the transient, the fleeting, the contingent' as against the other half of art, 'the eternal and immovable'. The fear ... is of not going fast enough, of letting the spectre escape before the synthesis has been extracted and taken possession of'. But all that rush leaves behind a great deal of human wreckage. 'The thousand uprooted lives' cannot be ignored.
>
> (ibid.: 175)

Reflecting on the complex allegories in Baudelaire's *Paris Spleen*, Harvey subtly yet powerfully alters his understanding of Hausmann, who becomes something more than capitalism's puppet-master:

> The towers from which the triangulation of Paris proceeded symbolized a new spatial perspective on the city as a whole, as did his attachment to the geometry of the straight line and the accuracy of levelling to engineer the flows of water and sewage. The science he put to work was exact, brilliant, and demanding; 'the dream' of Voltaire and Diderot had learned to calculate. But there was ample room for sentiment – from elaborate street furnishings (benches, gas lights, kiosks) and monuments and fountains

(like that in the place Saint-Michel) to the widespread planting of trees along the boulevards and the construction of gothic ghettoes in the parks, everything reimported romance into the details of a grand design that spelled out the twin ideals of Enlightenment rationality and imperial authority. The modernity that Hausman[n] created was powerfully rooted in tradition.

(ibid.: 178)

And gradually Harvey's awareness of the overdetermination of the politics of Paris during the Second Empire prompts him to reinscribe his earlier and somewhat stylized account of the geometry of the divided city within the fleeting encounters and multiple spheres that make up the geographies of everyday life:

Paris experienced a dramatic shift from the introverted, private and personalized urbanism of the July Monarchy to an extroverted, public and collectivized style of urbanism under the Second Empire. . . . Public investments were organized around private gain, and public spaces appropriated for private use; exteriors became interiors for the bourgeoisie, while panoramas, dioramas and photography brought the exterior into the interior. The boulevards, lit by gas lights, dazzling shop window displays, and cafés open to the street (an innovation of the Second Empire), became corridors of homage to the power of money and commodities, play spaces for the bourgeoisie. When Baudelaire's lover suggests the proprietor might send the ragged man and his children packing, it is the sense of proprietorship over public space that is really significant, rather than the all-too-familiar encounter with poverty. . . .

It was for this reason that the reoccupation of central Paris by the popular classes took on such symbolic importance. For it occurred in a context where the poor and the working class were being chased, in imagination as well as in fact, from the strategic spaces and even off the boulevards now viewed as bourgeois interiors. The more space was opened up physically, the more it had to be partitioned and closed off through social practice. Zola, writing in retrospect, presents as closed those same Parisian spaces that Flaubert had seen as open.

(ibid.: 204)

If this provides a gloss on my previous discussion of what Harvey describes elsewhere as Frédéric's 'casual penetration' of Parisian space, it still holds to capital and class as the primary axis through which power is inscribed in the city; but it would not, I think, be difficult to open up this text to quite other modalities of power. The complex distinctions between 'private' and 'public', 'interior' and 'exterior' that run through the passage I have just quoted were also enforced through a series of patriarchal codes whose spatialities were not a reflection of – were sometimes even orthogonal to – those produced through the grid of class relations (Pollock 1988; Wolff 1990).[22] In other words, one can transpose these partitions into different and irreducible registers (as contemporaries did them-selves) and read their coincidences, oppositions and transgressions as what Benjamin would have called a tense *constellation* of different human geogra-phies.[23] Indeed, Benjamin's twin exposés of the Arcades Project, which were both published as 'Paris, capital of the nineteenth century', prefigure the distinctions that Harvey makes in his dialectical rendering of the interiors and exteriors of Second Empire Paris.[24]

I hope it will be obvious that the two textual strategies I am trying to separate here do not correspond to an economic 'base' and a politico-cultural super-structure. Harvey's primary narrative is indeed structured by the basal logic of the capitalist mode of production, but this secondary text cannot be compressed into the superstructure. It stubbornly refuses to respect such a classical dualism. To be sure, the commodity remains at the centre of this 'other' historical geography, but its multiple meanings constantly threaten to escape the narrow-gauge wires of Harvey's circuit diagrams. 'Then, as now', he says at one point,

> the problem was to penetrate the veil of fetishism, to identify the complex of social relations concealed by the market exchange of things. . . . It takes experience and imagination to get behind the fetishism, and imagination is as much a product of interior needs as it is a reflection of external realities.
>
> (ibid.: 200)[25]

So it is. But I sometimes worry that a problematic like this can too readily marginalize material culture, pushing its 'surface forms' to one side in its haste to disclose the social relations so deeply inscribed in the process of commodifi-cation. One of Benjamin's most prescient insights was to establish an interest in cultural artefacts as a legitimate (though hardly self-sufficient) moment of a critical history (cf. Appadurai 1986). Still more important, if the social relations which are symbolized – objectified, naturalized and fetishized – in this way are not fully transparent, then neither can they be made visible from a single perspective: experience and imagination are not sutured around a single (masculine) subject-position. Harvey seems to concede as much when he agrees that 'each particular perspective tells its own particular truth'. And yet, he continues, these perspectives

> scarcely touch each other and they come together on the intellectual barri-cades with about the same frequency as urban uprisings like the Paris Commune. The intellectual fragmentations of academia appear as tragic reflections of the confusions of an urbanized consciousness; they reflect surface appearances, do little to elucidate inner meanings and connections, and do much to sustain the confusions by replicating them in learned terms.
>
> (1982: 263)

I am not so sure (or perhaps I am just confused); but I do think that any attempt to fold 'the complex of social relations' concealed by commodity fetishism into the envelope of a 'simple narrative' ought to be resisted. One way of doing so is by deliberately opening the text to multiple ways of knowing, refusing to round the edges and plane the differences into a single integrated setting. In this sense, Harvey's reference to the barricades can be turned against him. Ross (1988: 36) has suggested that their subversive character derived not only from the way in which they demarcated insurgent spaces of political action but also from the way in which they were constructed: 'Monumental ideals of formal perfection, duration or immortality, quality of material and integrity of design [were] replaced by a special kind of *bricolage* – the wrenching of everyday objects from their habitual context to be used in a radically different way.' As I now want to suggest, it was Benjamin's lasting achievement to have realized this 'tactical mission of the commonplace', as Ross calls it, in his own luminous investigations of nineteenth-century Paris.

PASSAGES: WALTER BENJAMIN AND THE ARCADES PROJECT

At the centre of this world of things stands the most dreamed-of of their objects, the city of Paris itself.

(Walter Benjamin, *Surrealism: The Last Snapshot of the European Intelligentsia*, 1978c: 182)

Benjamin was born in 1892, the son of prosperous German-Jewish upper-middle-class parents.[26] He spent his childhood in Berlin, a period and place recalled in a typically artful way in his *Berlin Chronicle* (1989: 3–60). His parents supported him through university, but in 1925, shortly after his committee refused to accept his *Habilitationschrift* – a post-doctoral requirement for a university teaching post – they lost most of their money and Benjamin was obliged to make ends meet through his writing. In 1926 he moved to Paris, where he embarked upon a series of notes which would eventually thread their way into a vast project woven around the Paris arcades or *passages*. This undertaking was to occupy him, discontinuously and in different forms, until his death in 1940: it became known as the *Passagenarbeit*.[27] It was during this first extended period in Paris that Benjamin fell under the spell of surrealism. Although he soon disavowed its immanent idealism – he eventually participated in the Collège de Sociologie, the circle of dissident surrealists founded just before the war by Bataille, Caillois and Leiris – he always credited Aragon's surrealist novel *Le paysan de Paris* as the most direct inspiration for the Arcades Project. By 1928, however, Benjamin was back in Berlin, contributing to literary journals and newspapers and writing and presenting radio programmes. There he witnessed the gradual collapse of the Weimar Republic until, in 1933, the Nazi seizure of power forced him to flee to France. He returned to Paris as an exile – 'condemned to a way of life closely resembling that of the emigré extras in Rick's Café in *Casablanca*' – but still managed to live off his writing. From 1935 his modest income was supplemented by a stipend from the Institute of Social Research. Although he had become keenly interested in the writings of the Frankfurt School, as the institute became known, the relation between his work and that of Adorno or Horkheimer (its principal architects) can perhaps be best described as one of creative tension: certainly, he was no disciple. As far as Adorno was concerned it was probably the other way round. His debates with Benjamin were an attempt 'to establish himself on an equal footing with this man whose disciple he had become in 1929' (Buck-Morss 1977: 165). In any event, if, as Arendt (1968: 11) remarks, Benjamin was the most peculiar Marxist ever to be associated with the Frankfurt School – itself hardly a bastion of orthodoxy – there can surely be no doubt that his later writings were saturated in the tonalities of historical materialism.[28]

As these remarks perhaps indicate, locating Benjamin within an intellectual landscape is far from straightforward. He has to be placed in relation to, if never entirely within, a Jewish intellectual tradition (Cabbalism) which conceived of the cosmos as an endless network of correspondences and symbolic connections; and yet as late as 1933 he could still remind his close friend Gershom Scholem of 'the abyss' of his lack of knowledge of the Cabbala. He was fascinated by surrealism and intensely sympathetic to its attempts to disrupt the conventional expectations of a bourgeois consciousness; and yet he despaired of the surrealists ever taking leave of 'the world of dreams' and jolting their audience awake. He was deeply interested in developing historical materialism, perhaps even to the point where it would assume a symbolist or surrealist form; and yet he remained uncomfortable at what he took to be its continued proximity to a bourgeois conception of reason and 'progress'.

These three co-ordinates can be made to intersect in different ways and, not surprisingly, the contemporary recuperation of Benjamin has allowed itself considerable licence. To Eagleton (1981: 56, 131), for example, Benjamin appears as 'an archaeologist *avant la lettre*' – where the *lettre* is plainly Foucault's – whose work prefigures many of the current motifs of deconstruction and post-structuralism.[29] Eagleton is not alone in reading Benjamin's work as an archaeology (of sorts), but for my present purposes the affinities that he claims with contemporary critical practice are of more moment.[30] They are further enriched by what Eagleton (1990: 326) also calls the 'dialectical impudence' of this 'Marxist rabbi': that is to say, Benjamin's attempt to conjure a revolutionary aesthetics *from the commodity form itself*. I think it possible to use this interpretation (in outline, at least) to show that although Benjamin came to place the deconstruction of commodity fetishism at the centre of his work, he proceeded in a radically different analytical direction from Harvey. One word of caution is necessary: the Arcades Project is a notoriously difficult textual terrain and what follows is only the roughest map of one possible route through it.[31]

Image and narrative: the enchantment of modernity

Unlike Harvey, Benjamin's conception of history was a profoundly tragic one: he did not see the project of modernity as a process of progressive 'disenchantment', in which Reason had gradually released humankind from the snares of myth and superstition.[32] If, as Harvey says, Marx was a child of the Enlightenment, then Benjamin's own childhood memories suggested that the modern world was a place of re-enchantment. This was not, I think, a nostalgic vision of an age of lost innocence, as it was for so many German intellectuals who mourned the encroachments of a modern and intensely rational *Zivilisation* upon a traditional and essentially spiritual *Kultur*, or even the neo-Romantic prospect of a revolutionary Utopia which animated Bloch and Lukács (cf. Löwy 1985). It was, rather, a highly distinctive way of thinking about commodity fetishism which decisively shaped the form in which Benjamin wanted to present his account. I will focus on two features in particular: Benjamin's fascination with the image and his objections to linear narrative.

In his early drafts, Benjamin proposed to read the Paris Arcades as a dream world – 'a dialectical faery scene' – furnished by objects which were simultaneously desired and commodified. Awakening would constitute the moment when 'the spell or illusion of reconciling a desire for fulfillment with a structure of exploitation and alienation [could] be broken' (Geyer-Ryan 1988: 67).[33] In his more developed sketches, which he worked on following his return to Paris, Benjamin both deepened and widened the project by representing the cultural landscape of nineteenth-century Paris (and, by extension, Europe) as a *phantasmagoria*. The phantasmagoria was a magic lantern which became popular in the early decades of the nineteenth century. Painted slides were illuminated in such a way that a succession of ghosts ('phantasms') was paraded before a startled audience. The phantasmagoria was no ordinary lantern, however, because it used back-projection to ensure that the audience remained largely unaware of the source of the image: its flickering creations thus appeared to be endowed 'with a spectral reality of their own' (Cohen 1989; see also Castle 1988; Crary 1990). Benjamin uses the phantasmagoria as an allegory of modern culture, which explains both his insistence on seeing commodity culture as a projection – not a reflection – of the economy, as its mediated (even mediatized) representation, and also his interest in the visual, optical, 'spectacular' inscriptions of modernity. Indeed, Benjamin was one of the earliest commentators to understand the centrality, the constitutive force, of the image within modernity. What he proposed to do, in effect, was to harness the latent energy of the modern image, to turn it back on itself and thereby use that image as 'a critique of reason' (Abbas 1989: 52; see also Benjamin 1968a).

This project coincided with Benjamin's critique of narrativity. He believed that one of the most powerful ways in which the modern world had become enchanted was through the operation of Reason itself. In its name, meta-narratives had been imposed upon human history to purge it of its specificities and present historical eventuation as a continuous, organic and homogeneous progression towards the present. It was not merely the 'progressive' conception of history to which Benjamin took exception – that this is reinforced by the use of a narrative form is self-evident[34] – but also its 'homogenization'. In a superbly sustained reading of Benjamin, Eagleton (1981: 28–9) suggests that he saw the circulation of the commodity as paradigmatic of this process:

> the commodity, which flaunts itself as a unique, heteroclite slice of matter, is in truth part of the very mechanism by which history becomes homogenized. As the signifier of mere abstract equivalence, the empty space through which one portion of labour-power exchanges with another, the commodity nonetheless disguises its virulent anti-materialism in a carnival of consumption. In the circulation of commodities, each presents to the other a mirror which reflects no more than its own mirroring; all that is new in this process is the very flash and dexterity with which mirrors are interchanged. . . . The exchange of commodities is at once smoothly continuous and an infinity of interruption: since each gesture of exchange is an exact repetition of the previous one, there can be no connection between them. It is for this reason that the time of the commodity is at once empty and homogeneous: its homogeneity is, precisely, the infinite self-identity of a pure recurrence which, since it has no power to modify, has no more body than a mirror-image. What binds history into plenitude is the exact symmetry of its repeated absences. It is because its non-happenings always happen in exactly the same way that it forms such an organic whole. Since the significance of the commodity is always elsewhere, in the social relations of production whose traces it has obliterated, it is freed . . . into polyvalence, smoothed to a surface that can receive the trace of any other commodity whatsoever. But since these other commodities exist only as traces of yet others, this polyvalence is perhaps better described as a structure of ambiguity – an ambiguity that for Benjamin is 'the figurative appearance of the dialectic, the law of the dialectic at a standstill'.

Benjamin sought to interrupt this process by calling into question its endless suppression of difference beneath repetition. What was distinctive about his attempt to do so was that it went beyond disclosure of the logic of capital *to assault the modalities of representation*. A concern with what Wolin (1982: 100) calls the 'image-character' of truth became a vital moment in Benjamin's work, therefore, 'for in this way he sought to confer equal rank to the spatial aspect of truth and thereby do justice to the moment of representation that is obscured once truth is viewed solely as a logical phenomenon'. In other words, Benjamin effectively 'spatialized' time, supplanting the narrative encoding of history through a textual practice that disrupted the historiographic chain in which moments were clipped together like magnets. In practice, this required him to reclaim the debris of history from the matrix of systematicity in which historiography had embedded it: to blast the fragments from their all-too-familiar, taken-for-granted and, as Benjamin would insist, *mythical* context and place them in a new, radically heterogeneous setting in which their integrities would not be fused into one (Geyer-Ryan 1988: 66–8; Jennings 1987: 51). This practice of montage was derived from the surrealists, of course, who used it to dislocate the boundaries between art and everyday life (cf. Clifford 1988: 117–51). Benjamin displayed much the same interest in the commonplace in his *Einbahnstrasse* (Benjamin 1979) – which Bloch saw as an attempt to give philosophy a surrealist form – and in the notes and drafts for the Arcades Project itself. His intention was 'to carry the montage principle over into history' and use it to present nineteenth-century Paris – that glittering world which all the world came to see – as 'a rubble heap of found objects': an image which he hoped would dispel the established vision of the imperial past (Frase 1990–1: 20; Frisby 1985: 215). In many ways montage was another version of that anti-hierarchical gesture that Ross found in the strategies of the Commune, but it is important to see that in Benjamin's hands its transgressive power was also used to challenge the idea of an Olympian meta-narrative through what Sieburth perceptively identifies as an attempt 'to inscribe citation and commentary, the text and its interpretation on the same plane of the page'.

> By situating its 'primary' and 'secondary' reflections on the same textual ... level, Benjamin's manuscript places into radical question the very possibility of metalanguage, that is, of a discourse that might somehow stand above, outside or beyond that of which it speaks.... [His handwritten pages] democratize the traditional hierarchies separating author from reader, original from copy, citation from commentary – like the passage itself, the text that thus emerges has no outsides. In this it resembles the polyphonic play of language which Bakhtin terms heteroglossia – different voices, different discourses refracting each other in dialogue.
>
> (Sieburth 1989: 33)[35]

These concerns struck at the heart of the traditional aesthetic paradigm, whose strictures refused to allow what Eagleton (1990: 330) calls 'the specificity of the detail' any genuine resistance to 'the organizing power of the totality'. Benjamin's way of working – of seeing? – was directed, in contrast, towards the construction of a *constellation* that would embody 'a stringent economy of the object which nevertheless refuses the allure of identity, allowing its constituents to light each other up in all their contradictoriness'.

These are not easy ideas to grapple with, but I hope it is now possible to glimpse the field in which Benjamin's fascination with the image and his critique of narrativity could be brought together. As Buck-Morss (1989: ix, 218) puts it in her own version of the book that Benjamin never wrote – which she presents as 'a picture book of philosophy' putting into practice Benjamin's 'dialectics of seeing' – he was convinced that

> What was needed was a visual, not a linear logic. The concepts were to be imagistically constructed, according to the cognitive principles of montage. Nineteenth-century objects were to be made visible as the origin of the

present, at the same time that every assumption of progress was to be scrupulously rejected.

The conceptual form that Benjamin eventually gave to this textual strategy was the construction of what he called the 'dialectical image'.

The dialectical image

I should say at once that the noun is as difficult as the adjective. Many commentators do indeed draw attention to Benjamin's preoccupation with the *visual* image: with that sense of the optical and spectacular that is preserved so emphatically in the title of Buck-Morss's (1989) account of the Arcades Project and which is captured, somewhat more loosely, in Jay's (1988) meditation on what he calls the 'scopic regimes of modernity'.[66] But others are more reluctant to conceive of the image in strictly visual terms. In his discussion of the dialectical image, for example, Jennings (1987) emphasizes not so much the optical implications of the concept as Benjamin's rehabilitation of allegory as a 'revelatory instrument' which by virtue of its 'brokenness', its resistance to mimesis, was 'the only form of language capable of resisting the allure of the commodity' (see also Spencer 1985). Eagleton (1990: 326–7) agrees that this break between signifier and signified at once mimed the circulation of the commodity and released 'a fresh polyvalence of meaning, as the allegorist grubs among the ruins of once integral meanings to permutate them in startling new ways'. But, like Buck-Morss (1989: 71, 170–6), he also suggests that in Benjamin's later work the sense in which the allegorical signifier could inscribe 'its own network of "magical" affinities across the space of an inscrutable history' was given a new inflection in the dialectical image itself. And if Benjamin was deliberately reviving allegorical emblematics in the Arcades Project, one might argue that its subversive inscriptions were reinforced in avowedly iconographic terms through a reading of photographs, exhibitions, architectural forms and the like. The filiations between the visual image and the linguistic image are likely to be extremely complicated, therefore, and I think one has to recognize that there may well be something essential to Benjamin's project that cannot be grasped through the visual image alone (see Nicholsen 1990).

Buck-Morss argues that Benjamin produced the dialectical image by wrenching the fragments of the past out of their usual context and placing them on the conceptual grid shown in Figure 15.1. She identifies the two axes in conventional Hegelian terms – consciousness and reality – but one could also suggest that the lower arc describes the trajectory of surrealism, the resolution of dream and reality into what Breton called 'surreality', while the upper arc describes the trajectory of historical materialism, the physical inscriptions of a political consciousness.[36] If this is an acceptable approximation then the two arcs can be folded over one another in a relation of extreme tension, as Benjamin sought to give surrealism a materialist form and to develop historical materialism

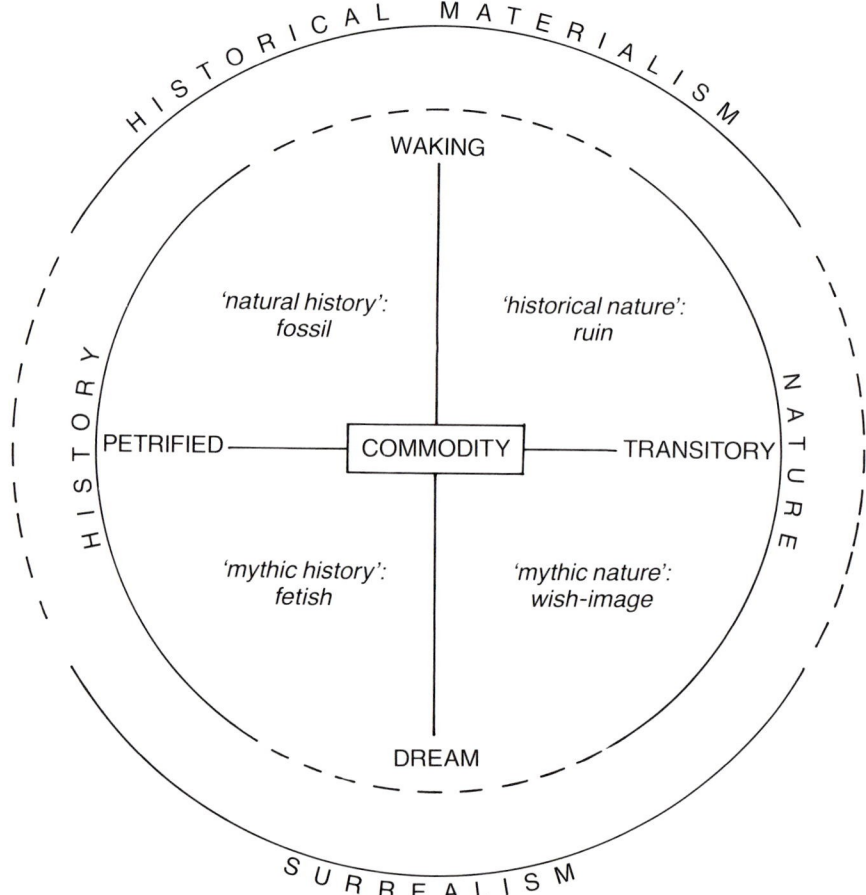

Figure 15.1 The conceptual grid of Benjamin's Arcades Project

Source: After Buck-Morss 1989

in a surrealist form. Another, equally tense relation can be produced by reading Benjamin through Adorno's eyes and seeing the left-hand arc as 'history' and the right-hand arc as 'nature'. To the two friends, these were mutually determining concepts constitutively involved in the formation of historical materialism. 'History and nature were not abstract "invariant concepts" but "arrange themselves around the concrete historical facticity", forming a constellation which released in the phenomena the moment of transitoriness which might break their mythical spell over the present' (Buck-Morss 1989: 57).[37]

Other interpretations of this grid are no doubt possible, but what matters is its dialectical structure: the way in which Benjamin sought to exploit the inter-penetrations of successive conceptual couples. At any rate, Buck-Morss (1989:

210–12) suggests that this screen of polarities may be regarded as the 'invisible, inner structure' of the Arcades Project.[38] Benjamin was not, I think, much interested in the fragments in themselves – he was certainly no empiricist – and neither did he concern himself with elaborating the grid in any systematic way. If I understand Buck-Morss properly, his hope was rather that when the fragments were placed on this conceptual grid a moment of 'profane illumination' would be generated. The dialectical image would appear, however fleetingly, at the null point of the grid, as what today we might think of as a sort of philosophico-political hologram. By this means, so Benjamin believed, the commodity (which had come to occupy a central place within the Arcades Project) would be seen as an historical object doubly constituted in the act of representation: *doubly* constituted because it was both displayed for sale (there and then) and disclosed in its duplicities (here and now).[39]

Benjamin achieved his emphasis on the *constitution* of the historical object *as* a dialectical image through the sheer force of the adjective.[40] The dialectical nature of the image, the profanity of its illumination, required a break with the conventional historiographic pietism that the present had to be seen in the light of a sepulchral past whose beams would usher historical eventuation into the future. Instead, Benjamin sought to bring about an explosion which would bring down the Dream House of History by forcing a discarded, forgotten, even repressed past into an unfamiliar, unreconciled constellation with the present. Like the surrealists, his intention was to shock: to stage a 'confrontation in which time is arrested to a compact monad, spatialized to a shimmering field of force, so that the political present may redeem an endangered moment of the past by wrenching it into illuminating correspondence with itself' (Eagleton 1990: 327). Unlike the surrealists, however, Benjamin's constellation was not an arbitrary construction: the commodity was not placed at the centre of the grid by chance. Indeed, Buck-Morss contends that Benjamin's sense of objective necessity distances his way of working from allegory and deconstruction too. She claims (1989: 241, 339) that the meaning of the allegorical image remains an expression of subjective intention and so 'is ultimately arbitrary' – in contrast to the dialectical image – and that the carnival of deconstruction reduces revolutionary criticism to an endless parade of new interpretations: 'Fashion masquerades as politics.' If Benjamin's project can be seen as a deconstruction of commodity fetishism, therefore, this has to be a deconstruction of a very different kind: one which somehow invests the text with new yet necessary meaning. But Benjamin's attempt to square this particular circle was not uniquely determined by the conceptual thematics of historical materialism (important though these were). It depended, crucially, on the present being understood as a moment of revolutionary possibility, on the notion of a revolutionary 'now time' or *Jetztzeit*, which acted as what Buck-Morss calls 'a lodestar for the assembly of historical fragments'. Without its 'power of alignment', she continues, 'the possibilities for reconstructing the past [would have been] infinite and arbitrary' (ibid.: 338–9; see also Habermas 1987: 11–16).[41] And yet the constellation was not a means of closure; 'necessity' was not transmuted into teleology. Benjamin's purpose was to prise open the texture of historical eventuation and create a space for revolutionary political action. The moment of *Jetztzeit* was supposed to contain within itself, as an immanent possibility, an interruption in the empty, homogeneous time of conventional history. Benjamin's belief was as complex as it was controversial, deriving as it did from both surrealism and Jewish mysticism, but a rough approximation of what he had in mind can be gained from his final thesis on the philosophy of history:

> the Jews were prohibited from investigating the future. The Torah and the prayers instruct them in remembrance, however. This stripped the future of its magic ... [but] does not imply ... that for the Jews the future turned into homogeneous, empty time. For every second of time was the strait gate through which the Messiah might enter.
>
> (Benjamin 1968c: 264)[42]

I have no wish to minimize the difficulties of Benjamin's strategy, which inhere in its mythical-Messianic genealogy and its radicalization of modern conceptions of temporality and historical consciousness.[43] Nevertheless, the (weak) Messianic power that Benjamin invoked was, I think, a distinctly worldly one.[44] In his eyes, within his 'materialist optics', the moment of profane illumination – the spark of the dialectical image – had the power to intervene in the directionality imposed upon human history by traditional, bourgeois narratives of progress. To put it still more concretely, conceiving of the 'history of the present' in this way (and by this means) was a way of empowering the emancipatory production of human geographies. It would surely be difficult to think of a more brilliant illustration of the politics of representation.

THE VEGA CAP: ALLAN PRED AND *FIN-DE-SIÈCLE* STOCKHOLM

A *Vega* cap finally tops his head.
(Allan Pred, *Lost Words and Lost Worlds: Modernity and the Language of Everyday Life in Late Nineteenth-Century Stockholm*, 1990a: 230)

Pred's account of Stockholm centres on the period between 1880 and 1900:

> A twenty year span during which modernity made its many guised entrance, during which consumption came to be marked by a whirl of quickly passing fashions and fads, during which commodity fetishism took hold, during which the circulation of money accelerated, during which the iron cage of bureaucratic regulation and surveillance dropped down over previously ungoverned details of daily existence, during which everyday life on the streets was characterized by restless and anonymous movement, by fleeting, fragmented impressions.
>
> (Pred 1990a: xiii)[45]

I want to show how Pred brings these restless geographies into focus by making a series of comparisons between his project and the reconstructions of Second Empire Paris offered by Harvey and Benjamin. I begin by situating his work in relation to 'mainstream' historical materialism. There are many points of contact between them, if sometimes at several removes, but Pred is clearly unconvinced by the closures of classical Marxism – in particular its silence over space – and so, like Harvey, he endorses the development of a materialist analysis of the production of space.[46] But it is equally obvious that Pred also wants to go beyond historico-geographical materialism: largely, I think, because he regards its spatial analytics as too unyielding to accommodate the contingencies of everyday life as they unfold in place (see Pred 1984, 1986). This sense of uncertainty has more than a theoretical valency, however, because Pred increasingly argues that it has textual implications too. He is particularly disconcerted by the imposition of a single meta-narrative on what he takes to be the multiple constellations of the past:

> There cannot be one grand history, one grand human geography, whose telling only awaits an appropriate metanarrative. Through their partici-pation in a multitude of practices and associated power relations, through their participation in a multitude of structuring processes, people make a plurality of histories and construct a plurality of human geographies.
>
> (Pred 1990b: 14)

My main concern will thus be the ways in which Pred elects to represent the concrete particularities of *fin-de-siècle* Stockholm, where the parallels with Harvey and Benjamin seem to me particularly instructive.

Materialism and modernity

Like Harvey, Pred registers the rising tide of commodification: 'the carousel of consumption and commodity fetishism – the merry-giddy-go-round and around and around of money in circulation ... [that] hailed the entrance of modernity in the capital cities of Europe' (1990a: 51). And he too sees the city pounded by waves of speculative investment and plunged in a maelstrom of creative destruction. 'From the early 1880s onward, when there was a high-tempo speculative investment of overaccumulated capital in the city's construction sector, Stockholm was more than figuratively a place where all that was solid melts into air' (ibid.: 122). The allusion is to Marx's description of modernity as a world in which 'all fixed, fast-frozen relations, with their train of ancient and venerable prejudices and opinions are swept away, all new-formed ones become antiquated before they can ossify', a world in which 'all that is solid melts into air' (Berman 1982: 21).[47] Where Harvey draws on Marx for a more formal rendering of these transformations, however, Pred offers only impressionistic outlines. He does not attempt to reconstruct the circuits of capital through which these pulsating changes were set in motion and brought so crashingly together. There is little sign of the theoretical notes that score Harvey's recital of Second Empire Paris, so to speak, only their echoes in the murmurs of everyday life.

Like many others working at the interface of social theory and human geography, Pred (e.g. 1990b: 228–34) invokes Marx's reminder in the *Eighteenth Brumaire of Louis Bonaparte* that 'men [and women] make their own history; but they do not make it just as they please; they do not make it under circumstances chosen for themselves, but under circumstances directly encountered, given and transmitted from the past'.[48] This is a deceptively simple formulation, however, and in working out its implications Pred borrows from traditions of social theory which, while they certainly have something in common with historical materialism, are none the less removed from most mainstream versions of Marxism. Among the most important are Bakhtin's discourse theory, Bourdieu's symbolic anthropology, Foucault's 'history of the present' and Giddens's structuration theory. This is a heterogeneous list, of course, but it is more than a collection of materials that just happen to be at hand. Pred strongly implies that there is an elective affinity between these various discourses: or, to put things somewhat differently, that their 'multiple voices' articulate a set of common concerns which reach beyond Marx's seminal discussions of praxis to become central to the contemporary reconstruction of critical social theory.[49] He thinks it possible to use the ideas assembled from these various sources in concert with ideas derived from human geography to inscribe propositions about social practice in time and space. His purpose is to illuminate the construction of historical geographies in ways which transcend – and transgress – the logic of capital etching its signature on successive landscapes of accumulation.

The urban morphologies that loom so large in the landscape of Harvey's Paris provide only a skeletal backdrop for Pred's Stockholm. Although he too connects 'creative destruction' to the spatial polarization of class relations, Pred is more interested in the ways in which this modern *grille* was imposed on the skeins of interaction which were spun, snapped and spliced together in the course of everyday life. These paths intersected in myriad ways which were not regulated by the residential lattice of class relations. The bourgeoisie and the working class may have lived in different parts of the city but their day-to-day activities brought them into constant, unavoidable and unpredictable contact with one another. If most members of Stockholm's bourgeoisie felt secure enough inside their houses, their vulnerability once they were outside the front door was almost palpable.

> Nowhere did many of the bourgeoisie sense more heightened apprehension or greater threat than amidst the welter of activities and bustling movement of the city's streets. There one might be cast adrift in a ... sea of pedestrian promiscuity, where the banker and the bum, the wholesaler and the whore, the retailer and the rag-picker, the respectable and the disrespectful, the high and the low, the clean and the dirty, flowed and jostled, side-by-side, over the same spaces.
>
> (1990a: 129)

Their characteristic response, as in Second Empire Paris, was to try to turn these savage 'exteriors' – the streets – into domesticated 'interiors', and it is hardly surprising that Pred should interpret the bourgeois politics of space in Stockholm as another version of 'Hausmannization' (ibid.: 133–4). But Pred is less interested in the logistics of this strategy – in its class filiations and geometries – than in its 'ideo-logistics', and in tracing its inscription (and subversion) in the language of everyday life he goes some considerable distance beyond Harvey's 'other' historical geography.

His enquiry depends on a close connection between 'lost words' and 'lost worlds'. To ordinary language philosophers the two are fused in the fluidity of language games, but Pred sutures them around constellations of power and practice embedded in place. 'To uncover lost words', he declares, 'is to lift the lid from a treasure chest of past social realities, to reveal fragments shimmering with the reflections of lost worlds of everyday life' (ibid.: 7–8). Moved by the same instincts that animated Benjamin's collector, Pred unpacks distinctive glossaries of production and consumption, of spatial orientation, of social distinction, and shows how these otherwise unpromising attics contain the discarded furniture of people's everyday lives.[50] His recovery of a 'popular geography' existing alongside, underneath and often in place of the official toponymy of the city is perhaps closest to the spirit of what Sieburth (1989: 15) describes as Benjamin's 'onomastic inventory'.[51] Benjamin was profoundly aware of the linguistic universe conserved in the city and he devoted one of the files of the *Passagenarbeit* (Konvolut P) to a collection-contemplation of the street names of Paris:

> All that remains given of the increasingly swift dissipation of perceptual worlds is 'nothing other than their names: *Passagen* .... The forces of perversion work deep within these names, which is why we maintain a world in the names of old streets.'
>
> (Menninghaus 1988: 311)

In much the same way, Pred treats the volatility of naming as an allegory of commodification. The 'ephemeral commodities and consumption practices' that punctuated the whirling world of modernity

> were also one with a wealth of lost words and meanings ... that either disappeared after decades or centuries of usage, or briefly appeared some time between 1880 and 1900 only to fade away during that period or shortly after the turn of the century.
>
> (ibid.: 53)

Benjamin believed that the inventory of these names not only revealed the elements of a 'mythical geography', a topographical domain charged with mythical significance, but also simultaneously preserved and prefigured what he called 'the habitus of a lived life' (Menninghaus 1988: 310–11; Sieburth 1989: 15). It is in this latter sense that Pred seeks to bring to light the undervocabulary, the subversive, rough-and-tumble street names and the scatalogical and unashamedly sexist transpositions which undermined and challenged the hegemonic inscriptions of the bourgeoisie on the symbolic landscape of modernizing Stockholm (1990a: 136–42).

Spaces of representation and everyday life

Although Pred refuses to submit the multiple historical geographies of Stockholm to the discipline of a singular narrative, this does not commit his enquiries to the prison of the parochial. He wants his empirical findings to speak to a series of larger questions, and his purpose is to use Stockholm to bring the wider world into the frame without ever losing sight/site of the city itself. I put it like that because the visual thematic is such an unusually prominent motif in a text which is so attentive to spoken language. Indeed, Pred sees the problem of representation in terms which clearly mirror Benjamin's own visual obsessions.

> modernity was characterized by an increasingly fragmented social life and rapid changes in impressions and relationships, especially in the streets. Particularly in the case of Stockholm, the experience of modernity was by no means limited to the restless, phantasmagoric movement of people on the city's principal thoroughfares. Social contacts around dwelling and workplace also frequently took on a fleeting, flickering quality because connections to both were so often short-lived. Those living or labouring beside a person, those people who one saw and interacted with on a daily basis, could disappear as quickly and suddenly as they appeared, could leave an unfilled void or be replaced by others who might also evaporate with the same rapidity.
>
> (ibid.: 186)

How, then, to convey this 'phantasmagoric' world? Let me consider two strategies which Pred pursues where the echoes of Benjamin are, I think, unmistakable.

First, and most generally, Pred deploys what, following Geertz, he calls 'blurred genres': a deliberate erasure of the finely etched line between the academic and the artistic. He insists that he is 'free to play the art-ificer, to play upon words'.

> The repetitive phrase-sentence as driving-pulsating sentence, as pound-pound-pummel sentence, straightforwardly strives to violate reception barriers, to hammer home a message; while the repetitive-phrase sentence as facet-rotating sentence, as prism-turning sentence, by putting a cubistic face on the subject or object around which it circles, strives to seduce, to be (mind's) eye-opening, to whisper sweet (and sour) newthings in the (mind's) ear. The use of poetic forms is not an end in itself, but
>> an attempt to exploit the physicality of the text,
>>> to exploit the landscape of the page....
>
> (ibid.: xv–xvi)

296

This is not an innocent strategy. Pred's purpose is to call attention to the multiple, labile meanings embedded in seemingly the most straightforward and stable of propositions.[52] If the conventional closures of sentences and texts can be opened up in this way then, by virtue of the embeddedness of language in the material flow of power and practice, the taken-for-granted categories of social life become equally vulnerable to displacement and deconstruction. 'The poetics of my textual strategy', Pred warns, 'are the politics of my textual strategy' (ibid.: xv). Other writers have made the same connection. Benjamin understood it very well, of course, though he was not very interested in typographical acrobatics and word-play, and Harvey is no less sensitive to the demands of dialectical modes of representation. I doubt that he would assent to the formal articulation of historical materialism and deconstruction proposed by some authors – still less to the claim that Marx and Derrida 'are on the same side' (Ryan 1982: 102) – but he has often acknowledged that one of the most intriguing (and far from incidental) aspects of Marx's use of language is its polysemy. 'Marx's words are like bats: one can see in them both birds and mice.'[53] For this reason, Harvey also prefers a non-linear mode of representation, although this refers more to the logic of the argument than to what Pred calls 'the landscape of the page', and he develops a relational style of theorizing which exploits the tensions within and between concepts. 'Different starting-points yield different perspectives,' he admits, 'and what appears as a secure conceptual apparatus from one vantage point turns out to be partial and one-sided from another' (Harvey 1985: xvi).[54] All of this may be granted, but it provides no ready-made solutions: and I have to say that Pred's 'poetic forms' do not always enlarge my horizon of meaning. There is no question that his experiments are serious, but somehow his traverse across the typographical high-wire often seems to end in a tangle of thoughts and a pile of words. This almost always happens whenever Pred is trying to convey a series of theoretical ideas which are inter-connected in complex ways, when he feels compelled to spatchcock several sentences into one and have their clauses skewer one another from multiple directions (see also Pred 1989).[55] I understand and sympathize with the intention; but I simply do not think the strategy works very well. Yet the images Pred conjures up through his more concrete – and, significantly, more graphical – modes of representation seem to me considerably more accomplished, and it is to these that I now turn.

Second, then, Pred convenes his separate accounts of the vocabularies used in different spheres of social life within a plenary reconstruction of a single docker's everyday path through the city. This is a telling way to drive language into the concrete, of course, and to reinforce the 'ordinariness' of its use; but it is also a way of suggesting a more general 'interweaving' of these heteroglossia within the fabric of everyday life (1990a: 203). Benjamin is often credited with much the same 'micrological' sensivity, the ability to see a large picture in the smallest of details.[56] As I have already indicated, such an emphasis on a 'way of seeing' is particularly appropriate, but here I want to draw attention to the graphical imagination at work in Benjamin's reflections on his childhood in Berlin:

> I have long, indeed for years, played with the idea of setting out the sphere of life – bios – graphically on a map.... I have evolved a system of signs, and on the gray background of such maps they would make a colourful show if I clearly marked in the houses of my friends and girl friends, the assembly halls of various collectives, from the 'debating chambers' of the Youth Movement to the gathering places of the Communist youth, the hotel and brothel rooms that I knew for one night, the decisive benches in the Tiergarten, the ways to different schools and the graves that I saw filled, the sites of prestigious cafés whose long-forgotten names daily crossed our lips, the tennis courts where empty apartment blocks stand today, and the halls emblazoned with gold and stucco that the terrors of dancing classes made almost the equal of gymnasiums.
>
> (Benjamin 1968b: 5)

The graphical impulse was not alien to Benjamin's reconstruction of Second Empire Paris either, but I make so much of this passage because it seems to evoke something of the same sensibility that is displayed in the diagrams of Hägerstrand's time-geography: a sensitivity to the micro-topographies of everyday life.[57] Pred (1984: 292) insists that it is essential to use Hägerstrand's graphical notation to represent the path of Sörmlands-Nisse, his imaginary docker (Figure 15.2), because such a mode of representation is able to capture simultaneities and conjunctures that 'can easily escape linear language'.[58] Again and again, he argues that diagrams of this sort have the unique capacity to make the structuration of social life seen/scene in the double sense of both making the processes *visible* and embedding them in *place*

But how is such a time-geography to be read? In one sense, I suggest, the line diagram depicts exactly that *danse macabre* which once so alarmed Hägerstrand's critics (and friends) – an alienated world of bodies in autonomic motion – and yet it is one which also captivated both Baudelaire and Benjamin. 'Baudelaire describes neither the Parisians nor their city,' wrote Benjamin.

> Forgoing such descriptions enables him to invoke the ones in the form of the other.... As Baudelaire looks at the plates in the anatomical works for sale on the dusty banks of the Seine, the mass of the departed takes the place of the singular skeletons on these pages. In the figures of the *danse macabre*, he sees a compact mass on the move.
>
> (Benjamin 1968e: 168)[59]

In much the same way, I think, the 'singular skeletons' of Hägerstrand's time-geography are able to evoke 'the mass of the departed'.[60] The bare bones of the line diagram reveal what one might even call a 'post-mortem' geography: a world in which, through the endless repetition of daily paths (time-space routines), the ever-same is reproduced as the ever-new. But this is not an objection to representing the world in this way, as I once thought, because – to recall Eagleton's

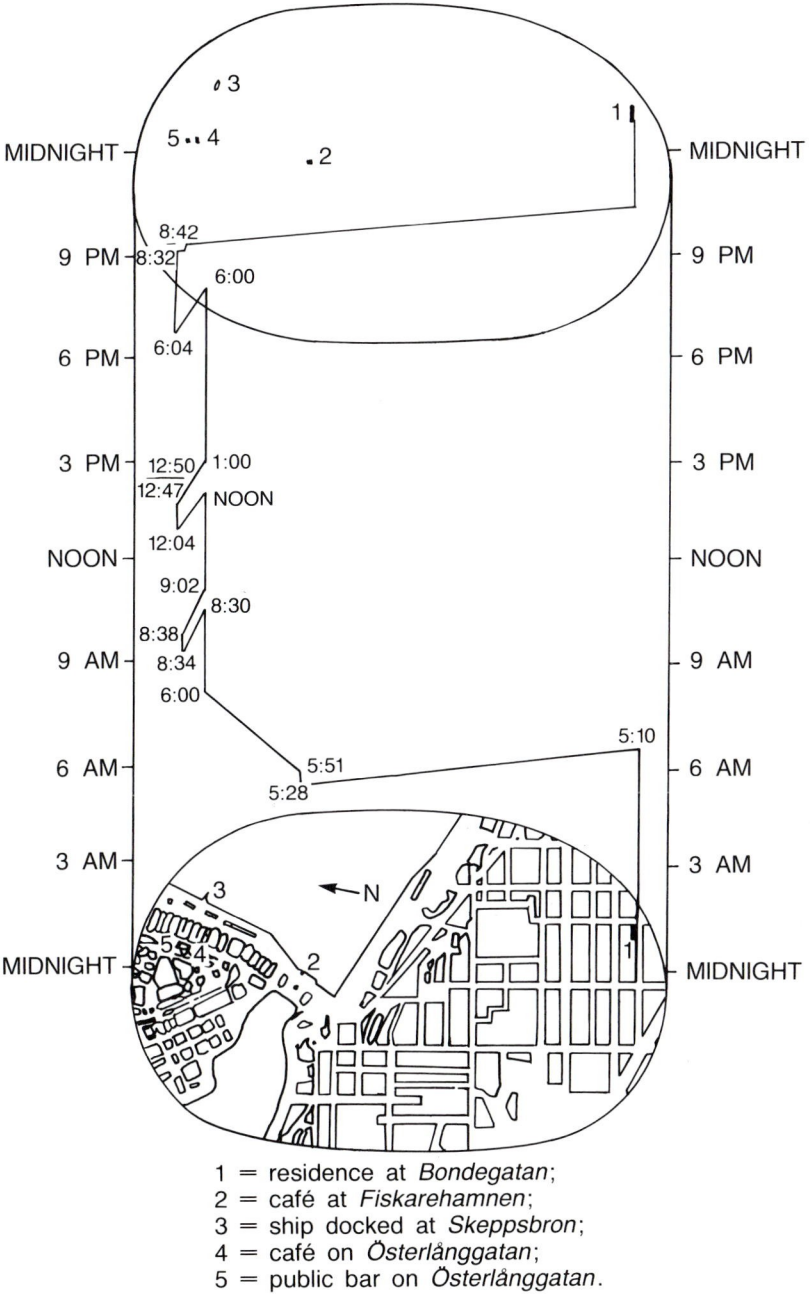

1 = residence at *Bondegatan*;
2 = café at *Fiskarehamnen*;
3 = ship docked at *Skeppsbron*;
4 = café on *Österlånggatan*;
5 = public bar on *Österlånggatan*.

Figure 15.2 Danse macabre? The daily path of Sörmlands-Nisse

Source: Pred 1990a

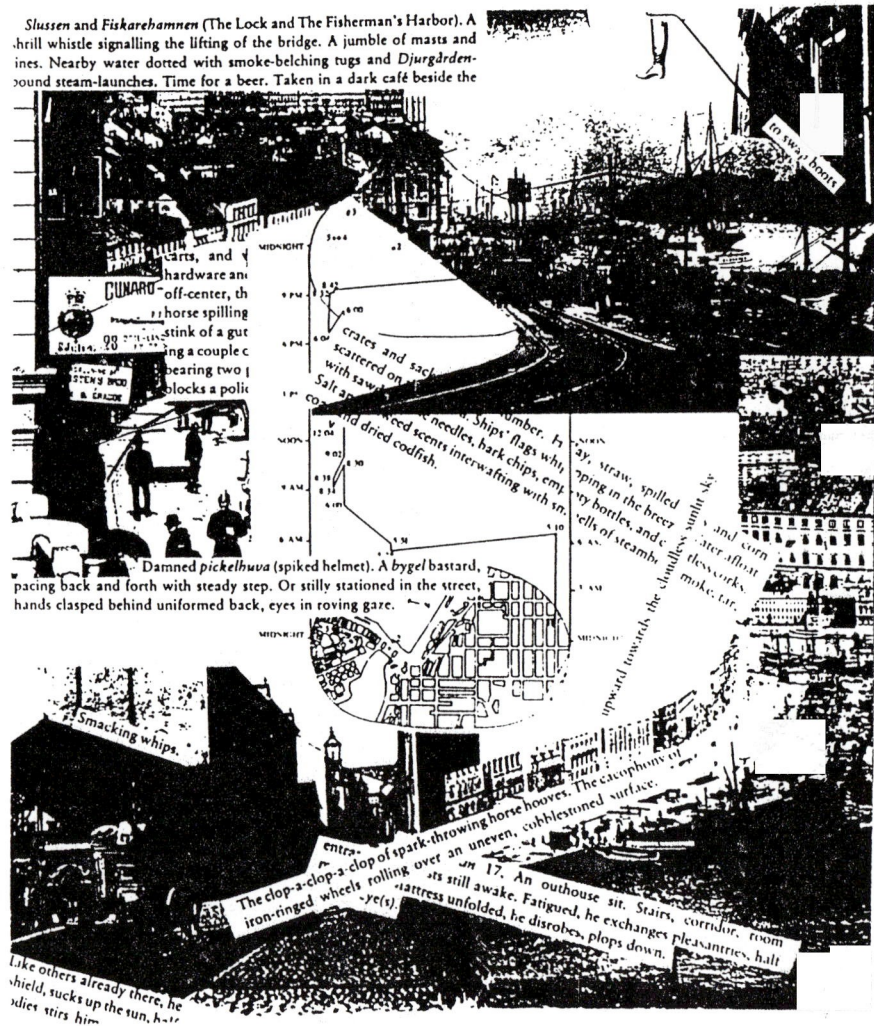

Plate 15.1 Montage? The time-geography of Sörmlands-Nisse

Source: Pred 1990a

discussion of Benjamin – *it allows the diagram to be read as an allegory of commodity circulation itself.* The figure of the docker then becomes doubly effective because so much of his life-world was anchored at 'pivotal points where goods circulated between the city and nonlocal worlds' (Pred 1990a: 200).[61]

But the diagram does not stand alone. Pred intends 'the imagery of this entire daily-path depiction [to be seen as] a . . . montage' (ibid.: 291). Again, the echoes of Benjamin are clear. In Plate 15.1 I have taken Pred at his word and brought together some of the photographs and word-pictures – the images – which frame and compose the text. It would be unduly fanciful to suppose that the result is a dialectical image, but it none the less discloses a field of extreme tensions. In quite another sense, therefore, once the implications of time-geography are internalized, then I think it is possible to achieve something like a moment of profane illumination: to grasp the multiple ways in which the docker made this world meaningful *to himself*, sometimes on terms which replicated the dominant order (xenophobia, patriarchy), to be sure, but sometimes on terms which were hostile to and even, on occasion, subversive of it. In my view, these passages contain some of the finest writing in the book (ibid.: 229–35). There is no extravagant experimentation, no fiddling around with composition, and yet the result is an astonishingly vivid evocation of *fin-de-siècle* Stockholm. Pred achieves this effect through a multi-layered series of overlapping images which assault the senses. The stench-drenched visit to the outhouse; the shopkeeper's signs swinging above the street; the smell of sour beer and stale cigar smoke issuing from the open café doors; the ships' flags whipping in the breeze; the thrumming of iron-ringed wheels over cobblestones; the eyes of the policemen, watching and waiting; the bodies dozing on the sun-warmed sacks of grain: all of these word-pictures hang flesh on the bare bones of the line diagram and bring it explosively to life in an intensity of experience which is quite alien to most conventional human geographies. In this sense, I suggest, Pred's redemption of 'the habitus of a lived life' depends as much on the meaning Benjamin gave to that phrase as it does on the those subsequently given to it by Bourdieu and Hägerstrand.

And yet this plenary representation, for all its power, does not quite achieve the force of a constellation. To be sure, Pred does not work with a screen of polarities, but his conceptual architecture is not at issue here. More to the point is his way of working. Imagine what would have happened had he constructed not one time-geography but several time-geographies: not just Sörmlands-Nisse's daily path to and from the docks, but also the path of one of the merchants he saw at breakfast through the open windows of the Skeppsbro Cellar, the path of the waitress who had to endure his lunchtime assaults and the path of one of the policemen patrolling the Österlånggaten. Presumably the word-pictures would have been different, both in composition and in detail: but what of the photographs which Pred uses to make the docker's daily path scene/seen? At present, clearly, one is tacitly invited to view these images through Sörmlands-Nisse's eyes, not simply to accompany him through the streets and into the bars but to enter into the fiction of seeing what he saw.[62] But how would one then make these same images scene/seen from *other* points of view? The merchant, the waitress and the policeman would have walked along at least some of the same streets, but did they see – could they possibly have seen – the same things? Would they have seen their surroundings in the ways in which they have been composed by the photographer? In raising questions like these I am seeking to make two points. On the one hand, photography (even 'documentary' photography) is not an innocent art. Pred is endlessly inventive in making sense of the textual archive, revealing the 'lost worlds' hidden within its seemingly silent glossaries; and yet, for all his visual sensibility, he does not subject the visual archive to the same process of interrogation (cf. Shapiro 1988; Tagg 1988). On the other hand, by convening a series of distinctive vocabularies within the life-world of one docker, suturing them around the figure of Sörmlands-Nisse, and in particular by passing off the point of view represented in these photographs as a 'plenary' perspective, accessible to and even shared by other participants in the making of Stockholm's multiple historical geographies, Pred risks losing the very heterogeneity which is contained within his own remarkable vision of the structuration of social life in time and space. These are not minor quibbles, because they suggest that Pred's poetics *could* be put at the service of a much more radical politics, one which is sensitive to difference without fragmenting politics in endless mutilations.

ARCHIVES AND ARCHAEOLOGIES

> It is surprising how long the problem of space took to emerge as a historico-political problem.
>
> (Michel Foucault, *Power/Knowledge*, 1980: 149)

I have used Benjamin to mediate between Harvey and Pred, but another figure has been wandering through these passages too: Michel Foucault. In several places the language I have used has been Foucauldian and in several others parallels could be drawn between Benjamin (in particular) and Foucault. I do not mean to imply that one can be assimilated to the other, of course, but there are two affinities between their projects which seem to me particularly telling. The first is their sense of historicity or, as I have called it here, of interventions in historical geography. Foucault (1989) once suggested that, as he used the term, 'archaeology' could almost mean 'description of the archive'. By archive, he continued, he meant

> the set ... of discourses actually pronounced; and this set of discourses is envisaged not only as a set of events which would have taken place once and for all and which would remain in abeyance, in the limbo or purgatory of history, but also as a set that continues to function, to be transformed through history, and to provide the possibility of appearing in other discourses.[63]

It would not be difficult to interleave this passage with Benjamin's theses on the philosophy of history and, indeed, with the labyrinthine construction of the Arcades Project itself. On such a reading, the discourses of the past are reactivated in the present – *our* present – not, as Thompson (1978: 234) once put it, in order to 'shake Swift by the hand' (or Walpole by the throat) but precisely to call such continuities into question: to disrupt the unilinear, progressive trajectory of historical eventuation and to substitute a much sharper, less self-assured 'history of the present'. Harvey's sensibilities are clearly much closer to Thompson's than are Pred's. We have much to learn from the struggles of the Commune, so Harvey would argue, not only because they can illuminate the tactics and strategies of an otherwise unremarked politics of space but also because they help to enlarge our own moral space: 'We are saying that these values, and not those other values, are the ones which make this history meaning-ful to us, and that these are the values which we intend to enlarge and sustain in our own present' (ibid.). I do not want this to be misunderstood: Benjamin and Foucault have a highly developed ethical sense too. But neither of them allows hindsight to become a privilege, and perhaps it is *that* privilege that we should also 'unlearn as our loss' (above, p. 277).

That visual thematic haunts Foucault as much as it does Benjamin. One commentator describes his way of working as an 'art of seeing', and another represents Foucault's archaeology – and I would add his genealogy – as a way of opening up what he calls 'visibilities':

> Visibilities are not forms of objects, nor even the forms that would show up under light, but rather forms of luminosity which are created by the light itself and allow a thing or object to exist only as a flash, sparkle or shimmer.
>
> (Deleuze 1988: 52; see also Rajchmann 1991)

The prose is elliptical, but Benjamin's heteroclite constellation seems to mine the same historico-geographical vein: in both cases 'archaeology' is not so much an excavation, bringing buried or hidden objects to the surface, as a way of showing the particular – anonymous, dispersed – practices and the particular – differ-entiated, hierarchized – spaces through which particular societies make particu-lar things visible. Deleuze argues that Foucault's archaeology also involves the recovery of a tense, chameleon-like struggle between the visible and the 'articulable' (fields of statements): and again, one can read a similar tension in the Arcades Project. But it is not so easy to translate these protocols into Harvey's Paris or Pred's Stockholm. Where Harvey draws on Benjamin, I think it *is* possible to glimpse elements of different – jostling, colliding – human geographies, but these are almost always submerged by his insistence on treating power as centred, marshalled, possessed. And one rarely hears in these pages the multiple voices that animated everyday life. Pred's archaeology of Stockholm is more attentive to the fissures and fractures of modern urbanism; he is acutely aware of the relations between the visible and the articulable, and his imbrication of 'lost worlds' (heterotopias?) and 'lost words' (heteroglossia) begins to suggest a much more complex human geography than most of us have dreamed of. All the same, its uneven edges and its awkward topographies are finally resolved in what I take to be a thoroughly modern (not modernist) dénouement. The time-geography of the solitary docker embodies a world whose different topographies and glossaries are at last made whole and centred around the inscriptions of power and practice in and on the individual. And yet, as I have tried to suggest, it is possible to use his own mode of representation to open this account to the uneven, unequal unfolding of multiple human geographies. That more remains to be done is not Pred's fault: it is Pred's challenge. It is also, I suggest, as much the challenge of modernism as it is of any postmodernism.

NOTES

1 Baudelaire's original essay was written in 1859–60.
2 For the co-ordinates of modernism, see Anderson (1984) and the elegant extension in Callinicos (1989).
3 Benjamin's essay was first published in 1936.
4 This much is made plain in Kern (1983) but the singular 'culture' of the book's title is surely misleading.
5 Entrikin (1990) portrays the disjuncture between space and place as diagnostic of, perhaps even constitutive of, the crisis of modernity.
6 Hence my disagreement with Soja (1989); see Gregory (1990). It seems to me that *high* modernism – by which I mean those shifts in cultural and intellectual registers which were installed after the *Second* World War – is a much more plausible candidate for Soja's thesis about the subordination of space in social thought. For it was during those decades, in the social sciences at any rate, that time was valorized (most obviously in the plenary trajectory of models of 'modernization') and space reduced to an isotropic plane (most obviously in the geometric lattices of spatial science). If such a reading can be sustained, then postmodernism can be seen as affecting a break with *high* modernism but simultaneously maintaining a deep affinity with an earlier, essentially *fin-de-siècle* modernism.
7 I take the term 'local knowledge' not directly from Geertz (as would usually be the case) but from Rouse (1987). In emphasizing the contextuality of knowledge, Rouse does not foreclose on its generalization: he merely insists that generalization is a (conditional) *achievement*, not a given.
8 The phrase is, of course, Foucault's but I do not intend to close my argument around the genealogy of the human sciences offered in Foucault (1979).
9 To avoid misunderstanding, I should say that I do not regard Thompson as an empiricist – and neither does he! – but I do think his attack on Althusser was spiteful and ill informed. There are several objections that may be made to structural Marxism, but there is also much to learn from Althusser's writings: and one of the lessons, in my view, concerns the constitutive function of theoretical work.
10 At one time Harris was much more suspicious of theory and, as I have suggested elsewhere, there are obvious affinities between Harris (1978) and Thompson (1978). But the 'theory' from which he distanced himself in the 1970s is emphatically not the 'theory' which he now commends in the 1990s.
11 For example: 'Who else but Europe could draw from its own traditions the insight, the energy, the courage of vision – everything that would be necessary to strip from the ... premises of a blind compulsion to system maintenance and system expansion their power to shape our mentality[?]': Habermas (1987: 367).
12 To be sure, the fact that Mitchell is able to achieve this effect says something about the *particular* traditions of European theory with which he works, and any more detailed discussion of ethnocentrism would have to offer a much more discriminating analysis than I am able to do here.
13 Important (and varied) considerations of the sexism of critical theory will be found in Benhabib and Cornell (1987), Diamond and Quinby (1988) and Fraser (1989).

14 Coming so long after the first signs of a feminist geography, it comes as a jolt – and it *should* come as a jolt – to read Deutsche (1991) and Massey (1991).

15 Here I am indebted to Bauman (1987).

16 '[T]heory has to be grasped in the place and the time out of which it emerges': Said (1984: 24). This means that working with social theory does not involve a double hermeneutic – spun recursively between participant and observer – but a *multiple* hermeneutic constituted through engagements between different historical geographies.

17 All page numbers given without further reference in this section are taken from Harvey (1985). I have borrowed 'The literary diver' from the epigraph that heads Harvey's essay. 'Paris is indeed an ocean,' says Balzac, so deep that 'there will always be something extraordinary, missed by the literary diver.'

18 Not only Lefebvre's own investigations of the Commune but also his more general theoretical work: see in particular Lefebvre (1991); this book was originally published in French in 1974.

19 In Benjamin's case, however, representations of femininity and sexuality have a complex genealogy and Rauch argues that their deconstruction functions as an allegory of (aesthetic) history which both unmasks the *appearance* of femininity as a projection of the male subject and marks the site of a radical inscription of the *corporeality* of the female body. See Rauch (1988). Other feminist critics have read Benjamin in other ways: see, for example, Buci-Glucksmann (1986). Although there are substantial differences between these two writers, neither of them treats Benjamin as being complicit in the contemporary fetishization of 'woman'.

20 On 'Liberty', for example, see the various readings offered by Agulhon (1981), Pointon (1990) and Silverman (1988).

21 Massey is in fact referring to the discussion of Frédéric in Harvey (1989: 263–4).

22 Benjamin's understanding of femininity was more nuanced than Wolff acknowledges: see Buci-Glucksmann (1986) and Rauch (1988).

23 According to Adorno, Benjamin thought of the constellation as 'a juxtaposed rather than integrated cluster of changing elements that resist reduction to a common denominator, essential core or generative first principle': Jay (1984a: 15).

24 Benjamin's 1935 exposé has been translated from German into English as 'Paris, capital of the nineteenth century' in Benjamin (1978a); the 1939 exposé was originally written in French and may now be found as 'Paris, capitale du XIXe siècle' in Benjamin (1989: 47–59). The dialectical structure of the exposés is clarified in Higonnet *et al.* (1984) but the commentary is insufficiently attentive to the differences between the two versions.

25 It is hardly necessary to point out that the imagery of 'penetrating the veil of fetishism' is freighted with many of the same assumptions to which I have already objected.

26 For fuller biographies see Roberts (1982) and Witte (1985).

27 Rolf Tiedemann, the editor of Benjamin's collected works, consistently refers to the Arcades Project – Volume V of the *Gesammelte Schriften* – as *Das Passagen-Werk* but several commentators have objected to this description for its implicit closure. It is, says Sieburth (1989: 26–7), 'an editorial invention that runs the risk of reifying an exploratory process of writing into an inert textual artifact. Benjamin, by contrast, describes his [project] not as a work but as an ongoing event, a peripatetic meditation or *flânerie* in which everything chanced upon en route becomes a potential direction his thoughts might take.'

28 For perceptive discussions of Benjamin's relation to historical materialism, see Buck-Morss (1977, 1981) and Lunn (1985).

29 Other commentators have read Benjamin's work as an archaeology of modernity

too. To Frisby (1985), for example, Benjamin was 'the archaeologist of modernity' who excavated 'three spatial, labyrinthine layers of reality' – the arcade, the city and the mythological underworld that lay beneath the city (represented by the catacombs) – in order to 'cut through yet another labyrith ... that of human consciousness' (pp. 210–11). See also Sagnol (1983).

30 For a sympathetic discussion of the relations between deconstruction and Benjamin's early writings, see Norris (1983).

31 For summaries, see Frisby (1985: 190–207), McCole (1985), Sieburth (1989) and Tiedemann (1988). A more extensive discussion will be found in Buck-Morss (1989).

32 Here I follow Jennings (1987). It is only fair to add that other commentators portray Benjamin in a somewhat more optimistic light: notably Habermas (1983) and Wolin (1982).

33 The emphasis on 'reawakening' indicates that Benjamin had already moved suffi-ciently far from surrealism to use Proust as his model rather than Aragon; the moment of awakening was analogous to the shock of recognition produced by Proust's *mémoire involontaire*. See Abbas (1990: 226–9) and Sieburth (1989: 18).

34 'Narrative was useless for his purpose because the notion of progress is built into its structure of continuity, the form whereby the later not only grows out of the earlier utterance but subsumes, surpasses or completes it in complexity and inclusivity': Frase (1990–1: 20–1). For a more general discussion, see White (1987).

35 This rebounds nicely against Richard Dennis's objections that Harvey's essay on Paris is an exercise in 'trading and speculating on the labour of others': Dennis (1987: 311).

36 The proximity of Hegel to historical materialism or surrealism is far from simple, of course. Jay's (1984b) description of Hegel as a thinker 'whose importance for the Western Marxist tradition was second only to that of Marx himself' (p. 53) is unexcep-tional: indeed, in what is often construed as one of the founding texts of Western Marxism the young Lukács announced his desire to achieve 'a Hegelianism more Hegelian than Hegel'. Nevertheless, as Jay's remarkable intellectual odyssey demonstrates, the reception of Hegel's ideas has been highly uneven. Many writers in the turbulent world of the 1920s and 1930s thought it impossible 'to see things whole' (p. 219) as Jay puts it, and the architects of the Frankfurt School (notably Horkheimer and Adorno) were particularly instrumental in sounding the retreat from Hegelian Marxism. Jay notes that the surrealists 'were among the first in France to recover the importance of Hegel for revolutionary thought' (p. 285): but even there he recognizes substantial differences between them which revolved, precisely, around their contrasting views of the prospects for a rationalist resolution of contradictions in some final synthesis.

37 For a fuller discussion, see Buck-Morss (1989: pp. 52–62).

38 I cannot imagine that 'structure' is being used here in any very technical sense. Buck-Morss's rendering of the Arcades Project undoubtedly risks imposing an imaginary coherence upon Benjamin's corpus, but I am at a loss to understand how Jameson can describe the constellation as 'virtually Althusserian *avant le lettre*; it also still retains something like a nostalgia for centeredness and for unified (if not necessarily organic) form': Jameson (1990: 244).

39 'Each field of the co-ordinates can ... be said to describe one aspect of the physiognomic appearance of the commodity, showing its contradictory "faces"': Buck-Morss (1989: 211). Buck-Morss devotes four of her chapters to discussions of the four fields, but I should emphasize that this grid is a *plenary* conception of the Arcades Project which was put into practice through *multiple* dialectical images: arcades and exhibitions, wax figures and mechanical dolls, the *flâneur* and the prosti-tute, and so on. For a more focused discussion, see Buck-Morss (1986).

40 The historical object is ... something constituted in the act of writing history or criticism. It is in fact identical with ... the dialectical image': Jennings (1987: 205).

41 It is for this reason that Ivornel (1986: 69–70) elects to write of 'Paris, capital of the popular front': as a way of 'explicitly [introducing] Benjamin's own present into the work of disengaging the past'. Ivornel is aware that such a title signposts only one constellation. 'It should not be forgotten that this moment and this place (Paris 1934–40) form the center of a vortex ... from which concentric circles expand to include the contemporary history of the whole of Europe.' Buck-Morss (1989: 8–43) attempts a similar genealogical mapping and when she identifies Naples, Moscow, Paris and Berlin as 'points of [Benjamin's] intellectual compass' (p. 40) she presumably prefigures her discussion of the 'lodestar' which orientates the construction of the constellation.

42 This essay was intended to serve as the preface to the Arcades Project; it was completed in the spring of 1940.

43 See Habermas (1983), Mosès (1989), Tiedemann (1989), Wohlfarth (1986) and Wolin (1982: 107–37). Buck-Morss (1989: 228–52) provides a characteristically creative reading of Benjamin's 'theology' in which she turns it into 'an axis of philosophical experience' and as such makes its function quite distinct from that of ' "religion" as part of the ideological superstructure'.

44 Cf. Wolin (1982: 264): 'Benjamin's relevance for historical materialism is to be found in [his] late attempt to secularize the notion of redemptive criticism.' Even so, Wolin's emphasis on the secular and my description of this Messianic power as 'worldly' do not mean that Benjamin's project is without important implications for political theology: see McCarthy (1991).

45 All page numbers given without further reference in this section are taken from Pred (1990a).

46 This is true in both a general and a specific sense. Most obviously, Pred constantly accentuates the materiality of social life – its 'material continuity' and 'physicality' – but there is another thread running through his writings in which he more obliquely positions his work in relation to historical materialism. This is usually accomplished with reference to particular traditions within Western Marxism, broadly conceived, and most concretely through those whose work has influenced social history in seminal ways: E.P. Thompson and Raymond Williams. See, for example, Pred (1981a).

47 Marx's original account will be found in the *Communist Manifesto*

48 The *Eighteenth Brumaire* has a much more direct and concrete relevance to Harvey's account of Second Empire Paris, of course, but he is also alert to its meta-theoretical implications. He commends the *Eighteenth Brumaire* as a model of the process of reflection that is indispensable for anyone seeking to negotiate 'the path between the historical and geographical grounding of experience and the rigors of theory construction': that is to say, 'the evaluation of experience, a summing up that can point in new directions, pose new problems and suggest fresh areas for historical and theoretical enquiry': Harvey (1985: xvi).

49 Pred (1981a: 5) once described this as an 'emerging consensus' in social theory whose roots supposedly lay in the marchlands between realism and Marxism. I am not sure that he would (or could) still hold to such a view.

50 The imagery is mine; in fact Benjamin talks about unpacking cases of books – 'what memories crowd in upon you!' he declares – but the *Passagenarbeit* suggests a more general rummaging around the accumulated debris of the past. The figure of the collector is none the less apposite, however, because Benjamin argues that 'for a true collector the whole background of an [object] adds up to a magic encyclopaedia whose quintessence is the fate of his object'. In other words, what matters is not the

object alone – Benjamin was dismissive of contemporary cultural history precisely because it fetishized the object – but rather the constellation of past and present fragments in which it is placed. See Benjamin (1968d); for a discussion see Abbas (1990: 231–2).

51 Onomastics is the study of the origin and forms of names of persons or places. 'As he charts the cultural topography of nineteenth-century Paris,' Sieburth (1989: 15) remarks, 'Benjamin finds himself entering into a landscape of proper nouns.' From its early drafts through to its later elaborations, he argues, the Arcades Project maintains 'as its basic structural and heuristic device the form of a *list*, a paratactic mapping of cultural traffic, a gazeteer of urban signs'.

52 Pred (1990c: 48) gives this strategy a gendered inflection and claims that he prefers to 'prioritize polyvalent, uncertainty accepting – largely female – modes of expression over univalent, certainty claiming – largely male – modes of expression'. I am not convinced by these claims, which seem (to me) to advance a series of strikingly absolutist propositions. They would be more plausible if they referred to 'masculine' and 'feminine' – which is how Cixous and Irigaray represent the contrast – but they would still be problematic.

53 The original phrase is Pareto's, but the most sustained discussion of its implications (and one to which Harvey returns again and again) is Ollman (1971: 3–71).

54 More particularly, Harvey seems drawn to the way in which Marx's concepts are continually breaking apart and coming together in new combinations. He tries to follow the same strategy in Harvey (1982: see xv–xvi), and in Harvey (1989: 51) he suggests that Benjamin worked in much the same way: 'Benjamin ... worked the idea of collage/montage to perfection, in order to capture the many-layered and fragmented relations between economy, politics and culture.'

55 The obvious parallel is with Olsson's dazzling linguistic experiments, but he somehow glides 'behind the back' of language without marking the surface of the text.

56 The term was originally Bloch's, but in praising Benjamin's 'sensitivity for individual detail' he was addressing more than an ability to recover overlooked and discarded fragments of the past. He claimed that 'Benjamin possessed an unequalled micrological – *philological* sensitivity' (emphasis added), so that his point was that Benjamin approached the world – and hence the city – as a *text* whose elements have been scattered and suppressed by the closures of conventional history. See Frisby (1985: 213–14).

57 I have provided a critical discussion of time-geography in Gregory (1985).

58 Harvey (1989: 211–12) agrees that the Hägerstrand model provides 'a useful [description] of how the daily life of individuals unfolds in space and time' but in itself, he objects, it tells us nothing about the production of the locational structures within which these time-space paths are spun nor about the relations of power and domination which are instantiated through them. Pred's investigations of Stockholm do not address the first of these objections – nor are they intended to – but they clearly meet the second by making the relations between power and practice focal to any developed version of time-geography: see Pred (1981b).

59 It was Anne Buttimer who told Hägerstrand that the world he depicted reminded her of a *danse macabre*: see Gregory (1985: 335).

60 In this connection I think it significant that Hägerstrand should have originally thought of his project as a 'population archaeology' and that he did not conceive of it in purely individualistic terms: see Gregory (1985: 305–6).

61 Like Flaubert's Frédéric, of course, his masculinity conferred a particular (though evidently class-bounded) mobility upon Sörmlands-Nisse: the generalized nexus of patriarchal relations was inscribed in his particular time-space path through the mediation of what Hägerstrand would call 'capability' and 'authority' constraints.

62 So, for example, plate 10 is not only '*Götagatan*, looking north' but also 'the view from Sörmlands-Nisse's eyes shortly after turning off from *Bondegatan*'.
63 The interview was originally published in 1969.

ACKNOWLEDGEMENTS

This essay was originally written for the Vegasymposium in Stockholm, April 1991, when the Anders Retzius medal was presented to Allan Pred by the Svenska Sällskapet för Anthropologi och Geografi. I am indebted to Gunnar Olsson and Allan Pred for their warm hospitality and to Michael Watts for his companionship. I am also grateful to Trevor Barnes, Dan Clayton, Robyn Dowling and Matt Sparke for their comments on the revised version of the text. It originally appeared in *Geografiska Annaler* 73B (1991: 17–44), and is reprinted here by permission of the editor.

REFERENCES

Abbas, A. (1989) 'On fascination: Walter Benjamin's images', *New German Critique* 48: 43–62.

—— (1990) 'Walter Benjamin's collector: the fate of modern experience', in A. Huyssen and D. Bathrick (eds) *Modernity and the Text: Revisions of German Modernism*, New York: Columbia University Press, 216–39.

Agulhon, M. (1981) *Marianne into Battle: Republican Imagery and Symbolism in France, 1789–1880*, Cambridge: Cambridge University Press.

Anderson, P. (1984) 'Modernity and revolution', *New Left Review* 144: 96–113.

Appadurai, A. (ed.) (1986) *The Social Life of Things: Commodities in Cultural Perspective*, Cambridge: Cambridge University Press.

Aragon, L. (1980) *Paris Peasant*, trans. S.W. Watson, London: Pan.

Arendt, H. (1968) 'Introduction', in W. Benjamin *Illuminations: Essays and Reflections*, ed. H. Arendt, New York: Schocken.

Barnes, J. (1989) *A History of the World in 10½ Chapters*, London: Jonathan Cape.

Baudelaire, C. (1984), 'The painter of modern life', in C. Baudelaire *The Painter of Modern Life and Other Essays*, London: Phaidon, 12–15.

Bauman, Z. (1987) *Legislators and Interpreters: On Modernity, Postmodernity and Intellectuals*, Cambridge: Polity.

Benhabib, S. and Cornell, D. (eds) (1987) *Feminism as Critique: Essays on the Politics of Gender in Late Capitalist Societies*, Cambridge: Polity.

Benjamin, W. (1968a) 'The work of art in the age of mechanical reproduction', in W. Benjamin *Illuminations: Essays and Reflections*, ed. H. Arendt, New York: Schocken, 217–51.

—— (1968b) 'The storyteller: reflections on the works of Nicolai Leskov', in W. Benjamin *Illuminations: Essays and Reflections*, ed. H. Arendt, New York: Schocken, 83–109.

—— (1968c) 'Theses on the philosophy of history', in W. Benjamin *Illuminations: Essays and Reflections*, ed. H. Arendt, Schocken, New York: 253–64.

—— (1968d) 'Unpacking my library', in W. Benjamin *Illuminations: Essays and Reflections*, ed. H. Arendt, New York: Schocken, 59–67.

—— (1968e) 'On some motifs in Baudelaire', in W. Benjamin *Illuminations: Essays and Reflections*, ed. H. Arendt, New York: Schocken, 155–200.

—— (1978a) 'Paris, capital of the nineteenth century', in 'A Berlin chronicle', in W. Benjamin *Reflections: Essays, Aphorisms, Autobiographical Writings*, ed. P. Demetz, New York: Schocken, 146–73.

—— (1978b) 'A Berlin chronicle', in W. Benjamin *Reflections: Essays, Aphorisms, Autobiographical Writings*, ed. P. Demetz, New York: Schocken, 1–60.

—— (1978c) 'Surrealism: the last snapshot of the European intelligentsia', in W. Benjamin *Reflections: Essays, Aphorisms, Autobiographical Writings*, ed. P. Demetz, New York: Schocken, 177–92.

—— (1979) 'One way street', in W. Benjamin *One Way Street and Other Writings*, London: New Left Books, 45–106.

—— (1989) *Paris, capitale de XIXe siècle: le livre des passages*, Paris: Editions du Cerf.

Berman, M. (1982) *All That Is Solid Melts into Air: The Experience of Modernity* London: Verso.

Buci-Glucksmann, C. (1986) 'Féminité et modernité: Walter Benjamin et l'utopie du féminin', in H. Wismann (ed.) *Walter Benjamin et Paris*, Paris: Editions du Cerf, 403–20.

Buck-Morss, S. (1977) *The Origin of Negative Dialectics: Theodor Adorno, Walter Benjamin and the Frankfurt Institute*, New York: Free Press.

—— (1981) 'Walter Benjamin: revolutionary writer', *New Left Review* 128: 50–75; 129: 77–95.

—— (1986) 'The *flâneur*, the sandwichman and the whore: the politics of loitering', *New German Critique* 39: 99–140.

—— (1989) *The Dialectics of Seeing: Walter Benjamin and the Arcades Project*, Cambridge, Mass.: MIT Press.

Callinicos, A. (1989) 'Modernism and capitalism', in A. Callinicos *Against Postmodernism: a Marxist Critique*, Cambridge: Polity, 29–61.

Castells, M. (1983) 'Cities and revolution: the Commune of Paris, 1871', in M. Castells *The City and the Grassroots: A Cross-Cultural Theory of Urban Social Movements*, Berkeley: University of California Press, 15–26.

Castle, T. (1988), 'Phantasmagoria: spectral technology and the metaphorics of modern reverie', *Critical Inquiry* 15: 26–61.

Christopherson, S. (1989) 'On being outside "The Project"', *Antipode* 21: 83–9.

Clifford, J. (1988) *The Predicament of Culture: Twentieth-Century Ethnography, Literature and Art*, Cambridge, Mass: Harvard University Press.

Cohen, M. (1989) 'Walter Benjamin's phantasmagoria', *New German Critique* 48: 87–107.

Connor, S. (1989) *Postmodernist Culture: An Introduction to Theories of the Contemporary*, Oxford: Blackwell.

Crary, J.C. (1990) *Techniques of the Observer: On Vision and Modernity in the Nineteenth Century*, Cambridge, Mass.: MIT Press.

Deleuze, G. (1988) *Foucault*, Minneapolis: University of Minnesota Press.

Demetz, P. (1989) 'Introduction', in W. Benjamin *Paris, capitale du XIXe siècle*, Paris: Editions du Cerf.

Dennis, R. (1987) 'Faith in the city?', *Journal of Historical Geography* 13: 310–16.

Deutsche, R. (1991) 'Boys town', *Environment and Planning D: Society and Space* 9: 5–30.

Diamond, I. and Quinby, L. (eds) (1988) *Feminism and Foucault: Reflections on Resistance*, Boston: Northeastern University Press.

Eagleton, T. (1981) *Walter Benjamin or, Towards a Revolutionary Criticism*. London: Verso.

—— (1990) 'The Marxist rabbi: Walter Benjamin', in T. Eagleton *The Ideology of the Aesthetic*, Oxford: Blackwell, 316–40.

Entrikin, J.N. (1990) *The Betweenness of Place: Towards a Geography of Modernity*, London: Macmillan.

Foucault, M. (1979) *Discipline and Punish: The Birth of the Prison*, trans. A. Sheridan, London: Penguin.

—— *Power/Knowledge*, Brighton: Harvester Press.

—— (1989) 'The archaeology of knowledge', in M. Foucault *Foucault Live: Interviews 1966–1984*, New York: Semiotext(e), 45–56.

Frase, B. (1990–1) 'Raising debris', *Hungry Mind Review* 16: 20–1.

Fraser, N. (1989) *Unruly Practices: Power, Discourse and Gender in Contemporary Social Theory*, Minneapolis: University of Minnesota Press.

Frisby, D. (1985), 'Walter Benjamin: the prehistory of modernity', in D. Frisby *Fragments of Modernity*, Cambridge: Polity, 187–265.

Geyer-Ryan, H. (1988) 'Counterfactual artefacts: Walter Benjamin's philosophy of history', in E. Timms and P. Collier (eds) *Visions and Blueprints: Avant-Garde Culture and Radical Politics in Early Twentieth-Century Europe*, Manchester: Manchester University Press, 66–79.

Gregory, D. (1985) 'Suspended animation: the stasis of diffusion theory', in D. Gregory and J. Urry (eds) *Social Relations and Spatial Structures*, London: Macmillan, 296–336.

—— (1990) '*Chinatown*, part three? Soja and the missing spaces of social theory', *Strategies: A Journal of Theory, Culture and Politics* 3: 40–104.

Habermas, J. (1983) 'Walter Benjamin: consciousness raising or rescuing [redemptive] critique', in J. Habermas *Philosophical-Political Profiles*, Cambridge, Mass.: MIT Press, 131–63.

—— (1987) *The Philosophical Discourse of Modernity*, Cambridge: Polity.

Harris, R.C. (1978) 'The historical mind and the practice of geography', in D. Ley and M. Samuels (eds) *Humanistic Geography: Prospects and Problems*, London: Croom Helm, 123–37.

—— (1991) 'Power, modernity and historical geography', *Annals of the Association of American Geographers* 81: 671–83.

Harvey, D. (1982) *The Limits to Capital*, Oxford: Blackwell.

—— (1985) 'Paris, 1850–1870', in D. Harvey *Consciousness and the Urban Experience*, Oxford: Blackwell 62–220.

—— (1989) *The Condition of Postmodernity: An Inquiry into the Origins of Cultural Change*, Oxford: Blackwell.

Higonnet, A., Higonnet, M. and Higonnet, P. (1984) 'Façades: Walter Benjamin's Paris', *Critical Inquiry* 10: 391–419.

Ivornel, P. (1986) 'Paris, capital of the popular front or the posthumous life of the nineteenth century', *New German Critique* 39: 61–84.

Jameson, F. (1990) *Late Marxism: Adorno, or the Persistence of the Dialectic*, London: Verso.

—— (1991) *Postmodernism, or the Cultural Logic of Late Capitalism*, Durham: Duke University Press.

Jay, M. (1984a) *Adorno*, Cambridge, Mass.: Harvard University Press.

—— (1984b) *Marxism and Totality: The Adventures of a Concept from Lukács to Habermas*, Cambridge: Polity.

—— (1988) 'Scopic regimes of modernity', in H. Foster (ed.) *Vision and Visuality*, Seattle: Bay Press, 3–23.

Jennings, M. (1987) *Dialectical Images: Walter Benjamin's Theory of Literary Criticism*, Ithaca: Cornell University Press.

Kern, S. (1983) 'The Cubist war', in S. Kern *The Culture of Time and Space, 1880–1918* Cambridge, Mass.: Harvard University Press, 287–312.

Lefebvre, H. (1991) *The Production of Space*, Oxford: Blackwell.

Löwy, M. (1985) 'Revolution against "progress": Walter Benjamin's romantic anarchism', *New Left Review* 152: 42–59.

Lunn, E. (1985) *Marxism and Modernism*, London: Verso.

McCarthy, T. (1991) 'Critical theory and political theology: the postulates of communicative action', in T. McCarthy *Ideas and Illusions: On Reconstruction and Decon-*

struction on Contemporary Critical Theory, Cambridge, Mass.: MIT Press, 200–15.

McCole, J. (1985) 'Benjamin's *Passagen-Werk*: a guide to the labyrinth', *Theory and Society* 14: 497–509.

Massey, D. (1991) 'Flexible sexism', *Environment and Planning D: Society and Space* 9: 31–58.

Menninghaus, W. (1988) 'Walter Benjamin's theory of myth', in G. Smith (ed.) *On Walter Benjamin: Critical Essays and Reflections*, Cambridge, Mass.: MIT Press, 292–325.

Mitchell, T. (1988) *Colonizing Egypt*, Cambridge: Cambridge University Press.

—— (1989) 'The world as exhibition', *Comparative Studies in Society and History* 3: 217–36.

Mosès, S. (1989) 'The theological-political model of history in the thought of Walter Benjamin', *History and Memory: Studies in Representation of the Past* 1: 1–33.

Nicholsen, S.W. (1990) Review of Susan Buck-Morss, *The Dialectics of Seeing*, *New German Critique* 51: 179–88.

Norris, C. (1982) 'Image and parable: readings of Walter Benjamin', in C. Norris *The Deconstructive Turn: Essays in the Rhetoric of Philosophy*, London: Methuen, 107–27.

Ollman, B. (1971) *Alienation: Marx's Conception of Man in Capitalist Society*, Cambridge: Cambridge University Press.

Pointon, M. (1990) *Liberty on the Barricades*: woman, politics and sexuality in Delacroix', in M. Pointon *Naked Authority: The Body in Western Painting, 1830–1908*, Cambridge: Cambridge University Press, 59–82.

Pollock, G. (1988) 'Modernity and the spaces of femininity', in G. Pollock *Vision and Difference: Femininity, Feminism and Histories of Art*, London: Routledge, 50–90.

Pred, A. (1981a) 'Social reproduction and the time-geography of everyday life', *Geografiska Annaler* 63B: 5–22.

—— (1981b) 'Power, everyday practice and the discipline of human geography', in A. Pred (ed.) *Space and Time in Geography: Essays Dedicated to Torsten Hägerstrand*, Lund Studies in Geography Series B, No. 48, Lund: C.W.K. Gleerup, 30–55.

—— (1984) 'Place as historically contingent process: structuration theory and the time-geography of becoming places', *Annals of the Association of American Geographers* 74: 279–97.

—— (1986) *Place, Practice and Structure: Social and Spatial Transformation in Southern Sweden 1750–1850*, Cambridge: Polity.

—— (1989) 'The locally spoken word and local struggles', *Environment and Planning D: Society and Space* 7: 211–33.

—— (1990a) *Lost Words and Lost Worlds: Modernity and the Language of Everyday Life in Late Nineteenth-Century Stockholm*, Cambridge: Cambridge University Press.

—— (1990b) *Making Histories and Constructing Human Geographies*, Boulder: Westview Press.

—— (1990c) 'In other wor(l)ds: fragmented and integrated observations on gendered languages, gendered spaces and local transformation', *Antipode* 22: 33–52.

Probyn, E. (1990) 'Travels in the postmodern: making sense of the local', in L.J. Nicholson (ed.) *Feminism/Postmodernism*, New York: Routledge, 176–89.

Rajchmann, J. (1991) 'Foucault's art of seeing', in J. Rajchmann *Philosophical Events: Essays of the '80s*, New York: Columbia University Press, 68–102.

Rauch, A. (1988) 'The *Traverspiel* of the prostituted body, or woman as allegory of modernity', *Cultural Critique* 10: 77–88.

Roberts, J. (1982) *Walter Benjamin*, London: Macmillan.

Ross, K. (1988) *The Emergence of Social Space: Rimbaud and the Paris Commune*, Minneapolis: University of Minnesota Press.

Rouse, J. (1987) *Knowledge and Power: Toward a Political Philosophy of Science*, Ithaca: Cornell University Press.

Ryan, M. (1982) 'The limits of *Capital*', in M. Ryan *Marxism and Deconstruction: A Critical Articulation*, Baltimore: Johns Hopkins University Press, 82–102.

Sagnol, M. (1983) 'La méthode archéologique de Walter Benjamin', *Les temps modernes* 40(444): 143–65.

Said, E. (1978) *Orientalism*, London: Routledge & Kegan Paul.

—— (1984) 'Travelling theory', in E. Said *The World, the Text and the Critic*, London: Faber, 226–47.

Shapiro, M. (1988) *The Politics of Representation: Writing Practices in Biography, Photography, and Policy Analysis*, Madison: University of Wisconsin Press.

Sieburth, R. (1989) 'Benjamin the scrivener', in G. Smith (ed.) *Benjamin: Philosophy, Aesthetics, History*, Chicago: University of Chicago Press, 13–37.

Silverman, K. (1988) 'Liberty, maternity, commodification', *New Formations* 5: 69–89.

Soja, E. (1989) *Postmodern Geographies: The Reassertion of Space in Critical Social Theory*, London: Verso.

Spencer, L. (1985) 'Allegory in the world of the commodity: the importance of *Central Park*', *New German Critique* 34: 59–77.

Spivak, G.C. (1988) *In Other Worlds: Essays in Cultural Politics*, New York: Routledge.

—— (1990) *The Post-Colonial Critic: Interviews, Strategies, Dialogues*, New York: Routledge, 95–112.

Tagg, J. (1988) *The Burden of Representation: Essays on Photographies and Histories*, London: Macmillan.

Thompson, E.P. (1978) 'The poverty of theory: or an orrery of errors', in E.P. Thompson *The Poverty of Theory and Other Essays*, London: Merlin, 192–406.

Tiedemann, R. (1988) 'Dialectics at a standstill: approaches to the *Passagen-Werk*', in G. Smith (ed.) *On Walter Benjamin: Critical Essays and Recollections*, Cambridge, Mass.: MIT Press, 210–27.

—— (1989) 'Historical materialism or political messianism? An interpretation of the theses "On the concept of History"', in G. Smith (ed.) *Benjamin: Philosophy, Aesthetics, History*, Chicago: University of Chicago Press, 175–209.

White, H. (1987) *The Content of the Form: Narrative Discourse and Historical Representation*, Baltimore: Johns Hopkins University Press.

Witte, B. (1985) *Walter Benjamin*, Berlin: Rowohlt.

Wohlfarth, I. (1986) 'Re-fusing theology: some first responses to Walter Benjamin's Arcades Project', *New German Critique* 39: 3–24.

Wolff, J. (1990) 'The invisible *flâneuse*: women and the literature of modernity', in J. Wolff *Feminine Sentences: Essays on Women and Culture*, Berkeley: University of California Press, 34–50.

Wolin, R. (1982) *Walter Benjamin: An Aesthetic of Redemption*, New York: Columbia University Press.

Young, R. (1990) *White Mythologies: Writing History and the West*, London: Routledge.

16

READING, COMMUNITY AND A SENSE OF PLACE

Brian Stock

This essay is conceived as a modest contribution to our understanding of relations between reading, the notion of community and a sense of place. More specifically, it is an attempt to trace the historical genesis of metaphors of reading and writing as they apply to the literary typology of places and to analyse some of the contexts in which they occur. My point of entry into the debate is the manner in which the issues arise in cultural geography; however, the illustrations also come from art history and ethnography. The assumption throughout is that the rules governing the transformation of the visual into the readable are not derivable from any single discipline: they are aspects of a more general mentality.[1] Also, I am not concerned with what images and words are in the brain – a problem which, at present, has no satisfactory solution[2] – but with known traditions of representing places in visual or verbal terms. It is within such histories that places are configured as readable or writable texts. The question I address is why this way of thinking seems appropriate.

Within geography, the success of reading and writing metaphors is unquestionable.[3] They have opened the subject to interpretative theory; and they have made it possible for historians, anthropologists and students of literature to take part in interpretative discussions using a shared vocabulary. However, the use of such expressions amounts to a restating of the priorities of geographical enquiry.[4] This has given rise to a certain amount of intellectual tension between 'observational' and 'linguistic' styles of interpretation: it has driven a wedge between the followers of Carl Sauer, who champion the recording of what is seen as an end in itself, and the sceptics of this position, for whom the nineteenth-century rise of observational techniques is itself a problematical issue.[5]

There are parallels to this debate in many of the disciplines that concern themselves with the visual. It is worth recalling at the outset that they are consequences of a change in outlook occasioned, in the early years of the twentieth century, by the appearance of relativity and quantum mechanics: one that was mediated in subsequent discussions through artistic and literary theories that gradually broke down beliefs in fixed standards, values, and canons of taste. Some three generations after the reorientation of physics away from Newtonian axioms, few figures in the field of science maintain that nature can be studied and understood as it 'really is'. 'This may have been my philosophy,' Einstein remarked to Werner Heisenberg, 'but it is nonsense all the same. It is never possible to introduce only observable quantities in a theory. It is the theory that decides what can be observed.'[6] As the notion of ocular objectivity has retreated from the sciences, so it has been modified in other disciplines that depend on the visual sense, including, of course, the visual arts themselves.

Language metaphors are statements about relativities of a different sort – those arising between speaker and hearer, writer and reader, or text and audience. In the study of art, architecture and geography, they appear to offer a way of sidestepping the problem of the observer. In this respect, they perform a useful destructive or, as some would have it, deconstructive service. The technique works best in the description of places in which two preconditions are fulfilled: a significant text is embedded in the place; and it can be recognized by the viewer as falling within a class of literary tropes that he or she already knows. Charlemagne's palace church at Aachen, for instance, deliberately built in imitation of San Vitale, Ravenna, symbolized for ninth-century visitors, as it does for us, the historic transfer of the empire from the Byzantine to the northern European world. The original viewers of the church had little trouble extracting a narrative of imperial hegemony, together with a political agenda, from the visual setting. Twentieth-century viewers are able to do so as well, provided that they have the required historical information and the necessary interpretative tools. There is nothing mysterious in the way in which such a monument, first seen with the eye, is later read as a text.

Let us call such a scene a 'landscape'. It is a place in which we can speak of what is seen as a genre of legible writing: we assume there is a grammar and syntax, a logic and rhetoric and a social, cultural or political context that is understandable through a literary typology. However, we must also bear in mind that using this vocabulary involves a number of assumptions about relations between what is seen and what is interpreted in linguistic terms. In what follows, I examine three of these, which can be listed summarily as follows:

1   If we speak of 'seeing' as 'reading', a metaphor is clearly involved. We have to ask how it operates. Not, it would appear, through substitution or comparison; nor do we find the 'literal' and 'figurative' poles of metaphor as traditionally conceived. The meaning of the two terms is arrived at by a type of 'interaction'. What we first focus upon, i.e. what we see, becomes the 'frame' for what we later read.[7]

2   However, we do not actually 'read' what we 'see'. If we did, there would be no need for metaphor. If we think of what we see as a reading, there must be something seen before we engage in metaphorical thinking. In addition to asking how the metaphor works, we have to ask what this is. It is clear that it cannot be the metaphor, i.e. 'reading': that would amount to circular thinking.

3   Metaphors of reading or writing work by analogy with literate experience.

We have to consider what they may mean, if anything, to two classes of non-literates who also visualize and understand landscapes. One consists of those who do not know how to read, for whom landscape may be a substitute for reading within a reading culture. The other includes all those who know how to read but who are denied access to writings, on religious, political or other grounds: for them, a landscape may deliberately conceal the message of a culturally relevant text. Moreover, in each of these cases, we have to ask what role memory plays. For all reading is remembering.

When issues like these are raised among those whose foundation discipline is literature, the discussion usually turns to one of two acknowledged aspects of the reading experience in contemporary literary debate, namely the text and the reader. If there is no text, it is argued, there is obviously nothing to read. But, if there is no reader, there is nothing to read either, since the text, as a web of meaning, is the reader's creation. If we transfer this way of thinking to land-scapes, it means that, where there is neither text nor reader, what we see is demoted to a mere physical location with no meaningful associations. On this view, there would seem to be nothing to say beyond what is said about the act of reading.

There is a serious weakness in this position, as regards both legible landscapes and texts more generally. This arises from taking only two variables into consideration. No notice is given of a third partner in any reading experience, the inherited concept of reading. In order to carry on the analysis of texts and readers we assume this to be a constant – something that is the same for all readers at all times and places. But, just as texts and reading practices change over time, so does our conceptualization of what is being done. It is not only what we are looking at that has a history; the way we talk about that is a product of history too.

During the fifteenth century, landscape, printing and silent reading came together in one such constellation of ideas. Silent reading antedated printing (Saenger 1982: 383–401), but the numbers of private, recreational and silent readers increased greatly when men and women could buy and own their own books. The rise in private readership took place during the period in which the idea of landscape was beginning to re-enter thinking about art, architecture and city design. The change in outlook was dramatic, and united landscape and the new style of reading into a single entity. In a typically medieval garden compendium, like Pietro de' Crescenzi's *Liber Ruralium Commodorum*, *c.* 1305, there are numerous flora, but they are not placed in an independent setting. Each plant or flower has its own meaning and its symbolic associations, and, to find those meanings, one reads the text, not the layouts of actual gardens. However, within a century and a half, in the gardens of Alberti, Michelozzi and Sangallo, plants, flowers and trees had all become parts of a larger entity – a fusion of design and natural space that we recognize as a landscape. We 'read' the floral arrangements of Renaissance architects because, like them, we can read the ancient works on which they were modelled. Printing made them available on a scale inconceivable during the manuscript age. For their erudite owners, the gardens were the visual equivalents of voiceless texts.

There is a faint echo of these changes in the evolution of the English word 'landscape' – although, I would stress, too much emphasis must not be placed on the sense of a single term.[8] The word is first employed by painters in England from about 1600 in order to distinguish scenes on land from those at sea and from portraits. By the middle of the seventeenth century, the observer's role is frequently enhanced: a landscape is what the eye can take in through one viewing, or what can be seen from a single perspective. By the turn of the eighteenth, the meaning shifts from the painting to the natural scenery itself or, significantly, to its depiction in words. From that point, it is difficult to separate what is seen in a landscape from the literary stylistics of observers of the scene. The interdependence is most vividly revealed in descriptions which clearly had no high culture exemplars, such as those recorded by Paul Carter for Botany Bay:

> The investment of the landscape with rhetorical interest reflected the explorer's linguistic practice, for the language of exploration was not the language of dictionaries, but the active, dialectical utterance characteristic of travelling. . . . Australian features might not resemble features elsewhere . . . but the difficulty was not beyond expression. . . . Often, in fact, the landscape was connected, not by invoking picturesque English scenes, but by following the direction of language itself, the way in which words became sentences, sentences conversations.
>
> (1988: 59)

Let me sum up what I have said to this point. When we use metaphors of reading, writing and texts to describe our experience of a landscape in the post-print age, we are partly reading what we see. But we are also seeing through eyes that are historically predisposed to read. We read, not because a text or a reader is present each time a landscape is viewed – although that may well be so – but because our notion of landscape contains within it an already conceptualized notion of the reading process. The metaphor is a part of the act.

Moreover, if readerly attitudes are implanted in notions of landscape at an early stage of reflection, then the debate within cultural geography, as well as art, architecture and the plastic arts, concerning the relative value of observation and literary context can be seen in a single perspective. The age of print is the logical starting-point for both the pro- and contra-observational positions. This is the period when the first accurate maps, guidebooks and itineraries are established – all by-products of the print revolution.[9] It is also the time when the first iconographic treatises on landscape appear.[10] The reader of landscapes, not surprisingly, is the humanist literary critic, the heir to the medieval commentator, who takes up the non-literal aspects of the 'book of nature', which are now secularized, even repaganized.[11] It is a short step to the eighteenth-century hermeneutics of the scene, in Goethe and Wincklemann, and to the beginnings of art criticism in mid-eighteenth-century France, which is concerned, Michael Fried emphasizes, with the 'self-absorption' of the beholder in the paintings of Greuze, Chardin, Van Loos and David – a problem, frequently configured by figures in paintings engaged in reflective reading.[12]

The position I am advocating can be generalized by distinguishing between 'reading' and 'rules for reading'. The two are evidently not the same.[13] When I read a piece of writing, I do not have before me a set of rules on 'how to read', any more than I have a grammar book in front of me when I speak. I just read. I do so because, having learned to read, I have the required knowledge lodged in my memory as a set of procedures. I can draw upon these whenever I wish; and I am aware that the mental record of the procedures endures, even though I may be unable to recall the specific texts I have read. Moreover, while I am reading, I cannot provide an answer, independent of my reading experience, to the question, 'How does one read?' I can only demonstrate *that* I am reading. Of course, if I am asked the question before or after I read, the situation is different. Then I can provide an answer; but I am not then reading. My reply would consist of my personal rules for reading – 'personal' because there is no way for others to tell whether my reply is an accurate account of what I am doing. It follows that, when I think about reading, I have in mind a *concept* of the reading process. This is something I can think with, for instance, when I view a landscape. But it is not the thinking that is reading.

Therefore, when I say, 'I read the landscape', what I really mean is, 'I understand the landscape by means of a conceptual instrument I call reading.' The concept is something I understand because I invented it. And it stays still, so to speak, while I use it. But, when I say, 'I read a text', what I mean is that I read it: but I cannot describe how I do this as a concept that is separate from the experience of reading. And this, obviously, never stays still while I am doing it.

There is, then, an element of concealment in all metaphors of reading and writing when they are applied to signifying places. The more I know about the metaphors, it would appear, the less I know about what it is in a landscape that is not read. I can illustrate this through recalling some facts about the psychology of reading, and by adding an example.

When I read a text, my impression is that my eyes move continuously across a line of text. But this is not the case. When I read or, for that matter, view a landscape, my eye moves in starts and stops, which experimentors call 'saccades' and 'fixations'. Learning takes place almost exlusively during the periods when the eye is at rest.[14] In other words, reading a text and viewing a landscape both involve illusions about visual perception. We distinguish the activities in a manner in which the brain does not.

In the case of the text, the illusion is one of regularity, whereas, in the case of a landscape, it is one of irregularity, even spontaneity. The reader of a written text may stop, start again, refocus or review selected passages: none the less, whenever reading takes place, it seems to follow a predictable pattern of perceptual cues. But a landscape does not appear to be scanned in this way.

Attention may gravitate to a part of the scene, a narrative within it, for instance, which can be understood linearly like a sequence of words. Yet, while we may 'read' some elements, we have the impression that we cannot read all of them; and, while there may be an order in the whole, we have the impression that it is not the reader's perceptual order. Even landscapes that embody narratives that the viewers are assumed to have read, such as those of Claude Lorrain, are unlike written texts in this respect. What the landscape tells us is that we cannot read all of it. Therefore, when we apply the regular pattern of our concept of reading to such landscapes we reverse the normal relationship between the parts and the whole. For we, as perceivers, and they, as objects perceived, use the concept of reading, which pertains to a part of them, as a metaphor governing the whole, whereas, in a written text, apart from the paper, pen and ink, everything that means something is given that meaning through linearly operating linguistic conventions: in this case, it is a concept of the whole that determines our understanding of the parts.[15]

The difference can be shown through a simple example. Suppose I want to see the Taj Mahal. I can go to Agra and see it; or, if I do not have the means, I can read about it and look at it in an illustrated guidebook. In each case, I have different visual experiences of the monument and, in some sense, both can be called readings. Also, in both cases, there is something that is not read, but which is completed by a readerly experience of the scene. In the book, what I see is a captioned picture, and what is completed is the narrative I read. In the observed reality, I cannot read everything that I see, only the part of the landscape that says architecturally, 'This is the tomb of Mumtaz Mahal, the wife of Shah Jehan, who died in 1629.' But this too is a kind of completion. For the statement is essential to my understanding of what I see: if I did not know it, and only appreciated the Taj in an architectural relationship with earlier mosques, I might assume that it is a place in which God dwells, when, as it was intended, it is the burial place of an emperor's wife, masquerading as a *locus sanctus*, since strictly speaking, Muslim law does not permit monuments to the dead.

Therefore, as it is viewed, the Taj is a statement in marble that deliberately incorporates an imperial narrative. Not knowing the text, I cannot reconstruct the architectural allusions in the place; but, using a guidebook alone, I cannot reconstruct the experience *within* its non-narrative setting on the Agra river, a setting made more dramatic by the absence of Shah Jehan's own tomb, which was intended for the other bank but never built. What is more, this is a fact that I can only know from my reading, since I cannot see a tomb that does not exist. And when I read about it, I experience something that does not happen when I look at the scene: this is the visual completion of the monument as intended *only* by the literary narrative. It does not matter, then, whether I look at the Taj, then read about it, or read about it, then look at it. In each case, something implicitly read is necessary for my understanding; and, while I am looking at the monument, or thinking about it visually, I conveniently forget about what is read. I see a readerly whole, but I have read the Taj only in part. To return to my point: metaphors of reading both reveal and conceal, and they do so through an 'interaction' that makes one element the framework of the other.

This is possible for a reason I did not emphasize in my earlier discussion of landscape, private reading and printing. Landscapes, in fact, are unlike silently read texts in a critical respect. Although the landscape *may* be read privately, it none the less remains a public statement, which, in the age of the printed book, the written text normally is not. If the analogy of the text is used, we must speak of a hybrid between the written text, accessible to the individual reader, and the oral text, accessible via the ear to many hearers. Bearing in mind that we are dealing with metaphorical analogies, the comparison between the landscape and the oral text is particularly helpful in accounting for the effects of visual spectacle, rather than visual narrative. Like the oral reading of an epic poem, the viewing of a landscape normally begins 'in the middle of things'; and it proceeds, as noted, not sequentially, but by the reiteration of selected motifs. The landscape is not read once, in the manner in which a text is scanned line by line as the eye moves down the page. It is read in several different ways, simultaneously: the view, perspective and position change, as we keep in mind the section we have just looked at and proceed to the next. Indeed, while reading a text demands an intensive tunnelling of our vision, viewing a landscape elicits some of the expressive, regulatory and monitoring functions of the 'gaze'.[16]

In the reading of a landscape, all is not read: a good deal is preread, as it is in the case of a text that one knows, all or in part. The 'reader' is somewhat like the performer of memorized lines. But those lines are not visible as a written text; and the recitation of the text has more in common with an oral reading than it does with a rapid, silent perusal of a piece of writing, which, as we understand it, is mainly a perceptual skill. Rather than speak of the viewing as a reading, one could perhaps speak of it as a 'visual ritual', since viewing a landscape is as much a ritual of the eye as it is the reading of a text: it is like a pilgrimage, but one which is suited to an age in which most journeys are made through books, since its itinerary lies within the eye's field of vision. The temple landscapes of South Asia are not read like newspaper articles: the visit to the site, the religious ceremonies and the embedded political messages are aspects of a public experience.[17] Indeed, before the text is read visually in the monument, it is known by oral means to everyone.

The analogy with oral tradition and ritual also makes historical sense, at least in the West. Landscape became fashionable during the period in which much of the symbolism, public ritual and allegory of the Middle Ages was losing its vitality. Oral literatures had been written down, in part as an act of preservation. Many rituals had deteriorated into formal procedures and had lost their legitimizing force. Collective sentiments, withdrawn from these areas, reappeared in the hermeneutic bonds that united readers of the same books – normally the Bible but, as time went on, poetic, dramatic and secular philosophical writings as well. But landscape could do something that the reading of books alone could not: it permitted the viewer to live in two worlds at once, that of the private reader and that of the public beholder.

Max Weber long ago drew attention to compensatory elements in moral behaviour during the Reformation. Individuals, he argued, deprived of the support systems of the Middle Ages, had no choice but to devote themselves unceasingly to innerworldly goals, in whose materiality they tried vainly to relocate their spiritual values. Some forms of life and thought of the preceding age were abandoned: more, one might add, were adapted, and transformed. The advance of science went hand in hand with the revival of pseudo-scientific ways of coping with the future, such as magic, astrology and witchcraft, as well as with new methods for measuring uncertainty, for instance those based on mathematical probability (Daston 1988). These were substitutes for the loss of faith in God's providential guidance of the world, for the erosion of the ontological relations between words and things, and for the destruction of a moral universe that saw good invariably triumphing over evil – a notion scorned by Machiavelli and Hobbes. In an age which was rapidly becoming aware that the methods for deriving meaning from nature and from texts were irreconcilable, some consolation could likewise be sought in landscape, which put forward an image of a nature in which men and women could locate themselves coherently in the scheme of things.

One can view this change in the short term, as affecting only the visual arts in the age of print, or, in the long term, as the final statement of a more protean development involving more general linguistic change. In the second scenario, the phenomenon of landscape, which echoes a new literacy and yet retains links with an old orality, reflects a tension that is felt throughout the earlier centuries of European education: the polarities are simply restated in terms understandable to an age fascinated with the visual for its own sake.[18] Broadly speaking, in regions in which established educational institutions were maintained, the medievals adapted and transformed existing traditions. This was the case in Italian cities like Naples, Rome and Venice, where the practice of notarial education was never abandoned. One literate tradition simply grew out of another. However, in regions of oral culture, that is, where Irish, Scandinavian or Germanic institutions prevailed, a different type of development took place. Oral laws, customs and modes of thought persisted; and writing was normally introduced through a process of acculturation, i.e. by imitating already existing verbal institutions. Between the sixth and thirteenth centuries, the presence of scribes, parchment and scriptoria in north-west Europe did not necessarily mean that oral traditions were abandoned: it only signified that someone wanted them preserved. *Beowulf* and the *Chanson de Roland*, let us not forget, are examples of works recorded, not created, by writing.

It is worth suggesting, at least as a hypothesis, that these two pathways to modern literate institutions are also expressed in different attitudes towards the visual elements in landscape at the end of the Middle Ages. The one sees landscape as a means of presenting a narrative in visual terms: it is the visual retelling of a story that the viewer already knows, a mythological tale or an episode from the Bible. The other sees landscape chiefly as a means of representing an individual's view of reality: it is, as Svetlana Alpers says, like a conversation that is taking place before the viewer's eyes (1983: introduction). It is likewise an artistic expression of the rise of scientific techniques of observations, which offers an unnarratized view of reality. Of course, the two types of landscape are rarely if ever found in a pure state, just as, during the Middle Ages and the Renaissance, the written is never found without the oral. Yet the two linguistic and visual traditions can be distinguished: the literate and narrative possibilities of the visual are most widely exploited in Italy, where written traditions had a virtually unbroken continuity, while oral tradition, together with the placing of lifelikeness above narrative, allegory or emblem, is exemplified in the north, particularly in the Netherlands, in the paintings of artists like Vermeer, Van Eyck and Rembrandt.

There is another way to make this point: not to look at the emergence of landscape after 1400, but to try to explain its absence before. The presence of landscape in later ancient art and architecture is well attested – at Pompei, in Roman villas in North Africa and in the descriptions of the Greek romances. Rome was deliberately laid out as a civic monument legible in its art, design and inscriptions.[19] There were also *loco sancta*, to which Varro devoted three of his sixteen books on religious antiquities. These places were made sacred by the rituals enacted there, which involved ceremonies and local myths. The sacred landscape was conceived as a 'text', notes Hubert Cancik (1985–6: 260), one that was 'composed by natural, artificial, and religious signs according to rules'.[20] Moreover, the deities were not located in writings: they were present in officially sanctioned places, to which appropriate observances had periodically to be made.

Christianity attempted to put an end to this sort of thinking, which was regarded as a type of superstition. The notion that deity could inhabit a place, or that the place could be read as a text invested with deity, was emphatically rejected by the Greek and Latin fathers, as it was before them by Jewish and afterwards by Muslim thinkers.[21] In contrast to pagan views, the Church asserted that the word of God could be put anywhere: but there was only one location in which the text of this place could be found – the Bible. It was not the physical site that mattered, but the 'spirit' that dwelt in the sacred writings – a spirit which, if those writings were read properly, could also be located in each reader.

Such intellectualist views could hardly be expected to appeal to the largely illiterate masses of Hellenistic cities who found themselves, suddenly but irrevocably, faced with a new state religion early in the fourth century. Christianity's announcement of the death of spiritualized localities turned out to be premature. True, in the West, Gregory the Great asserted in a famous statement that the pictures on the walls of churches comprised a type of text directed towards the unlettered,[22] and St Basil even spoke of a parallel between narrative, which is understood through the ear, and painting, which is taken in silently through 'imitation' – a readerly and pictorial type of mimesis.[23] But the associations of place resurfaced in numerous ways: in the cults of local saints, in the worship of relics and in pilgrimages to holy places. So the ancient landscape did not really disappear: it was transformed into two separate traditions in accordance with the linguistic development and religious ideals of the early Middle Ages. The embedding of texts in places was demoted to a popular religious tradition, suitable, it was thought, for non-literates. The dissociation of reading from place, and the setting up of mental narratives – which, St Benedict reminded his monks, were to be lived as well as read – was the hallmark of the lettered individual, who communicated in Latin.

It is clear, then, that metaphors of reading cannot be applied uncritically to visual experiences. But can they be applied at all? Sceptics will say no – based on the sort of evidence I have so far considered. However, I wish to conclude this brief look at the subject by replying in the affirmative. The evidence comes from anthropology: its consideration brings me to the third of my initial points of departure; it also suggests a different type of contrast between reading and landscape.

In the post-Reformation, secularizing world, in which education made slow but perceptible progress, reading was normally looked upon as a liberating, democratizing, even enlightening instrument of change. However, in many of the archaic societies discovered in the wake of European expansion, just the opposite was the case. Access to key writings was not democratic. The critical knowledge was restricted to a tiny minority, a priestly, clerical or poetic caste. Selective non-literacy was, and is, an effective instrument of political control. In such societies, then as now, reading does not promote cultural literacy, but rather constrains and channels it. To the degree that readings are represented in landscapes, places offer an equally effective means of perpetuating the power of a chosen few. Ancient cultures were the first to recognize that reading is not an innocent bystander in the unfolding of societal narratives. They are part of the rewriting of those narratives.[24]

Yet the public dimension of landscape, together with many readerly qualities, is alive and well in many such societies. Understanding its nature merely requires that we reverse the Western tendency to put the text before the landscape. Three peoples whose different approaches to the symbolism of place have been recorded are the Puluwatese of Micronesia, the Iatmul of New Guinea and the Mandaya of Mindanao in the Philippines.[25] In each culture, the equivalent of 'landscape' is the spatial location of clues to knowledge based on oral memory. Objects and localities have both a physical and a symbolic significance within the community. The manner in which the clues are selected, organized and cate-gorized in space varies with each. But, in all cases, Aram Yengoyan emphasizes, 'the connection between them is neither linear nor causal.... Mnemonic structures are not simply linear registers of spatial and temporal events; they must be interpreted as a means by which all cultural information is maintained and transmitted in a society' (1985: 158, cf. 160). Reading, therefore, as we know it, with its implicitly linear system of ordering facts, is not an appropriate analogy for what takes place, except at the level of oral narrative. Nor is the textual metaphor helpful in explaining the guidelines for recall.

Among the Mandaya, oral memory is interlocked with history and geography. The knowledge is transmitted to the community by skilled orators, both men and women, whose repertory ranges from riddles and proverbs through songs, folktales and morality stories to epic accounts of past deeds. The mnemonic devices of the oral sages depend on a knowledge of plants, or of the places in which significant events have taken place, and these in turn are intertwined with the 'genealogy' of property rights in the places where mangos, bananas and other fruits are grown. Cultural continuity is provided by history, but this history, transmitted by word of mouth from one generation to the next, is intimately involved with the real and symbolic events that have taken place on the sites. Thus, place offers a vast system of environmental clues to history, clues which, as structures in memory, 'are anterior to behaviour' (Yengoyan 1985: 174).

To take another example, a different sort of relationship between oral memory, narrative and landscape exists in aboriginal Australia. Ronald Berndt has shown that, in the northern sector of the western desert,

> any piece of country ... is defined by the presence and activities of a mythic being, either by himself or herself, or in conjunction with other characters of mythic significance. In a sense, land not only locates but also identifies a particular deity: the one is the necessary condition of the other.
>
> (1985–6: 267, cf. 273)

The connection between the mythical deities, their narratives and the landscape is made in the concept of the 'dreaming', which enshrines the eternally present qualities of these beings as well as providing physical clues to the creation of the world and countless generations of Aborigines:

> Mythic personages journeyed across the country, had adventures, inter-acted with others, and were eventually metamorphosed as tangible, visible aspects of nature – rocks, trees, waterholes and so on.... Wherever they went ... they left part of their own spiritual essence. This ... remains at specified places today, and is accessible to Aborigines *if* the circumstances are congenial and *if* the occasion so demands.
>
> (Berndt 1983: 15)

The instinct to implant significance in localities may well be universal. But a larger sampling of cases than can be given here would also illustrate that texts and landscapes produce similar but not identical kinds of communities. In the case of the text, all that is needed is the writing, someone to interpret it and an audience. The moment the listeners (or readers) hold a common view of what has been told to them, and are prepared to act on it, they comprise a 'textual community'. Speech acts then become real acts: a structure of discourse begins to live a life of its own narrative possibilities.[26] A place cannot become such a community, as John Agnew has recently reminded us.[27] For when a community is created by a piece of writing – whether the text is read or transmitted orally – the writing is consciously present only as long as the community of understanding does not exist. The moment the community comes into being, this textual presence begins to fade from consciousness, until, at the moment of action, it disappears altogether. The landscape is not capable of a similar vanishing act: it remains 'out there', a part of lived reality. It is both its own text and, in part, its own reading: it is able to refer back to itself, as does any text, while making a public statement, visible to all. It thereby achieves a double hold on its beholders, which is perhaps one reason why landscapes can inspire communities in oral cultures when no written texts exist.

NOTES

1　For an examination of the issues in two different periods, see the introductions to Brilliant (1984) and Bryson (1981). A more recent review is found in Fried (1990: 3–46), where, on pp. 293–294, n. 14, there is a useful summary of articles on the question of the implied viewer in the light of the analysis of *Les Meninas* by Foucault (1973: 3–16), a contemporary *locus classicus*.

2　Goodman (1976: 37–9) states the case in favour of interdependence on logical grounds, a view shared by many cognitive scientists; see Johnson-Laird (1983: 146–66). The opposite position is supported by the experiments of Kosslyn; see (1980: chs 5–6).

3　For a comprehensive review and bibliography of relations between interpretative theory and cultural geography, see Duncan (1990: 3–24), on which I draw here.

4　For two discussions of the issues in differing historical contexts, see Brilliant (1984: introduction, 'Sight reading') and Bryson (1981: introduction).

5　For a recent review of the issues, see Crary (1990: chs 1, 4, 5), with a full bibliography pp. 150–62.

6　I draw these statements from Heisenberg (1989: 9–10; originally published as 'Tradition in science', a lecture delivered on 24 April 1973 at the Smithsonian Institution).

7　I draw on Black (1962: 27–45), substituting the visual components of the scene for 'word' and 'sentence' in the definition of 'focus' and 'frame', p. 28. See, as well, Black's reaffirmation of his position (1990: 58–62).

8　See *The Oxford English Dictionary* (2nd edn, Oxford, 1989, Vol. 8, pp. 628–9).

9　Cf. Eisenstein (1983: 195–9), who overstates the connections between the printing and the growth of science; for a corrective, based on a wide survey of English manuscript evidence, see Voigts (1989: 349–51).

10　For an outline, see Cosgrove, in Cosgrove and Daniels (1988: 1–8).

11　Thereby transforming *ekphrasis*; for a discussion, see the outline in Curtius (1953: 183–202), which is, however, in need of serious revision.

12　See, in particular, Fried (1980: 8–70), especially the discussion of Greuze, Chardin, Van Loo, Vien and Eustache Le Sueur, as well as the criticism of Du Bos, Coypel, Diderot and Grimm, pp. 71–105, and the analysis of landscapes pp. 118–45.

13　At a linguistic level, one of the earliest thinkers to observe this was St Augustine; this is acknowledged by Johnson-Laird (1983: 205–6). For a way into this problem through the analysis of pictorial realism, see Bryson (1981: 7–13).

14　For a recent summary, see Pollatsek and Rayner (1990), from which this statement draws on p. 417. The pioneering work was done by Buswell; see the review of Kolers (1976). A general survey of the issues is found in Haber and Hershenson (1981:

ch. 19); for recent experimental results on word recognition, see Humphreys *et al.* (1990: 546–47). Bryson (1981: 20 ff.) proposes to analyse reading using the linguistic terms 'syntagm' and 'paradigm', derived from Saussure via Barthes; but it is clear to all in the field, although they may disagree on much else, that the understanding of spoken and written language is not the same; for a brief statement, see Kolers (1985: 405–11).

15 A point emphasized by Wollheim (1980: 69–70, cf. 87–91). For an informed discussion of the linguistic possibilities, see Sanford and Garrod (1981: chs 1–4).

16 Kendon (1967), where there is a good review of a wide range of previous studies, including Sartre, Simmel and Erving Goffman, pp. 22–4; for distinctions I employ, see pp. 52–7; cf. Rutter (1984: chs 2–3).

17 For a discussion, see Duncan (1990) and literature cited therein.

18 Summarizing Stock (1983: ch. 1).

19 Scholarship is ably summarized in Corbier (1987: 27–30).

20 On inscriptions, see Corbier (1987).

21 E.g. Augustine, *Confessions*, 7.1.

22 For a review of the issues and a full bibliography, see Kessler (1985: 75–6).

23 For a recent review of the issues, see Belting (1990).

24 Cf. Clifford and Marcus (1986: 1–26).

25 This paragraph is based on Hage (1978) and Yengoyan (1985). For corroboration based on a study of the Vai of West Africa, see Scribner and Cole (1978); and, for a more general survey, linking oral memory and narrative, see Colby and Cole (1973: 69–91), as well as Heath (1983: *passim*).

26 Cf. Heath (1983: 149–57), on oral stories within contemporary communities; on reading as a social activity, see pp. 196–201.

27 For a review of the issues and a full bibliography in the social sciences, see Agnew (1989).

ACKNOWLEDGEMENTS

This is a revised version of a talk delivered at the symposium on New Directions in Cultural Geography, Syracuse University, on 24 March 1989. I am grateful for the symposium organizers, James and Nancy Duncan, for the kind invitation to engage in an interdisciplinary dialogue, as well as to James Duncan for many helpful comments and bibliography. I should also like to express my gratitude to Herbert Kessler, Maruja Jackman and Mrs G.G.R. Harris.

REFERENCES

Agnew, J.A. (1989) 'The devaluation of place in social science', in J.A. Agnew and J.S. Duncan (eds) *The Power of Place: Bringing Together Geographical and Sociological Imaginations*, Boston: Unwin Hyman, 9–29.

Alpers, S. (1983) *The Art of Describing: Dutch Art in the Seventeenth Century*, Chicago: University of Chicago Press.

Belting, H. (1990) *Bild und Kunst: Eine Geschichte des Bildes vor dem Zeitalter der Kunst*, Munich: C.H. Beck.

Berndt, R.M. (1983) 'Images of God in aboriginal Australia', *Visible Religion* 2: 14–39.

—— (1985–6) 'Identification of deity through land: an Australian aboriginal view', *Visible Religion* 4–5: 266–75.

Black, M. (1962) *Models and Metaphors: Studies in Language and Philosophy*, Ithaca: Cornell University Press.

—— (1990) *Perplexities: Rational Choice, the Prisoner's Dilemma, Metaphor, Poetic Ambiguity, and Other Puzzles*, Ithaca: Cornell University Press.

Brilliant, R. (1984) *Visual Narratives: Storytelling in Etruscan and Roman Art*, Ithaca: Cornell University Press.

Bryson, (1981) *Word and Image: French Painting of the Ancien Régime*, Cambridge: Cambridge University Press.

Cancik, H. (1985–6) 'Rome as a sacred landscape', *Visible Religion* 4–5: 250–65.

Carter, P. (1988) *The Road to Botany Bay: An Exploration of Landscape and History*, New York: Knopf.

Clifford, J. and Marcus, G.E. (eds) (1986) *Writing Culture: The Poetics and Politics of Ethnography*, Berkeley: University of California Press.

Colby, B. and Cole, M. (1973) 'Culture, memory, and narrative', in R. Horton and R. Finnegan (eds) *Modes of Thought: Essays on Thinking in Western and Non-Western Societies*, London: Faber, 69–91.

Corbier, M. (1987) 'L'écriture dans l'espace public romain', *L'urbs: espace urbain et histoire (1er siècle av. J.-C.–Ille siècle ap. J.-C.)*, Collection de l'Ecole française de Rome Vol. 98, Rome: Ecole française de Rome, 27–60.

Cosgrove, P. and Daniels, S. (eds) (1988) *The Iconography of Landscape: Essays on the Symbolic Representation, Design and Use of Past Environments*, Cambridge: Cambridge University Press.

Crary, J. (1990) *Techniques of the Observer: On Vision and Modernity in the Nineteenth Century*, Cambridge, Mass.: MIT Press.

Curtius, E.R. (1953) *European Literature and the Latin Middle Ages*, trans. W.R. Trask, Bollingen Series 36, New York: Pantheon.

Daston, L. (1988) *Classical Probability in the Enlightenment*, Princeton: Princeton University Press.

Duncan, J.S. (1990) *The City as Text: The Politics of Landscape Interpretation in the Kandyan Kingdom*, Cambridge: Cambridge University Press.

Eisenstein, E. (1983) *The Printing Revolution in Early Modern Europe*, Cambridge: Cambridge University Press.

Foucault, M. (1973) *The Order of Things*, New York: Pantheon.

Fried, M. (1980) *Absorption and Theatricality: Painting and Beholder in the Age of Diderot*, Berkeley: University of California Press.

—— (1990) *Courbet's Realism*, Chicago: University of Chicago Press.

Goodman, N. (1976) *The Languages of Art*, Indianapolis: Hackett.

Haber, R.N. and Hershenson, M. (1981) *The Psychology of Visual Perception*, 2nd edn, New York: Holt, Rinehart & Winston.

Hage, P. (1978) 'Speculations on Puluwatese mnemonic structure', *Oceania* 49: 81–95.

Heath, S.B. (1983) *Ways with Words: Language, Life and Work in Communities and Classrooms*, Cambridge: Cambridge University Press.

Heisenberg, W. (1989) *Encounters with Einstein and Other Essays on People, Places, and Particles*, Princeton: Princeton University Press.

Humphreys, G.W., Evett, L.J. and Quinlan, P.J. (1990) 'Orthographic processing in visual word identification', *Cognitive Psychology* 22: 517–60.

Johnson-Laird, P.N. (1983) *Mental Models: Towards a Cognitive Science of Language, Inference, and Consciousness*, Cambridge, Mass.: Harvard University Press.

Kendon, A. (1967) 'Some functions of gaze direction in social interaction', *Acta Psychologica* 26: 1–47.

Kessler, H. (1985) 'Pictorial narrative and church mission in sixth-century Gaul', *Studies in the History of Art* 16: 75–91.

Kolers, P. (1976) 'The role of eye movements in reading', in R.A. Monty and J.W. Senders (eds) *Eye Movements and Psychological Processes*, Hillsdale: Erlbaum.

—— (1985) 'Phonology in reading', in D.R. Olson, N. Torrance and A. Hildyard (eds)

Literacy, Language, and Learning: The Nature and Consequences of Reading and Writing, Cambridge: Cambridge University Press, 405–11.

Kossyln, S.M. (1980) *Image and Mind*, Cambridge, Mass.: Harvard University Press.

Ortony, A. (ed.) (1979) *Metaphor and Thought*, Cambridge: Cambridge University Press.

Pollatsek, A. and Rayner, K. (1990) 'Reading', in Michael J. Posner (ed.) *Foundations of Cognitive Science*, Cambridge, Mass.: MIT Press, 401–36.

Rutter, D.R. (1984) *Looking and Seeing: The Role of Visual Communication in Social Interaction*, New York: Wiley.

Saenger, P. (1982) 'Silent reading: its impact on late medieval script and society', *Viator* 13: 367–414.

Sandford, A.J. and Garrod, S.C. (1981) *Understanding Written Language: Explorations of Comprehension beyond the Sentence*, Chichester and New York: Wiley.

Scribner, S. and Cole, M. (1978) 'Literacy without schooling: testing for intellectual effects', *Harvard Educational Review* 48: 448–61.

Stock, B. (1983) *The Implications of Literacy: Written Language and Models of Interpretation in the Eleventh and Twelfth Centuries*, Princeton: Princeton University Press.

Voigts, L.E. (1989) 'Scientific and medical books', in J. Griffiths and D.A. Pearsall (eds) *Book Production and Publishing in Britain 1375–1475*, Cambridge: Cambridge University Press, 345–402.

Wollheim, R. (1980) *Art and its Objects*, 2nd edn, with six supplementary essays, Cambridge: Cambridge University Press.

Yengoyan, A.A. (1985) 'Memory, myth, and history: traditional agriculture and structure in Mandaya society', in K.L. Hutterer, A.T. Rambo and G. Lovelace (eds) *Cultural Values and Human Ecology in Southeast Asia*, Michigan Papers on South and Southeast Asia No. 27, Ann Arbor, Michigan: Center for South and Southeast Asian Studies, University of Michigan.

17

EPILOGUE

David Ley and James Duncan

If the chapters in this book hold a single objective, it is the view that landscapes and places are constructed by knowledgeable agents who find themselves inevitably caught up in a web of circumstances – economic, social, cultural and political – usually not of their own choosing. Every landscape is thereby a synthesis of charisma and context, a text which may be read to reveal the force of dominant ideas and prevailing practices, as well as the idiosyncrasies of a particular author. Within cultural geography, while this synthesis has been recognized, analytical priority has for the most part been given to a particular view of the cultural context. In contrast other authors have prioritized the role of the charismatic individual. We are arguing here for a charting of a middle ground between the poles of collectivism and individualism, whereby neither individual nor context is privileged but both are dialectically related in the making of geographies. Empirical studies which may serve as guideposts for such a venture have been published in recent years by a number of cultural geographers (Cosgrove 1984; Cosgrove and Daniels 1988; Daniels 1992; Duncan 1990; Ley 1987).

At the same time, landscapes are read as well as made, and the act of reading is not an unproblematic act either. Each reader, like each author, brings a past biography and present intentions to a text, so that the meaning of a place or a landscape may well be unstable, a multiple reality for the diverse groups who produce readings of it. A very particular reader of a place or a landscape is the social scientist, who brings a distinctive biography and set of concerns of his or her own. Consequently, any presentation of a landscape, whether popular or scholarly, is best thought of as a *representation*, that is, a construction that is contingent, partial and unfinished. It is, as ethnographers have suggested for their own accounts, in the very literal sense a fiction, a fabrication that depends in part upon the position of the interpreter (Clifford 1988; Fernandez 1991; Fox 1991; Tyler 1987). The insights from interrogating the positional categories that shape a text are too rich to be ignored – for example, the pervasive and pernicious effects of Eurocentrism (Amin 1989; Said 1979; Stratton 1990) or androcentrism (Colomina 1992; Mohanty *et al.* 1991; Trinh 1989). Yet as we have seen in earlier chapters, this is a dangerous realization, for with it the solid ground of interpretation itself becomes highly unstable. The door is flung wide open for a potentially lawless relativism to immobilize interpretative work, as deconstruction upon deconstruction creates an endless deferral of signification.

Yet at the same time this is a replay in a new guise of the old philosophical game of presuppositional criticism. Each philosophical system cannot be erected without its predicates, its logical positions. So it is with cultural relativism. Because positionality cannot be 'solved', let alone removed, for it is an inevitable property of the writer as a valuing subject, it must always be acknowledged as a constitutive element of any representation. With it, as a necessary rider, should come a certain modesty in the claims an interpretation can offer, for appeals to reflexivity must always assume the likelihood that positionality is taken for granted and not fully transparent for the authorial subject.

These insights are not entirely novel. Certainly they have been discussed in other fields.[1] But also in part these issues were alluded to by humanistic geographers like David Lowenthal (1961) and Yi-Fu Tuan (1971, 1982) and, though rarely explicitly, have been acted upon by some authors in historical geography (especially Harris 1991; also Lowenthal 1985). But the view of the contributors to this volume is that past cultural geography, in particular, has rarely exploited the richness, or addressed the authorial challenges, of an interpretation of either place or landscape. Too often, interpretation has been thin rather than thick, with an attention to artefacts and distributions that has suspended a whole range of questions to do with *process*; the constellation of economic interests, power relations, cultural predispositions and social differentiation which together constitute the character of a place. In an attention to process, cultural geography would be following the precedent of earlier work in the discipline that has repeatedly found the preoccupation with pattern or morphology an incomplete intellectual agenda; so geomorphology has moved from description of landforms to process geomorphology, urban geography from urban morphology to an examination of constitutive processes, and social geography from maps of spatial patterns to a concern with social processes. To interrogate form is also to engage theory, and so it is that contemporary cultural geography has sought inspiration from a new corpus of cultural theory, including literary theory (Cohen 1989), art theory (Ferguson *et al.* 1990), film theory (Burgin 1982) and the interconnected arguments of postmodernism, feminism, post-structuralism and post-colonialism (Chaudhuri and Strobel 1992; Nicholson 1990; Parker *et al.* 1992).

So too cultural geography has not been nearly attentive enough to the problems of authorship, to the task often summarily dismissed as 'writing up'. Again it is ethnographers who have become keenly aware of these major problems of representation (Clifford 1988; Geertz 1988). There are now several interesting experiments in new forms of writing by geographers (Olsson 1980; Pred 1990; Western 1992) which reveal just how substantial are the challenges in transcending the power/knowledge relations of conventional narrative.

Half of the essays in this book were prepared in first draft for a symposium at

Syracuse University in 1989, with the title of New Directions in Cultural Geography. The first stirrings of new approaches to cultural geography had occurred a year or two earlier, but what has become evident in the short period since the symposium has been an explosive growth of interest in this field, expressed by Jackson's (1989) innovative textbook inspired by the Birmingham school of cultural studies, the seminal initiative of the Social and Cultural Geography Specialty Group of the Institute of British Geographers (Philo 1991), the special double issue of *Environment and Planning D: Society and Space* (1992) addressing postmodernism and poststructuralism, packed sessions at the annual conferences of American and Canadian geographers, and the launch in 1993 of *Ecumene*, a new journal of cultural geography. This surge of interest within geography repeats a broader trend in the humanities and social sciences where 'a cultural turn' in the past decade has attracted an immense amount of intellectual energy. An explanation of this turn to culture is an interesting task, but instead, in concluding this volume, it does seem worth offering a few brief comments on the directions the cultural turn is taking in geography.

While the preceeding essays are part of this cultural turn, they do not exhaust it. But they do share with other recent work a premise that categories are socially constructed, a premise that has been growing in strength in human geography over the past twenty years. To assert that reality is socially constructed is to interrogate the taken-for-granted categories of everyday life and of intellectual endeavours alike. It is a perspective that is vitally concerned with the positioning of the author, with such contingencies as class, race, gender, nationality and political persuasion that shape an outlook and colour an interpretation. The constructionist perspective is concerned not only with the author but also with his or her authority, the power which is always confered upon the person with the authority to categorize. This power shapes the character of the 'Other', those people and places that fall outside the privileged domains of power/knowledge (Driver 1992).

Beyond this common point of departure the literature rapidly diverges. The inflection of this book has been one in which place and landscape, established geographical categories, have assumed a central position. The potential fruitfulness of geographical imaginations should not be understated, especially now, in an interdisciplinary literature in cultural studies which finds considerable purchase in geographical metaphors (Mohanty 1991; Probyn 1990).[2] Yet, surprisingly, some recent work in cultural geography has only a fleeting association with place or landscape. In part this stems from a pursuit of cultural process that has sometimes led to long detours from recognizably geographical problems. Another reason is an over-preoccupation with theory or philosophy which is scarcely accountable to the empirical realm. This is reminiscent of ultimately unprofitable earlier debates in economic theory and social theory. Some recent explorations of the formation of the self, for example, seem far detached from conventional geographical issues. Yet if the allure of psychoanalysis was complemented by earlier engagements with social psychology, geographical

331

connections would be far more evident. For example, social phenomenology and symbolic interactionism, emanating originally from Mead and Dewey, have been usefully employed by social and cultural geographers to interpret the mutual construction of place and identity (Duncan 1978; Jackson and Smith 1984; Ley 1977; Smith 1984, 1988). To a theoretical argument that sees the self becoming defined through interaction, the dialogue with place itself, with rhythms and routines of distinctive locales, frames the outlines of a shared identity (compare Western 1992).

In making space for culture in a discipline where social and cultural relations have either been 'black-boxed' as in traditional cultural geography or been regarded too unproblematically as outcomes of environmental or economic processes, it is perhaps not surprising that some recent work has celebrated the autonomy of the cultural perhaps too enthusiastically. While there are few authors who would still treat the cultural as a simple projection of the economic, none the less the economy remains as a key context of social and cultural relations. The fact that the art world is also the art market confirms the interconnected nature of analytical realms even in the heart of high culture (Clifford 1988; Karp and Lavine 1991; Price 1989). Recent ethnographic arguments have made the same point that local ethnographies should be located in broader economies and polities (Atkinson 1990; Marcus and Fischer 1986).

Whereas much has been made of the cultural turn within the social sciences over the past decade, perhaps what we have witnessed is more of a convergence between what were analytically treated as semi-autonomous realms of culture, politics and economy, than a cultural turn. For while it is unquestionable that specialists interested in economic and political questions have increasingly become cognizant of the cultural dimension of the issues they study, so equally students of culture have realized that one cannot profitably consider the cultural apart from the political and the economic.[3] And yet having said this, we are not proposing a single monolithic social science nor even a single way of doing geography. Rather we would argue for analytical differences both within social science and within specific fields such as geography. For while we recognize that the cultural is not an autonomous realm, neither would we conclude because of this that we are all cultural geographers now. What then does it mean to say that we practise *cultural geography*? First, it means that we prioritize culture within our scholarship, that is, we focus upon how the cultural as a signification system interpenetrates the economic and political systems within a social order. Second, we prioritize the geographic within our interpretations, that is, we focus upon the cultural dimensions of landscape, place or space. If we fail to do the former then we are some other kind of geographer, but if we fail to do the latter then we are not geographers at all. At this particular moment, above all, to concede both these positions is to concede some exceedingly promising intellectual territory.

NOTES

1   A most interesting recent interdisciplinary debate on this issue is found in Collini (1992).
2   For a discussion of geographical metaphors, see Bondi and Domosh (1992).
3   One should not be naïve here either, for this convergence has been more exceptional than we might imply. Consider, for example, the ultimately reductionist view of culture from two political economists who have sought to broaden their arguments (Harvey 1989; Wallerstein 1991).

REFERENCES

Amin, S. (1989) *Eurocentrism*, New York: Monthly Review Press.
Atkinson, P. (1990) *The Ethnographic Imagination: Textual Constructions of Reality*, London: Routledge.
Bondi, L. and Domosh, M. (1992) 'Other figures in other places: on feminism, post-modernism and geography', *Environment and Planning D: Society and Space* 10: 199–213.
Burgin, V. (ed.) (1982) *Thinking Photography*, London: Macmillan.
Chaudhuri, N. and Strobel, M. (eds) (1992) *Western Women and Imperialism: Complicity and Resistance*, Bloomington: Indiana University Press.
Clifford, J. (1988) *The Predicament of Culture: Twentieth-Century Ethnography, Literature and Art*, Cambridge, Mass.: Harvard University Press.
Cohen, R. (ed.) (1989) *The Future of Literary Theory*, London: Routledge.
Collini, S. (ed.) (1992) *Interpretation and Overinterpretation: Umberto Eco with Richard Rorty, Jonathan Culler and Christine Brook-Rose*, Cambridge: Cambridge University Press.
Colomina, B. (ed.) (1992) *Sexuality and Space*, Princeton: Princeton Papers on Architecture.
Cosgrove, D. (1984) *Social Formation and Symbolic Landscape*, Beckenham: Croom Helm.
Cosgrove, D. and Daniels, S. (eds) (1988) *The Iconography of Landscape: Essays on the Symbolic Representation, Design and Use of Past Environments*, Cambridge: Cambridge University Press.
Daniels, S. (1992) 'The implications of industry: Turner and Leeds', in T. Barnes and J.S. Duncan (eds) *Writing Worlds: Discourse, Text and Metaphor in the Representation of Landscape*, London: Routledge, 38–49.
Driver, F. (1992) 'Geography's empire; histories of geographical knowledge', *Environment and Planning D: Society and Space* 10: 23–40.
Duncan, J.S. (1978) 'The social construction of unreality: an interactionist approach to the tourist's cognition of environment', in D. Ley and M. Samuels (eds) *Humanistic Geography*, Chicago: Maaroufa, 269–82.
——— (1990) *The City as Text: The Politics of Landscape Interpretation in the Kandyan Kingdom*, Cambridge: Cambridge University Press.
Ferguson, R., Olander, W., Tucker, M., and Fiss, K. (eds) (1990) *Discourses: Conversations in Postmodern Art and Culture*, Cambridge, Mass.: MIT Press.
Fernandez, J.W. (ed.) (1991) *Beyond Metaphor: The Theory of Tropes in Anthropology*, Stanford: Stanford University Press.
Fox, R.G. (1991) *Recapturing Anthropology: Working in the Present*, Santa Fe: School of American Research Press.
Geertz, C. (1988) *Works and Lives*, Stanford: Stanford University Press.
Harris, R.C. (1991) 'Power, modernity and historical geography', *Annals of the Associ-*

ation of American Geographers 81: 671–83.

Harvey, D. (1989) *The Condition of Postmodernity: An Inquiry into the Origins of Cultural Change*, Oxford: Blackwell.

Jackson, P. (1989) *Maps of Meaning: An Introduction to Cultural Geography*, London: Unwin Hyman.

Jackson, P. and Smith, S.J. (1984) *Exploring Social Geography*, London: Allen & Unwin.

Karp, I. and Lavine, S. (eds) (1991) *Exhibiting Culture: The Poetics and Politics of Museum Display*, Washington, DC: Smithsonian Institute Press.

Ley, D. (1977) 'Social geography and the taken-for-granted world', *Transactions of the Institute of British Geographers* (new series) 2: 498–512.

—— (1987) 'Styles of the times: liberal and neoconservative landscapes in inner Vancouver, 1968–1986', *Journal of Historical Geography* 13: 40–56.

Lowenthal, D. (1961) 'Geography, experience and imagination: toward a geographical epistemology', *Annals of the Association of American Geographers* 51: 241–60.

—— (1985) *The Past Is a Foreign Country*, Cambridge: Cambridge University Press.

Marcus, G.E. and Fischer, M.M.J. (1986) *Anthropology as Cultural Critique: An Experimental Moment in the Human Sciences*, Chicago: University of Chicago Press.

Mohanty, C.T. (1991) 'Introduction: cartographies of struggle, third world women and the politics of feminism', in C.T. Mohanty, A. Russo and L. Torres (eds) *Third World Women and the Politics of Feminism*, Bloomington: Indiana University Press, 1–47.

Mohanty, C.T., Russo, A. and Torres, L. (eds) (1991) *Third World Women and the Politics of Feminism*, Bloomington: Indiana University Press.

Nicholson, L.J. (ed.) (1990) *Feminism/Postmodernism*, New York: Routledge.

Olsson, G. (1980) *Birds in Egg: Eggs in Bird*, London: Pion.

Parker, A., Russo, M., Sommer, D. and Yaeger, P. (eds) (1992) *Nationalisms and Sexualities*, New York: Routledge.

Philo, C. (ed.) (1991) *New Words, New Worlds: Reconceptualizing Social and Cultural Geography*, Aberystwyth: Cambrian Printers.

Pred, A. (1990) *Lost Words and Lost Worlds: Modernity and the Language of Everyday Life in Late Nineteenth-Century Stockholm*, Cambridge: Cambridge University Press.

Price, S. (1989) *Primitive Art in Civilized Places*, Chicago: University of Chicago Press.

Probyn, E. (1990) 'Travels in the postmodern: making sense of the local', in L.J. Nicholson (ed.) *Feminism/Postmodernism*, New York: Routledge, 176–89.

Said, E. (1979) *Orientalism*, New York: Vintage.

Smith, S.J. (1984) 'Practicing humanistic geography', *Annals of the Association of American Geographers* 74: 353–74.

—— (1988) 'Constructing local knowledge: the analysis of self in everyday life', in J. Eyles and D.M. Smith (eds) *Qualitative Methods in Human Geography*, Oxford: Polity.

Stratton, J. (1990) *Writing Sites: A Genealogy of the Postmodern World*, Ann Arbor: University of Michigan Press.

Trinh, M.-H. (1989) *Woman, Native, Other*, Bloomington: Indiana University Press.

Tuan, Y.-F. (1971) 'Geography, phenomenology, and the study of human nature', *The Canadian Geographer* 15: 181–92.

—— (1982) *Segmented Worlds and the Self*, Minneapolis: University of Minnesota Press.

Tyler, S. (1987) *The Unspeakable: Discourse, Dialogue, and Rhetoric in the Postmodern World*, Madison: University of Wisconsin Press.

Wallerstein, I. (1991) *Geopolitics and Geoculture*, Cambridge: Cambridge University Press.

Western, J.C. (1992) *A Passage to England*, Minneapolis: University of Minnesota Press.

INDEX